AF361720

Indoor Air Quality

Indoor Air Quality

Editor

Dikaia E. Saraga

MDPI • Basel • Beijing • Wuhan • Barcelona • Belgrade • Manchester • Tokyo • Cluj • Tianjin

Editor
Dikaia E. Saraga
National Centre for Scientific Research "Demokritos"
Greece

Editorial Office
MDPI
St. Alban-Anlage 66
4052 Basel, Switzerland

This is a reprint of articles from the Special Issue published online in the open access journal *Applied Sciences* (ISSN 2076-3417) (available at: https://www.mdpi.com/journal/applsci/special_issues/Indoor_Quality).

For citation purposes, cite each article independently as indicated on the article page online and as indicated below:

LastName, A.A.; LastName, B.B.; LastName, C.C. Article Title. *Journal Name* **Year**, *Volume Number*, Page Range.

ISBN 978-3-03943-703-0 (Hbk)
ISBN 978-3-03943-704-7 (PDF)

Cover image courtesy of Dikaia E. Saraga.

Contents

About the Editor

Dikaia E. Saraga is an Associate Researcher at the Institute of Nuclear & Radiological Sciences and Technology, Energy & Safety, NCSR "Demokritos", Greece. She holds a BSc in Physics (University of Athens, Greece), an MSc in Environmental Physics (University of Athens, Greece) and a PhD in Indoor Air Pollution (University of West Macedonia, Greece). Her full operating skills include, among others: indoor/outdoor field air quality measurements, application of reference methods for PM and VOC determination, ionic and OC/EC analysis; receptor modeling application for source apportionment studies. Her research interests also include: air pollution and climate change mitigation, population exposure, air quality sensors, big data treatment. She has participated in more than 20 European (Horizon 20202, FP6, FP7, Life+) and national research projects in the field of air quality. She is a member of the management board of COST17136 INDAIRPOLLNET action and a member of the FAIRMODE group. She is a recognized reviewer of 18 scientific journals and author of 44 publications in peer-reviewed journals, and more than 80 conference proceedings and scientific reports.

 *applied sciences*

Editorial

Special Issue on Indoor Air Quality

Dikaia E. Saraga

National Center for Scientific Research "Demokritos", Aghia Paraskevi, 15310 Athens, Greece; dsaraga@ipta.demokritos.gr

Received: 8 February 2020; Accepted: 11 February 2020; Published: 22 February 2020

1. Introduction

It is a fact that people in developed countries spend almost 90% of their time indoors, where they experience their greatest exposures. However, regulation of air pollution focuses on outdoor air, as indoor environment is less well-characterized and recognized as a potential location for exposure to air pollution.

What makes indoor air intrinsically more interesting than outdoor air from a scientific point of view? Some sources are undoubtedly uniquely building-related (e.g., cleaning agents, emissions from building materials and personal care products), while some contaminant dynamics operate only in buildings (e.g., the distribution of particles and gases by mechanical ventilation systems and photochemical reactions, and the infiltration of soil gases). Besides this, air pollutant concentrations are often higher indoors than outdoors, particularly following activities such as cleaning and cooking (with a greater source strength indoors than outdoor on a per area basis), while it has already been proven that many indoor air pollutants are harmful to human health.

Another issue is ventilation. While indoor microenvironments are a microcosm of most urban settings, the effective air exchange and renewal in buildings is much lower than outdoors, even in urban areas. This concern is amplified by the fact that energy efficiency measures, driven by climate change awareness, have made modern buildings more airtight, further degrading the quality of indoor air. Therefore, a person is significantly more likely to inhale a harmful chemical molecule if it is emitted indoors rather than outdoors.

Monitoring of indoor air pollutants in a spatio-temporal basis is challenging. A key element is the access to local (i.e., indoor residential, workplace, or public building) exposure measurements. Unfortunately, the high cost and complexity of most current air pollutant monitors results in a lack of detailed spatial and temporal resolution. Therefore, individuals of vulnerable groups (children, pregnant, elderly, and sick people) have little insight into their personal exposure levels. This becomes significant in cases of hyper-local variations and short-term pollution events such as instant indoor activity (e.g., cooking, smoking, and dust resuspension). Advances in sensor miniaturization have encouraged the development of small, inexpensive devices capable of estimating pollutant concentrations. This new class of sensors presents new possibilities for indoor exposure monitoring.

Furthermore, indoor air chemistry models typically account for air exchange with outdoors through ventilation, deposition on indoor surfaces, and photochemical reactions. Surface chemistry on furnishings, building materials, and human bodies is becoming increasingly recognized as being of crucial importance.

In light of the above, this Special issue on 'Indoor Air Quality' was introduced to collect latest research and address challenging issues in the areas of the triptych: Indoor environment quality monitoring, indoor air modeling, and exposure to indoor air pollution.

2. From Indoor Environment Quality Perception to Indoor Air Quality Monitoring and Control

In this Special Issue, 24 papers were submitted, and 16 were accepted for publication (67% acceptance rate). 63% of the published studies originate from Europe, while 37% of them were conducted in Asiatic countries. Various topics have been addressed in the contributed articles: Indoor air quality (IAQ) monitoring and modelling, occupants' comfort related to indoor environment parameters as well as innovative techniques for IAQ monitoring and improvement. When looking back, it can be concluded that the majority of the studies can be distinguished into two main groups. The first one refers to occupants' perception for the quality of the indoor environment as well as their comfort inside a building. The second group focuses on new techniques of monitoring and controlling the parameters determining the quality of indoor air. Finally, a quite smaller group includes studies performed in indoor environments of special characteristics.

To be more specific, nine papers discuss the issues of IAQ perception and control as well as thermal comfort. The study of M. Cho and M. Kim, [1] titled *'Residents' Perceptions of and Response Behaviors to Particulate Matter—A Case Study in Seoul, Korea'* aimed at understanding the perception of 171 people in Seoul for indoor air quality based on domestic particulate matter levels. In a European-scale study titled *'Personal Control of the Indoor Environment in Offices: Relations with Building Characteristics, Influence on Occupant Perception and Reported Symptoms Related to the Building—The Officair Project'*, Sakellaris I. et al. [2] focused on revealing the complex relationship between office employees' control over various indoor environment parameters and their comfort, health and productivity. 7441 occupants of 167 recently built or retrofitted office buildings in eight European countries participated in an online survey about personal/health/work data as well as physical/psycho-social information. In another study titled *'An Investigation of the Effects of Changes in the Indoor Ambient Temperature on Arousal Level, Thermal Comfort, and Physiological Indices'*, Gwak J. et al. [3] aimed to design a thermal environment that improves both the arousal level and thermal comfort of the occupants. To this end, they investigated the relationships between the physiological indices, subjective evaluation values, and task performance under several conditions of changes in the indoor ambient temperature. The study *'Probability of Abnormal Indoor Air Exposure Categories Compared with Occupants' Symptoms, Health Information, and Psychosocial Work Environment'* authored by Tähtinen K. et al. [4] aimed at (i) evaluating the relation between the four-level categorized probability of abnormal indoor air exposure and employees' work environment-related symptoms, group-level health information, and psychosocial work environment, (ii) assessing the relation between ventilation system deficiencies and employees' work environment-related symptoms and evaluating the impact of prolonged IAQ problem solution processes on perceived IAQ. The study *'Combined Model for IAQ Assessment: Part 1—Morphology of the Model and Selection of Substantial Air Quality Impact Sub-Models'* of Piasecki M. and Kostyrko K.B. [5] provided an overview of models defining occupants' comfort and satisfaction with IAQ. Specifically, subcomponents of three potential IAQ models were classified according to their application potential: IAQ quality index, IAQ comfort index, and an overall health and comfort index. The authors provide a method for using the combined IAQ index to determine the indoor environmental quality index, IEQ and a practical case study which provides IAQ and IEQ model implementation for a large office building assessment. The study titled *'An Accident Model with Considering Physical Processes for Indoor Environment Safety'* authored by Yang Z. et al. [6] also deals with thermal comfort in an indoor environment. In particular, authors presented an extension of Systems-Theoretic Accident Model and Process (STAMP) while considering physical processes in an indoor environment as temperature changes. Thermal comfort was also the subject of the study of Cheng X. et al. [7] titled *'A Contactless Measuring Method of Skin Temperature based on the Skin Sensitivity Index and Deep Learning'*. In this study, a skin sensitivity index was proposed to describe individual sensitivity of thermal comfort, and the index was combined with skin images for deep learning network training. Also, a novel contactless measuring algorithm (NISDL) based on SSI was proposed, with two different frameworks of NISDL having been designed for real-time thermal comfort measurements. Finally, a deep learning algorithm without SSI was also generated and trained. Two more studies have worked on the evaluation of the indoor environment quality while focusing on a specific population group: Children and teenagers. The study of Mainka A. and Zajusz-Zubek E., [8] named *'Keeping Doors Closed as One Reason for Fatigue in Teenagers—A Case Study'* investigated the variability of CO_2 concentration in naturally ventilated bedrooms of teenagers in Polland, by correlating bedroom door opening during the night with CO_2 concentration and thermal comfort. In the study *'Cooking/Window Opening and Associated Increases of Indoor PM2.5 and NO$_2$ Concentrations of Children's Houses in Kaohsiung, Taiwan'*, Yen Y. et al. [9] attempted to assess the influence of window opening and cooking activity to measured air pollutants levels in 60 children homes in an industrial city in Taiwan.

The second sub-group of papers published in this Special Issue, includes five studies which feature the introduction of new developments in technology and computational science to the field of indoor environment monitoring and control. The study of Kim S. et al. [10] titled *'Evaluation of Performance of Inexpensive Laser Based PM2.5 Sensor Monitors for Typical Indoor and Outdoor Hotspots of South Korea'* presents the results of the evaluation of a low-cost real-time PM monitor under indoor testing with common $PM_{2.5}$ sources of Korea (frying pork in a pan or smoking). In another study, *'Real-Time Monitoring of Indoor Air Quality with Internet of Things-Based E-Nose'*, authored by Tastan M. [11] and Gokozan H., an 'e-nose', a real-time mobile air quality monitoring system with various air parameters such as CO_2, CO, PM10, NO_2 temperature and humidity was presented and evaluated. The proposed e-nose is produced with an open source, low cost, easy installation and do-it-yourself approach. An environmental quality solution based on IoT to supervise Laboratory Environmental Conditions (LEC) named *iAQ+* was introduced in the paper titled *'An Internet of Things-Based Environmental Quality Management System to Supervise the Indoor Laboratory Conditions'* authored be Marques C. and Pitarma R [12]. This low-cost wireless solution for indoor environment quality supervision incorporates mobile computing technologies for data consulting, easy installation, significant notifications for enhanced living conditions, and laboratory activities. Further, Cheng X. et al. [7] study (presented in the previous paragraph) belongs to this sub-group as it included training of deep learning network. Finally, the study *'A Promising Technological Approach to Improve Indoor Air Quality'* authored by Maggos T. et al. [13] presents an innovative paint material which exhibits intense photocatalytic activity under direct and diffused visible light for the degradation of air pollutants, suitable for indoor use. This innovative photo-paint was tested under laboratory and real scale conditions.

Last but not least, three papers published in this Special Issue include studies of air quality in indoor environments of special characteristics. The first one, *'Study of Passive Adjustment Performance of Tubular Space in Subway Station Building Complexes'* by Li J. et al. [14] focused on the various tubular space forms in subway station building complexes with the scope of proposing an improvement of the indoor environment in terms of comfort and energy consumption. The second one, *'Application of Airborne Microorganism Indexes in Offices, Gyms, and Libraries'* authored by Grisoli P. et al. [15] quantified the levels of microorganisms present in the air in different places such as offices, gyms, and libraries. The third one, *'How Working Tasks Influence Biocontamination in an Animal Facility'* authored by Marcelloni M. et al. [16] aimed to determine what factors could be associated with a high level of exposure to biological agents in an animal facility, through the measuring and characterization of airborne fungi, bacteria, endotoxin, $(1,3)$-β-D-glucan and animal allergens.

3. Conclusions

As a final point, the papers in this Special Issue have pinpointed two thematic areas of IAQ that researchers currently focus on, and basically, answer two questions: (i) How do people perceive the quality of air inside their home, office, school etc.? and (ii) what are the state of the art tools (both instrumentational and computational) to monitor, control and improve indoor air quality? While this Special Issue has been closed, further research towards these directions is expected in the very near future. There are still several challenging research questions to be answered. Manuscripts addressing challenging future research for Indoor Air Quality are invited in the second volume, named 'New Challenges for Indoor Air Quality' launched by Applied Sciences, MDPI.

Author Contributions: D.E.S. performed the papers' review and wrote the editorial. The author has read and agreed to the published version of the manuscript.

Acknowledgments: This issue would not be possible without the contributions of various talented authors, hardworking and professional reviewers, and dedicated editorial team of Applied Sciences. Congratulations to all authors—no matter what the final decisions of the submitted manuscripts were, the feedback, comments, and suggestions from the reviewers and editors which substantially helped the authors to improve their papers. Finally, I place on record my gratitude to the editorial team of Applied Sciences, and special thanks to Daria Shi, Assistant Editor from MDPI Branch Office, Beijing.

Conflicts of Interest: The authors declare no conflict of interest.

References

1. Cho, M.; Kim, M. Residents' perceptions of and response behaviors to particulate matter—A case study in Seoul, Korea. *Appl. Sci.* **2019**, *9*, 3660. [CrossRef]

2. Sakellaris, I.; Saraga, D.; Mandin, C.; de Kluizenaar, Y.; Fossati, S.; Spinazzè, A.; Cattaneo, A.; Szigeti, T.; Mihucz, V.; de Oliveira Fernandes, E.; et al. Personal control of the indoor environment in offices: Relations with building characteristics, influence on occupant perception and reported symptoms related to the building—The officair project. *Appl. Sci.* **2019**, *9*, 3227. [CrossRef]

3. Gwak, J.; Shino, M.; Ueda, K.; Kamata, M. An investigation of the effects of changes in the indoor ambient temperature on arousal level, thermal comfort, and physiological indices. *Appl. Sci.* **2019**, *9*, 899. [CrossRef]

4. Tähtinen, K.; Lappalainen, S.; Karvala, K.; Lahtinen, M.; Salonen, H. Probability of abnormal indoor air exposure categories compared with occupants' symptoms, health information, and psychosocial work environment. *Appl. Sci.* **2019**, *9*, 99. [CrossRef]

5. Piasecki, M.; Kostyrko, K. Combined model for IAQ assessment: Part 1—Morphology of the model and selection of substantial air quality impact sub-models. *Appl. Sci.* **2019**, *9*, 3918. [CrossRef]

6. Yang, Z.; Lim, Y.; Tan, Y. An accident model with considering physical processes for indoor environment safety. *Appl. Sci.* **2019**, *9*, 4732. [CrossRef]

7. Cheng, X.; Yang, B.; Tan, K.; Isaksson, E.; Li, L.; Hedman, A.; Olofsson, T.; Li, H. A Contactless measuring method of skin temperature based on the skin sensitivity index and deep learning. *Appl. Sci.* **2019**, *9*, 1375. [CrossRef]

8. Mainka, A.; Zajusz-Zubek, E. Keeping doors closed as one reason for fatigue in teenagers—A case study. *Appl. Sci.* **2019**, *9*, 3533. [CrossRef]

9. Yen, Y.; Yang, C.; Mena, K.; Cheng, Y.; Chen, P. Cooking/window opening and associated increases of indoor $PM2.5$ and NO_2 concentrations of children's houses in kaohsiung, Taiwan. *Appl. Sci.* **2019**, *9*, 4306. [CrossRef]

10. Kim, S.; Park, S.; Lee, J. Evaluation of performance of inexpensive laser based PM2.5 sensor monitors for typical indoor and outdoor hotspots of south Korea. *Appl. Sci.* **2019**, *9*, 1947. [CrossRef]

11. Taştan, M.; Gökozan, H. Real-time monitoring of indoor air quality with internet of things-based E-nose. *Appl. Sci.* **2019**, *9*, 3435. [CrossRef]

12. Marques, G.; Pitarma, R. An internet of things-based environmental quality management system to supervise the indoor laboratory conditions. *Appl. Sci.* **2019**, *9*, 438. [CrossRef]

13. Maggos, T.; Binas, V.; Siaperas, V.; Terzopoulos, A.; Panagopoulos, P.; Kiriakidis, G. A promising technological approach to improve indoor air quality. *Appl. Sci.* **2019**, *9*, 4837. [CrossRef]

14. Li, J.; Lu, S.; Wang, Q.; Tian, S.; Jin, Y. Study of passive adjustment performance of tubular space in subway station building complexes. *Appl. Sci.* **2019**, *9*, 834. [CrossRef]

15. Grisoli, P.; Albertoni, M.; Rodolfi, M. Application of airborne microorganism indexes in offices, gyms, and libraries. *Appl. Sci.* **2019**, *9*, 1101. [CrossRef]

16. Marcelloni, A.M.; Chiominto, A.; Di Renzi, S.; Melis, P.; Wirz, A.; Riviello, M.C.; Massari, S.; Sisto, R.; D'Ovidio, M.; Paba, E. How working tasks influence biocontamination in an animal facility. *Appl. Sci.* **2019**, *9*, 2216. [CrossRef]

Article

Residents' Perceptions of and Response Behaviors to Particulate Matter—A Case Study in Seoul, Korea

Myung Eun Cho and Mi Jeong Kim *

School of Architecture, Hanyang University, 222 Wangsimni-ro, Seongdong-gu, Seoul 04763, Korea
* Correspondence: mijeongkim@hanyang.ac.kr; Tel.: +82-2-2220-1249

Received: 31 May 2019; Accepted: 27 June 2019; Published: 4 September 2019

Abstract: This study is interested in understanding the particulate matter perceptions and response behaviors of residents. The purpose of this study was to identify indoor air quality along with the response behaviors of residents in Seoul, to ascertain whether there is a difference in behaviors when particulate matter is present, according to the characteristics of residents and to grasp the nature of this difference. A questionnaire survey of 171 respondents was conducted. The questionnaire measured the indoor air quality perceived by residents, the health symptoms caused by particulate matter, residents' response behaviors to particulate matter and the psychological attributes affecting those response behaviors. Residents of Seoul were divided into college students in their twenties, male workers in their thirties and forties and female housewives in their thirties and forties. The data were calibrated by SPSS 23 using a one-way analysis of variance (ANOVA) and multiple regression analyses. The results show that most people found particulate matter to be an important problem but were unable to do sufficient mitigation action to prevent its presence. Residents showed greater psychological stress resulting in difficulty going out than physical symptoms. The most influential factor on response behaviors was psychological attributes. Participants were aware of the risks of particulate matter but believed it to be generated by external factors; thus, they felt powerless to do anything about it, which proved to be an obstacle to response behaviors.

Keywords: particulate matter; perception; response behavior; psychological attribute

---

1. Introduction

In recent years, particulate matter (PM) has emerged as a big problem in Korea. According to the Organization for Economic Cooperation and Development (OECD)'s annual report on the concentration of ultra-particulate matter in countries by 2017, the mean population was exposed to $PM_{2.5}$ and with pollution at 25.1 $\mu g/m^3$, Korea was the second worst of the member countries [1], with a level twice as high as the average OECD member countries (12.5 $\mu g/m^3$) and 2.5 times higher than the World Health Organization (WHO)'s annual average recommended concentration (10 $\mu g/m^3$). Based on Korea's PM forecast, the number of "bad" (36–75 $\mu g/m^3$) and "very bad" (more than 76 $\mu g/m^3$) days in metropolitan areas increased from 62 in 2015 to 77 in 2018 [2]. In early March 2019, Korea experienced the most severe PM situation. In Seoul, an 8-day ultra-particulate matter warning ($PM_{2.5}$ with a time-averaged concentration of more than 75 $\mu g/m^3$ for 2 h) and 2 days (March 5 and 6) with an alert level (an average $PM_{2.5}$ of more than 150 $\mu g/m^3$ for 2 h) [3]. As a result, the PM levels became hazardous to health.

PM is a WHO Level 1 carcinogen that has negative effects on health, contributing to cardiovascular and respiratory diseases [4,5]. Choe and Lee [6] investigated the effect of particulate emissions on specific diseases in Seoul and found that the number of hospitalizations for various respiratory diseases increased as the amount of ultra-PM increased. Korea's increase in PM is related to rapid economic growth. Large cities, such as Seoul, have high levels of energy use resulting from the concentration of population and economic activity and their direct emission of air pollutants is high. Further, since its geographical location is on the mid-latitude westerly wind area, seasonal influx of PM from neighboring China also affects the increase of PM in Korea [7].

According to the survey data on the perception of environmental problems among Koreans aged 13 and over, conducted by the Korea National Statistical Office (KNSO), 82.5% of respondents experience anxiety about PM [8]. Kim et al. [9] observed that Koreans regard PM as the most serious social risk factor. Because national concern about PM has been increasing, the government introduced a comprehensive plan for PM in 2017. In 2018, it attempted to reduce PM emissions by enforcing a Special Act on Particulate Matter in major cities across the country, including the capital region [10]. The Ministry of Environment, in consideration of atmospheric environmental standards and health effects, produced a PM forecasting system, which presents the levels of PM as well as countermeasures [11]. However, despite the various risk indicators for PM and notwithstanding the government measures, the residents of Seoul are notably passive in protecting individuals and society from PM despite viewing it as a threat [12]. To prevent and reduce the PM generated by anthropogenic rather than natural factors, public efforts must be accompanied by measures at the national level. Without ensuring that residents understand PM, it is predicted that reduction measures will be ineffective. Therefore, this study aims to identify levels of awareness of PM, recognition of indoor air quality, symptoms of PM exposure experienced by residents and coping behavior in relation to PM in Seoul. Specific research questions are as follows:

First, how do residents perceive the indoor air quality, how do they feel PM symptoms and how do they behave in response to PM?

Second, do the different characteristics of residents produce any variations in behavior in response to PM?

Third, what are the impediments to the proper responses to PM by residents?

In this study, we have sought to understand the perceptions of PM and the subsequent response behaviors of residents, as well as to identify the causes of these behaviors. It is critical to understand and solve the barriers to public engagement to avoid the worst consequences of PM. The results of this study are expected to be used as basic data for effective governmental measures to reduce PM.

2. Status and Risks of Particulate Matter in Korea

2.1. Characteristics of Particulate Matter Generation in Seoul

To identify the characteristics of PM generation in Seoul, we investigated the annual average concentration of PM along with its highest levels of concentration. As shown in Figure 1, the average concentrations of PM_{10} and $PM_{2.5}$ in Seoul were calculated using Seoul's atmospheric environment information from the past 10 years. From 2009 to 2018, the average annual concentration of PM_{10} in Seoul was 53.8 $\mu g/m^3$ and the average annual concentration of $PM_{2.5}$ was 27.2 $\mu g/m^3$ [3]. The average annual concentration of PM_{10} decreased from 76 $\mu g/m^3$ in 2002 to 41 $\mu g/m^3$ in 2012 but this decline has since slowed. Since the government enacted the Special Act on the Improvement of the Air Quality in the Seoul Metropolitan Area and established and performed its Basic Plan for the Management of Air Quality in the Seoul Metropolitan Area in 2003, air pollution as well as PM concentrations have been reduced [7]. However, no further improvement has occurred since 2012.

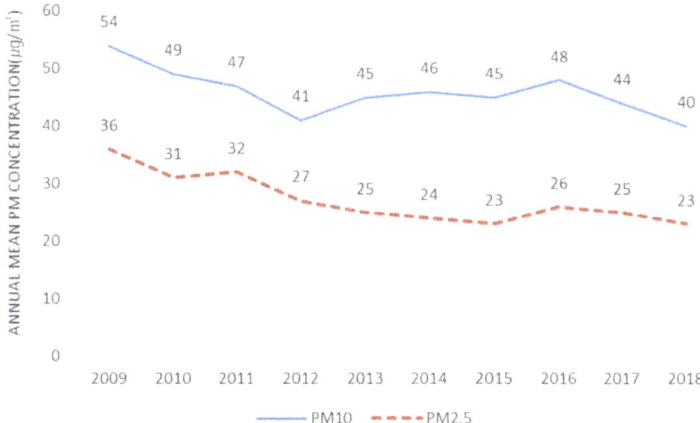

Figure 1. Annual mean particulate matter (PM) concentration in Seoul (2009–2018).

The results of a survey of "bad" and "very bad" days exceeding a $PM_{2.5}$ of 35 $\mu g/m^3$ for an average of 24 h showed that 44 days in 2015, 73 days in 2016, 64 days in 2017 and 61 days annually on average were recorded as "bad" and "very bad". This means that the number of days when the concentration of PM was significant is very large. This can be seen by comparing the number of PM warnings and days of alarm in Seoul. Figure 2 shows the number of PM and ultra-PM warning days using Seoul's atmospheric environment information. If there is a PM_{10} of 150 $\mu g/m^3$ or a $PM_{2.5}$ of 75 $\mu g/m^3$ for more than 2 h, a warning is issued. If there is a PM_{10} of 300 $\mu g/m^3$ or a $PM_{2.5}$ of 150 $\mu g/m^3$ for more than 2 h, an alarm is issued. In 2018, there were 17 PM warning days, 1 PM alarm day and 7 ultra-PM warning days. Seoul residents were exposed to extremely high dust concentrations on these days.

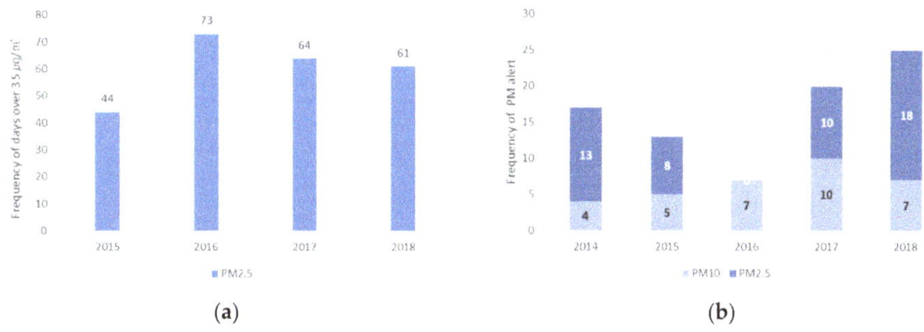

Figure 2. High PM concentration days: (**a**) days exceeding a $PM_{2.5}$ of 35 $\mu g/m^3$ on average for 24 h; (**b**) PM and ultra-PM warning and alarm days (Source: Seoul Atmospheric Environment Information).

This result is closely related to the monsoon season experienced in the geographical location of Korea. Korea's winter atmospheric circulation is affected by the northwestern winds associated with the winter monsoon in East Asia and the location of the barometer over the Korean peninsula. Inflow from China and Mongolia in winter greatly affects the concentration of PM in Korea [13]. Figure 3 shows the changes in the concentrations of PM and ultra-PM from March 2018 to February 2019 [3]. It shows that the levels of PM are high in winter and spring and low in summer. Specifically, they are lowest in September and highest in January. The atmospheric environmental standards for domestic PM are below 50 $\mu g/m^3$ on average for 24 h and below 25 $\mu g/m^3$ on average per year. Korea's levels exceed these standard values in all seasons except summer.

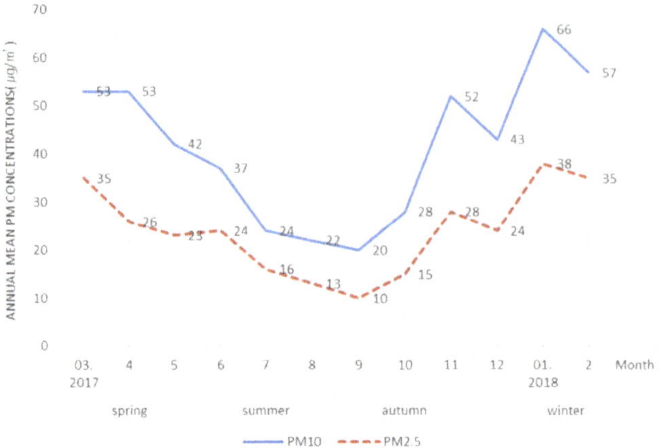

Figure 3. Seasonal dust concentrations in Seoul.

The air quality in Korea deteriorates considerably as the number of days with seasonally high concentrations of PM increase [14].

2.2. Domestic Particulate Matter Standards and Health Protection

The Ministry of Environment [2] proposed enhanced environmental standards for PM in 2018, taking into consideration national health effects, international standards, pollution status and achievability. Table 1 shows the WHO recommendations along with the standards in other regions. The concentration of PM in the domestic environment of higher than 50 µg/m³ far exceeds the WHO's recommended level of an annual average PM_{10} of 20 µg/m³ and is much greater than that of other regions. In addition, the rate of achieving the annual PM environmental standard was around 60% in 2015 and the rate of achieving the 24 h environmental standard for $PM_{2.5}$ and PM_{10} was very low at 4% and 10.7% respectively.

Table 1. Comparison of air quality standards for PM (Source: Ministry of Environment).

Category	Standard Time	PM Standards				Standard Achievement Rate (2015)
		Korea	WHO	USA	EU	
$PM_{2.5}$ (µg/m³)	yearly	15	10	12, 15	25	65.0%
	24 h	35	25	35	-	4.0%
PM_{10} (µg/m³)	yearly	50	20	-	40	65.6%
	24 h	100	50	150	50	10.7%

Considering public health, the Ministry of Environment has been conducting PM forecasting from February 2014 and ultra-PM forecasting with a warning system from January 2015 [2]. The PM forecasting system provides forecasts of the PM concentration four times a day for the present day, the following day and the day after that. The PM forecast is graded in four stages: "good", "normal", "bad" and "very bad" (Table 2). The PM warning system works to promptly inform the public when a high concentration of PM occurs and to reduce the damage. It is issued when the air quality is harmful to health. Air pollution alarms are classified into two stages—warning and alarm—as shown in Table 3. At the time of a PM alarm, there are eight countermeasures for people to take: remaining indoors; wearing a health mask; reducing external activities; washing the body after returning home; drinking water and eating fruits and vegetables; undertaking ventilation and indoor water cleaning; managing indoor air quality; and restricting air pollution inducing activities.

Table 2. Particulate matter forecast grade (2018).

PM Concentration (μg/m^3, Average for 24 h)	Good	Normal	Bad	Very Bad
PM$_{10}$	0–30	31–80	81–150	over 151
PM$_{2.5}$	0–15	16–35	36–75	over 76

Table 3. Particulate matter warning and alarm issuing grade (2018).

Category	Warning Issuing	Alarm Issuing
PM$_{10}$	PM$_{10}$ hourly average concentration over 150 μg/m^3 for 2 h	PM$_{10}$ hourly average concentration over 300 μg/m^3 for 2 h
PM$_{2.5}$	PM$_{2.5}$ hourly average concentration over 75 μg/m^3 for 2 h	PM$_{2.5}$ hourly average concentration over 150 μg/m^3 for 2 h

3. Materials and Methods

3.1. Participants and Questionnaire Design

This study surveyed the residents of Seoul. To identify differences in the response to PM associated with gender and age, respondents were divided into three groups. A total of 171 respondents were used for the analysis. The groups were 20-year-old college students (N = 70, gender = 32 male and 38 female, mean age = 21.88, SD = 2.33), 30- to 40-year-old male workers (N = 51, mean age = 41.11, SD = 5.45) and 30- to 40-year-old housewives (N = 50, mean age = 37.06, SD = 4.33). The questionnaire comprised five main parts. The first part included questions about the quality of indoor air perceived by residents in the home. The second part addressed the health of the residents in terms of objective symptoms, subjective symptoms and health behaviors concerning the symptoms. The third section elicited the residents' responses to PM, dividing them into mitigating behavior, adaptive behavior and behavior intentions. The fourth part sought to ascertain the psychological causes of interference with the response behaviors of the residents to PM and the final content consisted of questions investigating the residents' overall knowledge of PM.

3.2. Measures

3.2.1. Measuring Response Behaviors to Particulate Matter

Mitigating behavior and adaptive behavior are countermeasures to deal with risk. Swart and Raes [15] define "mitigation" as an anthropogenic intervention to reduce the sources of air pollution, whereas "adaptation" is an adjustment in natural or human systems in response to climatic stimuli. Mitigation is a way to reduce the cause of hazards associated with PM. Usually, the benefits of mitigating behavior are not seen in the short term, so it is considered a long-term countermeasure [16]. Mitigating behavior is a personal effort to reduce the generation of PM, which includes "using public transportation to reduce atmospheric gas generation", "not using electricity and heating to restrain unnecessary energy use", "using kitchen utensils that generate less harmful gas" and other similar measures. Adaptation refers to controlling the damage caused by PM.

Adaptive behavior can reduce the risks associated with PM by prophylactic and post-exposure measures that minimize the negative effects of PM [17]. In contrast to mitigating behavior, from which long-term effects arise, the effects of adaptive behavior are immediate and are characterized by the matching of subject and beneficiary [18]. Adaptive behavior is action to prevent the damage caused to individuals by high concentrations of dust. Behavioral intentions signify an individual's specific willingness to act in response to PM and include two positive intentions. To achieve sustainable development policies, it is necessary to integrate mitigating actions that directly reduce the concentration of PM and adaptive prevention actions that reduce the impact of the existing dust risk. In this study, mitigating behavior, adaptive behavior and the behavioral intentions of Seoul residents are explored, as shown in Table 4.

Table 4. Responsive behaviors to particulate matter.

Question: What do you do when Particulate Matter Occurs?	
Mitigating behavior	MB1 Using public transportation to reduce atmospheric gas generation MB2 Refraining from using electricity and heating for the reduction of unnecessary energy use MB3 Using kitchen appliances that generate less harmful gas
Adaptive behavior	AB1 Checking the levels of particulate matter concentration every day AB2 Wearing a particulate matter mask on high particulate matter density days AB3 Refraining from going out when the concentration of particulate matter is high AB4 Turning on an air purifiers to reduce particulate matter concentration AB5 Washing the whole body thoroughly in running water after returning home AB6 Wiping dust from the floor with a damp cloth and undertaking indoor water cleaning
Behavioral intentions	BI1 Being ready to suffer immediate damage or inconvenience to reduce particulate matter concentrations. BI2 Being willing to participate in actions to reduce particulate matter BI3 Not feeling any urgency to change my behavior to prevent particulate matter pollution BI4 Thinking that changes in my behavior do not affect particulate matter reduction BI5 Not knowing what concrete action may be taken to reduce particulate matter

3.2.2. Measuring Psychological Attributes that Hinder Reactions to Particulate Matter

Gilfford [19] observed that most people find environmental sustainability to be an important issue but that psychological barriers prevent them from engaging in sufficient actions to address it. These psychological barriers impede the behavioral choices that would facilitate mitigation, adaptation and environmental sustainability [20]. Risk perception is a factor that affects human attitudes and behavioral intentions and that is very important in making individual decisions about a behavior [21]. Brewer et al. [22] explained three dimensions of risk perception. The first is the likelihood that one will be harmed by the hazard; the second is susceptibility, referring to an individual's constitutional vulnerability to a hazard; and the third is severity, indicating the extent of harm a hazard may cause. Psychological distance describes how individuals participate in future events [23]. The perceived distance of events indicates how they are mentally construed. As the perceived distance increases, events are interpreted as more abstract, decontextualized and conceived in generalized terms. When events become closer, they use more specific, contextualized and detailed features [24]. The psychological distance for PM was evaluated using the four distance domains of geography, temporality, socialization and awareness. In addition, this research investigated various barriers increasing the perceived concern of the public, such as information distrust, externalizing responsibility and uncertainty about the causes of PM pollution. In this study, to understand the effects of response behaviors to PM, the psychological attributes of the residents were classified into nine concepts within three domains: "risk perception", "psychological distance", and "perceived concerns" (Table 5).

Table 5. Psychological attributes.

Question: What Are your Thoughts in Response to Particulate Matter?		
Risk perception	Likelihood	1. Contact with particulate matter is a health hazard 2. If I continue to be exposed to particulate matter, I will be damaged in a few years
	Susceptibility	3. Particulate matter is an important problem for me 4. I am more affected by the risk of particulate matter than other risks
	Severity	5. The risks of particulate matter for health are very serious 6. Even short-term contact with particulate matter can increase the likelihood of cancer and early death

Table 5. *Cont.*

		Question: What Are your Thoughts in Response to Particulate Matter?
Psychological distance	Geographic and Temporal	7. The particulate matter condition is more serious in Korea than in other countries 8. The risk of particulate matter is present very often or constantly
	Social	9. Particulate matter will have a significant impact on me and my family 10. Particulate matter is sure to be a serious social problem
	Awareness	11. I am very concerned about particulate matter 12. I am interested in issues related to particulate matter
Perceived concerns	Information distrust	13. I cannot trust the information about particulate matter 14. The risk of particulate matter is greatly exaggerated
	Externalizing responsibility	15. The problem of particulate matter is beyond my ability to solve 16. The solution to the excess of particulate matter should be provided by the government rather than individuals
	Uncertainty	17. The damage caused by particulate matter is unclear 18. It is difficult to measure health damage arising from particulate matter 19. Particulate matter does not affect me right now

3.3. Methodology

This research used a questionnaire to collect its data; the survey was conducted online in March 2019 and the data were analyzed using IBM's SPSS Statistics Program 23. The question concerning residents' perceptions of indoor air quality comprised positive and negative language (the semantic differential method). We distinguished between positive vocabulary and negative vocabulary over five levels, asking residents to assign one level to each question according to their perceptions. The vocabulary used to measure the indoor air quality of residents included "bad", "good", "stuffy", "refreshed", "unpleasant", and "comfortable". A five-point Likert scale was used to assess the symptoms, behaviors and psychological responses of residents, ranging from "very unlikely" (1 point), to "unlikely" (2 points), to "average" (3 points), to likely (4 points), to "very likely" (5 points). Questions measuring residents' knowledge of PM could be answered as "right" (○) or "wrong" (X) and the "right" answers were added to produce a score.

An ANOVA was used for the statistical analysis to determine whether there were differences in behavior as a response to PM, according to the characteristics of residents. Participants were divided into three groups based on their genders and ages: college students in their twenties, male workers in their thirties and forties and female housewives in their thirties and forties. In addition to the response behavior, we compared differences between the groups in terms of environmental exposures to PM, perceived symptoms and psychological attributes. Further, post-hoc tests were conducted to analyze variations between groups. Multiple regression analysis was conducted to identify the factors influencing response behavior.

4. Results

4.1. Perceived Indoor Air Quality and Overall Satisfaction

The perceived indoor air quality and the satisfaction of residents in Seoul were measured using a semantic differential approach. Residents assessed their perceptions of indoor air quality by comparing two opposing pairs of vocabulary in five steps. For example, the evaluation score for the first question on indoor air quality was divided into five levels, spanning two opposing experiences: "a lot of particulate matter" and "no particulate matter". The more PM respondents perceived, the closer to 1 point they scored and the less PM they observed, the closer they scored to 5 points.

A total of 171 residents were analyzed and the results are shown in Table 6. Respondents' answers to the four questions about the quality of indoor air produced an average score of approximately three (2.93–3.25) points. This score comprises the median value of 1 and 5, indicating that the residents of

Seoul understand the indoor air quality to be normal. The evaluations of overall indoor air quality also averaged 3.21 points. However, the average score for the question, "Do you think seriously about or are you interested in the quality of indoor air where you live?" was 4.17 points (where 1 point indicated "not at all" and 5 points denoted "very much"), implying that respondents were interested in the air quality of their living spaces.

Table 6. Perceptions of and satisfaction with air quality.

Perceptions of			Mean	Std. Deviation
	1 Point	**5 Points**		
Indoor air quality	very bad	very good	3.26	0.91
	a lot of particulate matter	little particulate matter	2.94	0.98
	stuffy	refreshed	2.93	1.02
	unpleasant	comfortable	3.25	0.92
Satisfaction	dissatisfaction	satisfaction	3.21	1.10
	no thought or interest	much thought and interest	4.17	1.10

Two phrases corresponded to 1 point and 5 points respectively and between these two phrases were expressions valued at 2 points (signifying the perception of slightly more PM), 4 points (signifying the perception of slightly less PM) and 3 points (signifying the perception of an average amount of PM).

4.2. Particulate Matter Environmental Exposures and Perceived Symptoms

When measuring the outdoor activity times of respondents, it was found that they stayed outdoors for an average of 4.54 h (SD = 4.31) per day and the analysis of variance showed a statistically significant difference ($p < 0.01$) depending on the subject group. The results are shown in Table 7. The group of 20 college students and the group of 30- to 40-year-old male workers remained outside for an average of 5 h a day—specifically 5.12 h and 5.07 h respectively—while the group of 30- to 40-year-old housewives stayed outdoors for 3 h a day on average.

Table 7. Average daily external activity time.

Group 1		Group 2		Group 3		F (p-Value)
Mean	**Std Deviation**	**Mean**	**Std Deviation**	**Mean**	**Std Deviation**	
5.12 h	3.43	5.07 h	5.49	3.20 h	3.81	3.54 (0.031)

Group 1: 20-year-old college students; Group 2: 30- to 40-year-old male workers; Group 3: 30- to 40-year-old housewives.

The physical symptoms and health problems experienced by residents as a result of PM were verified through objectively judged symptoms, such as cough, allergy and headache, as well as subjectively judged psychological symptoms, such as anxiety and stress. In addition, the extent of any actual treatment at the hospital because of these symptoms was investigated. The respondents' answers ranged from "not at all" (1 point) to "very strongly" (5 points) and the results are shown in Table 8.

The average score for the response, "Thinking that particulate matter has a negative effect on my health" was 4.39, the highest and the average score for the response "Thinking that particulate matter causes problems in my life" was 4.00, the second highest. Symptoms such as respiratory problems and complications with nose, skin and eyes rated 3.69 and 3.66 respectively, revealing that the average scores for the objective symptoms of residents with health problems arising from PM were normal. These findings indicate that the psychological symptoms (Mean = 4.39) were more significant than the objective symptoms (Mean = 3.75); further, the incidence of direct treatment in hospitals (M = 3.15) for such symptoms was not significant (Figure 4). However, ANOVA analysis showed statistically significant differences according to subject groups. A Tukey post-hoc test was conducted to analyze the differences between the groups and the results showed that the perceived symptoms were greater among the group of housewives in their thirties and forties than in the group of college students in their twenties or in the group of male workers in their thirties and forties. Notably, the differences in psychological symptoms were pronounced among these groups (Table 8).

Table 8. Health-related symptoms experienced by residents as a consequence of particulate matter.

	Perceived	Mean				F (*p*-Value)
		Group 1	Group 2	Group 3	Total	
Objective symptom	Suffering from cough, hoarseness, bronchitis, respiratory distress and difficulty breathing	3.54	3.50	4.08	3.69	3.33 (0.038 *)
	Experiencing clogged nose, rhinitis, sneezing, allergy and atopic symptoms, such as itchy or swollen skin or dry eyes	3.51	3.43	4.10	3.66	3.76 (0.025 *)
	Suffering from headache, arthritis and other pain	2.61	2.92	3.38	2.92	4.38 (0.014 *)
	Experiencing the deterioration of an already present illness	2.42	2.88	3.38	2.84	7.57 (0.001 **)
Subjective symptom	Suffering increased stress and fatigue	3.34	3.50	4.14	3.62	7.23 (0.001 **)
	Feeling worried and anxious	3.18	3.60	4.30	3.63	12.84 (0.000 ***)
	Thinking that particulate matter has a negative effect on my health	4.21	4.27	4.78	**4.39**	8.55 (0.000 ***)
	Thinking that particulate matter causes problems in my life	3.71	3.68	4.72	**4.00**	17.55 (0.000 ***)
	Thinking that particulate matter reduces my performance and concentration	3.20	3.31	4.02	3.47	6.80 (0.001 **)
Remedy	I have recently been to the hospital for any of the above symptoms	1.94	2.15	3.00	2.31	8.96 (0.000 ***)
	I have recently been taking medication for some of the above symptoms	2.12	2.31	3.04	2.45	6.19 (0.003 **)
	I have recently taken a break at home because of some of the above symptoms	2.51	2.39	3.42	2.74	7.80 (0.001 **)

* $p < 0.05$, ** $p < 0.01$, *** $p < 0.001$. Group 1: 20-year-old college students; Group 2: 30- to 40-year-old male workers; Group 3: 30- to 40-year-old housewives.

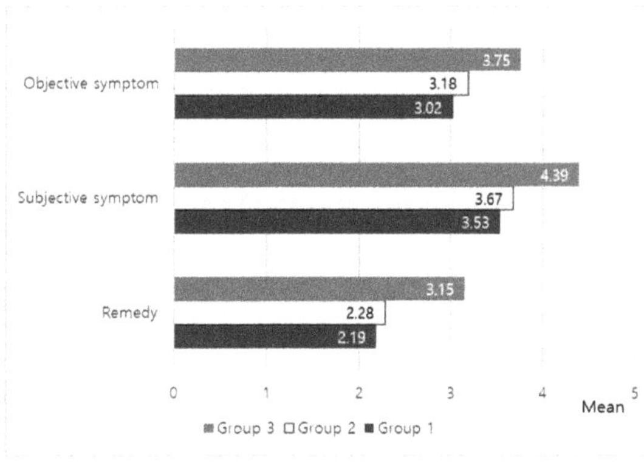

Figure 4. Physical symptoms of particulate matter experienced by residents.

4.3. Behavioral Changes in Response to Particulate Matter

Residents' response behaviors to the occurrence of PM were analyzed and the results indicate that adaptive behaviors (M = 3.66, SD = 0.95) controlling the damage caused by PM were more prevalent than mitigating behaviors (M = 2.98, SD = 1.03) to reduce the fundamental causes of PM, as shown in Table 9. The most common PM adaptive response behaviors were "AB1: Checking the PM concentration every day" (4.00 points) and "AB5: Washing your whole body thoroughly in running water after returning home" (3.78 points). Regarding behavioral intentions, the most common PM response behaviors were "BI2: Being willing to participate in actions to reduce particulate matter" (4.00 points) and "BI1: Being ready to experience immediate damage or inconvenience to reduce the particulate matter concentration" (3.69 points). For other responses, scores ranging from 2.50 to 3.50 were obtained, indicating that residents' responses to PM were not significant. However, there were statistically significant differences among the groups (Table 9). In particular, the group of 30- to 40-year-old housewives showed higher scores for adaptive behavior and behavioral intentions than the other groups (Figure 5).

Table 9. Residents' response behaviors to particulate matter.

Behavior Type		Mean				F (p-Value)
		Group 1	Group 2	Group 3	Total	
Mitigating behavior	MB1	4.04	2.82	2.60	3.25	19.63 (0.000 ***)
	MB2	2.80	3.00	3.00	2.91	0.49 (0.613)
	MB3	2.55	2.94	3.12	2.83	2.91 (0.057)
Adaptive behavior	AB1	3.55	3.80	4.84	**4.00**	19.63 (0.000 ***)
	AB2	2.77	3.35	4.28	3.38	20.36 (0.000 ***)
	AB3	2.82	3.41	4.32	3.43	19.53 (0.000 ***)
	AB4	2.77	3.92	4.32	3.56	19.03 (0.000 ***)
	AB5	3.48	3.76	4.24	3.78	5.55 (0.005 **)
	AB6	2.91	3.35	3.98	3.35	13.37 (0.000 ***)
Behavioral intentions	BI1	3.48	3.45	4.24	3.69	12.11 (0.000 ***)
	BI2	3.91	3.86	4.28	**4.00**	3.72 (0.026 *)
	BI3 [1]	−2.72	−2.82	−1.82	−2.49	12.44 (0.000 ***)
	BI4 [1]	−2.54	−2.82	−2.12	−2.50	4.88 (0.009 **)
	BI5 [1]	−3.05	−2.94	−2.86	−2.96	0.42 (0.657)

* $p < 0.05$, ** $p < 0.01$, *** $p < 0.001$, [1] Reverse scored for negative perceptions. Group 1: 20-year-old college students; Group 2: 30- to 40-year-old male workers; Group 3: 30- to 40-year-old housewives.

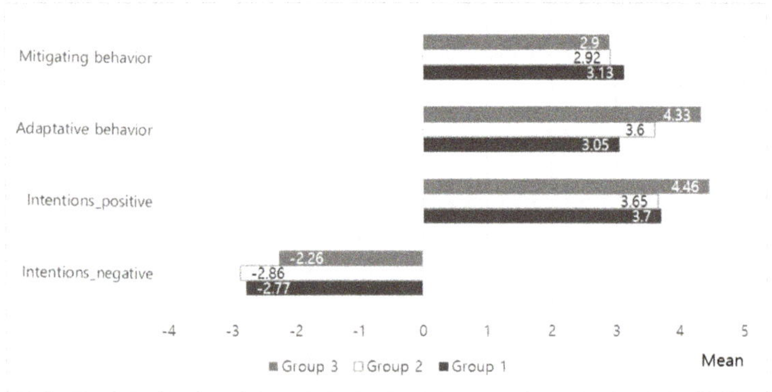

Figure 5. Residents' response behaviors to particulate matter.

4.4. Barriers to Mild Behavioral Changes as a Response to Particulate Matter

We investigated the residents' understandings of PM and their psychological causes as obstacles to behavioral changes to reduce and prevent PM.

4.4.1. Lack of Knowledge

The items indicating the residents' knowledge of PM consisted of 10 authentic scales (true/false) based on data from experts and the Ministry of Environment. The respondents selected "correct" or "incorrect" in answer to each item and answers of "correct" were treated as single points, producing a total score between 0 and 10 points. Table 10 shows the results for the 171 respondents. The average knowledge of PM was 6.56 (SD = 1.44) and the results of the analysis are shown in Figure 6. Knowledge of PM did not vary among the groups.

Table 10. Residents' knowledge of particulate matter.

Knowledge	Number of Correct Answers (%)
1. The influx of particulate matter is mainly caused by artificial actors, such as boilers, automobiles and power generation facilities	138 (80.7)
2. More than 80% of particulate matter is from abroad, as a result of the yellow dust and smog from China	39 (22.8)
3. Preliminary reduction measures against emissions of particulate matter are implemented nationwide and are targeted based on particulate matter concentration	43 (25.1)
4. Ultra-particulate matter comprises very thin, small particles but causes deterioration of visibility in places where the flow of air is stagnant, creating obstacles to traffic and navigation	118 (69.0)
5. When there is a particulate matter alarm, schools can prohibit outdoor classes, adjust the times of travel to and from school and temporarily shut down	151 (88.3)
6. Domestic particulate matter concentrations are similar to those in major cities of other OECD countries, such as New York and London	141 (82.5)
7. Even when exposure occurs for only a short time, particulate matter can penetrate directly into the alveoli, resulting in asthma, lung disease and even death	119 (69.6)
8. The mask is a quasi-drug certified by the Food and Drug Administration that can only prevent exposure to particulate matter if the product is marked "KF94" or "KF80"	133 (77.8)
9. Generally, high concentrations of particulate matter occur in spring and summer	89 (52.0)
10. It is helpful to eat fruits and vegetables that are rich in water and vitamin C to combat the effects of high dust concentrations	152 (88.9)

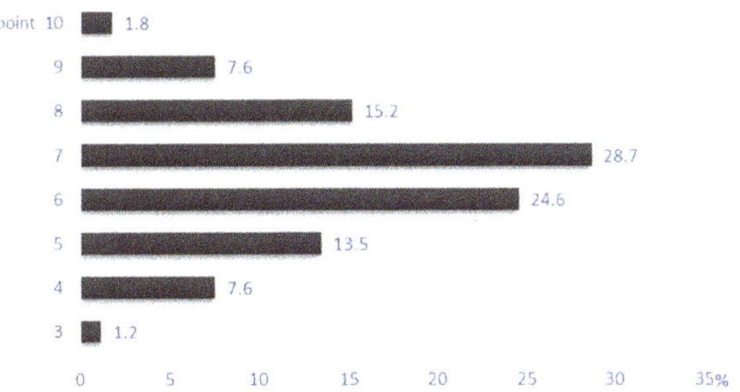

Figure 6. Residents' total knowledge of particulate matter.

4.4.2. Psychological Attributes of Particulate Matter

The results of the survey of the psychological influences on behavioral responses to PM show that risk perception (Mean = 4.01, SD = 0.76) and psychological distance (Mean = 4.32, SD = 0.73) achieved a higher than average score of 4.0 points (where 1 point = "not at all" and 5 points = "very much"). Table 11 shows the results of the analysis. Question 2 related to likelihood when considering residents' risk perception—"If you continue to be exposed to particulate matter, you will be affected in a few years" (Item 4.5)—as did Question 1, "Contact with particulate matter is harmful to health" (4.35 points)'; both of these questions received the highest scores. The score for the residents' perception of the psychological distance with which they regarded PM was higher than that for their risk perception. All questions concerning psychological distance were rated at greater than 4.0 and among them, the scores for Question 10, relating to the social implications of PM—"particulate matter is sure to be a serious social problem" (4.57 points)—and for Question 8, pertaining to the geographic and temporal significance of PM—"the risk of particulate matter is present very often or consistently" (4.47 points)—were the highest. Questions about perceived concerns, such as information distrust, the externalizing of responsibility and uncertainty, were reverse scored. Among them, Question 16, relating to the externalizing of responsibility—"The solution to the excess of particulate matter should be provided by the government rather than individuals" (−4.35 points)—and Question 15—"The problem of particulate matter is beyond my ability to solve" (−4.02 points)—received the highest scores. There was a statistically significant difference between the groups (Table 11). The group of housewives aged in their thirties and forties showed higher scores for risk perception and psychological distance than the other groups (Figure 7).

Table 11. Psychological influences on residents' response behaviors.

Psychological Attributes			Mean				F (p-Value)
			Group 1	Group 2	Group 3	Total	
Risk perception	Likelihood	1	4.08	4.33	4.74	4.35	8.58 (0.000 ***)
		2	4.44	4.29	4.82	**4.50**	6.22 (0.002 **)
	Susceptibility	3	4.10	4.19	4.62	4.28	5.11 (0.007 **)
		4	2.84	3.07	3.68	3.15	6.70 (0.002 **)
	Severity	5	4.10	4.15	4.58	4.27	4.69 (0.010 *)
		6	3.34	3.41	3.86	3.51	3.10 (0.048 *)
Psychological distance	Geographic and temporal	7	4.18	4.19	4.60	4.30	2.99 (0.053)
		8	4.41	4.25	4.80	4.47	8.13 (0.000 ***)
	Social	9	3.92	4.19	4.70	4.23	11.01 (0.000 ***)
		10	4.44	4.43	4.92	**4.57**	10.55 (0.000 ***)
	Awareness	11	4.02	4.13	4.76	4.27	9.96 (0.000 ***)
		12	3.70	4.00	4.60	4.05	11.86 (0.000 ***)
Perceived concerns [1]	Information distrust	13	−2.97	−3.62	−3.80	−3.40	10.84 (0.000 ***)
		14	−2.40	−2.43	−1.52	−2.15	13.65 (0.000 ***)
	Externalizing responsibility	15	−4.04	−3.96	−4.08	**−4.02**	0.30 (0.741)
		16	−3.98	−4.52	−4.70	**−4.35**	11.91 (0.000 ***)
	Uncertainty	17	−2.68	−2.88	−2.78	−2.77	0.36 (0.692)
		18	−2.97	−3.01	−2.88	−2.95	0.15 (0.857)
		19	−2.40	−2.50	−1.88	−2.28	4.08 (0.019 *)

* $p < 0.05$, ** $p < 0.01$, *** $p < 0.001$, [1] Reverse scored for negative perceptions. Group 1: 20-year-old college students; Group 2: 30- to 40-year-old male workers; Group 3: 30- to 40-year-old housewives.

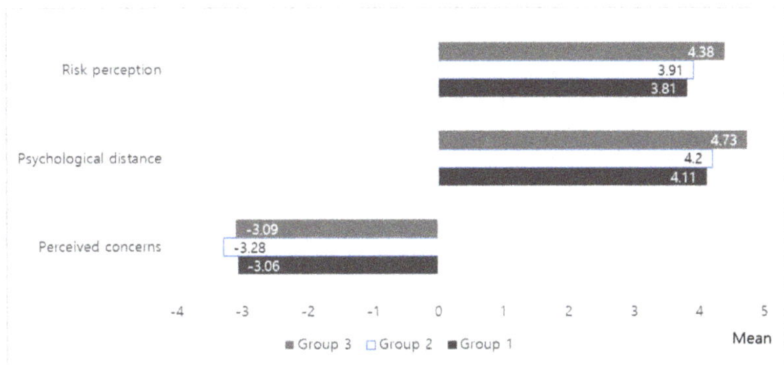

Figure 7. Residents' psychological influences.

4.5. Factors Affecting Response Behaviors to Paticulate Matter

A multiple regression analysis was conducted to determine the factors with the greatest effect on residents' response behaviors. Knowledge of PM, psychological attributes and PM-related symptoms were selected as factors influencing response behaviors and the analytical result was statistically significant ($p < 0.05$; F = 25.306, p-value = 0.000). Table 12 shows the evaluation of the contribution and statistical significance of the individual independent variables. Psychological attributes were the most influential factor on residents' responses to PM, followed by PM-related symptoms.

Table 12. Multiple regression analysis of the response behaviors to particulate matter.

Variable	B	Std Error	Beta	*t* Value	*p*-Value
Constant	8.103	3.577		2.265	0.025
Knowledge of PM	0.630	0.443	0.095	1.424	0.156
Psychological attributes	0.373	0.082	0.375	4.555	0.000
PM-related symptoms	0.191	0.070	0.219	2.737	0.007

Note: R = 0.55, R^2 = 0.31, Adj. R^2 = 0.30.

5. Discussion and Conclusions

In Seoul, it is inadvisable to go out on any day, except in summer, as a result of the PM concentration and its effect on human health. When the WHO standard is applied, the concentration of PM in Seoul is revealed to be remarkably high and cases of excessive concentration exceeding the daily average atmospheric environment standard are also frequent. The government has been pursuing various policies to reduce PM concentration but the concentration of PM has not improved to the extent that people recognize any change. Research has shown that most people believe PM to be an important problem but do not undertake enough mitigating behaviors to prevent its occurrence. Reliable research should be preceded by an analysis of cause and effect before the preparation of PM countermeasures. We sought to understand residents' perceptions of PM, their response behaviors and the psychological causes of these behaviors; the main findings of this study are as follows.

First, to identify the perceived indoor air quality of the residents, the questionnaire was distributed at a time when the concentration of PM had been significant throughout the year; thus, residents' interest in indoor air quality was great. However, despite high outdoor concentrations of PM, the indoor air quality was not considered poor or dissatisfactory. The symptoms that residents attributed to PM were not physical, such as allergies or bronchitis. They more commonly experienced psychological stress or anxiety, feeling that PM had a negative effect on health and was interfering more frequently with daily life, in that residents were going out less often to avoid it.

Second, the problem with the PM-related behaviors of residents in Seoul was that many people did not take actual action despite their willingness to do so. Mitigating behavior to try to minimize PM generation was less common than adaptive behavior to protect individuals from the risks of PM. This may have been because of the nature of mitigating behavior, which often entails loss or inconvenience. Even if the risks associated with PM were largely recognized, it would be necessary to compensate for the loss when the habitual behavior of individuals was abandoned in favor of following the recommended behavior. Unlike preventive actions to protect individuals from PM, compensation for mitigating behavior is only likely to be achieved in the distant future and even then, such compensation would provide a social rather than an individual benefit. This suggests that more persuasive strategies should be employed to stimulate changes in individual behaviors for collective interests.

Third, the psychological characteristics of individuals can either hinder or promote action for environmental sustainability. The greater the perceived risk and the more psychological that risk, the more likely it is that an active response will occur. Analysis of the psychological factors affecting residents in Seoul revealed that they perceived the risk of PM to be very close (in terms of psychological distance). However, despite revealing significant levels of risk perception and psychological distance, the results of this study did not show active response behaviors. Attention must be paid to the externalizing of responsibility among the psychological causes of this lack of action. This study found distrust of and uncertainty about PM information to be largely unrecognized as causes of apathy; on the contrary, understanding PM saturation to be externally produced was the main reason for inaction. If external factors are emphasized as the cause of the problem, it is possible that individuals perceive the risk as uncontrollable, which may nullify their will to respond. The results of this study also suggest that residents' psychological barriers have a negative effect on their response behaviors for the reduction of PM.

Fourth, the results of the correlation analysis show that the most influential factor in PM-related behavior was psychological attributes but that physical symptoms also affect response behaviors to PM. This result is meaningful in that it indicates desirable directions for PM policy. Based on the analysis of psychological factors, it was found that, because PM is also produced by individuals, active prevention is likely to occur if the importance of personal responses is conveyed. While knowledge of PM was not found to significantly affect response behaviors, residents in Seoul showed an average score of 6.56 out of 10 for PM knowledge. Lack of knowledge and understanding of the PM problem hinders appropriate response behavior. It is necessary to provide accurate and correct PM education and continuously available reliable information to enhance the knowledge of PM. In addition,

There was significant variation in residents' perceptions, response behaviors and psychological attributes concerning PM based on gender and age. Conversely, the time during which residents undertook external activities exposed to PM did not significantly affect PM-related symptoms or behavior. In contrast to college students in their twenties and male workers in their thirties and forties who remained outdoors for 5 h on average, women in their thirties and forties who stayed outdoors for only 3 h on average responded more sensitively to PM-related symptoms and their commitment to PM reduction and mitigating actions was much greater. Similarly, the results of the psychological factor analysis for PM showed that the group of women in their thirties and forties were more aware of the risks than the other groups and felt those risks to be closer psychologically. However, it is the younger generations who will become the subjects of society in future and will need to face these problems and solve them. There is an urgent need for plans to increase awareness of the risks and the will of the public to act.

Government efforts to reduce PM have focused primarily on expensive communication campaigns. However, the results of this study suggest that the mere encouragement of attitudinal change is not effective; public engagement in terms of accurate PM knowledge, health education and the study of various personal characteristics and psychological causes is required.

This study identified residents' thoughts, response behaviors and psychological factors relating to PM and showed that personal characteristics, cognition and emotions affect behavioral intentions. Unlike previous studies, the significance of this research is that it seeks to understand residents' behaviors in a multidimensional way. Psychological factors are expected to be used effectively to motivate individuals to protect themselves from the risks of PM exposure. The residents of Seoul involved in this study were well aware of the risks of PM but felt that its generation and any possible solutions were external and therefore beyond their control. This situation is likely to prevent public action to protect individuals and the society. It is recommended that we shift our focus from risk and attention to participation and action. It is expected that a proper direction for efficient PM reduction can be established using these results.

Author Contributions: Conceptualization, M.E.C. and M.J.K.; methodology, M.E.C. and M.J.K.; formal analysis, M.E.C.; investigation, M.E.C.; data curation, M.J.K.; writing—original draft preparation, M.E.C.; writing—review and editing, M.J.K.; supervision, M.J.K.; funding acquisition, M.J.K.

Funding: This research was supported by a grant (19AUDP-B127891-03) from the Architecture & Urban Development Research Program funded by the Ministry of Land, Infrastructure and Transport of the Korean government.

Conflicts of Interest: The authors declare no conflict of interest.

References

1. OECD. OECD Statistics. 2019. Available online: https://stats.oecd.org/index.aspx?queryid=72722 (accessed on 24 March 2019).
2. Ministry of Environment. Particulate Matter Status. 2019. Available online: http://www.me.go.kr/cleanair/sub02.do#2 (accessed on 12 March 2019).
3. Seoul Metropolitan Government. Seoul Atmospheric Environment Information. 2019. Available online: http://cleanair.seoul.go.kr/air_pollution.htm?method=average (accessed on 29 April 2019).
4. Sacks, J.D.; Stanek, L.W.; Luben, T.J.; Johns, D.O.; Buckley, B.J.; Brown, J.S.; Ross, M. Particulate matter—Induced health effects: Who is susceptible? *Environ. Health Perspect.* **2011**, *119*, 446–454. [CrossRef] [PubMed]
5. Pope, C.A., III; Burnett, R.T.; Thun, M.J.; Calle, E.E.; Krewski, D.; Ito, K.; Thurston, G.D. Lung cancer, cardiopulmonary mortality, and long-term exposure to fine particulate air pollution. *JAMA* **2002**, *287*, 1132–1141. [CrossRef] [PubMed]
6. Choe, J.I.; Lee, Y.S. A study on the impact of PM2.5 emissions on respiratory diseases. *Korea Environ. Policy Adm. Soc.* **2015**, *23*, 155–172. [CrossRef]
7. Lee, H.J.; Jeong, Y.M.; Kim, S.T.; Lee, W.S. Atmospheric circulation patterns associated with particulate matter over south Korea and their future projection. *J. Clim. Chang. Res.* **2018**, *9*, 423–433. [CrossRef]
8. Korean Statistical Information Service. Recognition of Environmental Problems (PM Inflow). 2019. Available online: http://kosis.kr/statisticsList/statisticsListIndex.do? (accessed on 1 April 2019).
9. Kim, Y.; Lee, H.; Kim, H.; Moon, H. Exploring message strategies for encouraging coping behaviors against particulate matter. *Korea J. Commun. Inf.* **2018**, *92*, 7–44.
10. Hwang, I.C. Particulate matter management policy of Seoul: Achievements and limitations. *Korea Assoc. Policy Stud.* **2018**, *27*, 27–51.
11. Han, H.; Jung, C.H.; Kim, H.S.; Kim, Y.P. The revisit on the PM10 reduction policy in Korea. *J. Korea Environ. Policy Adm.* **2017**, *25*, 49–79.
12. Yang, W.H. Changes in air pollutant concentrations due to climate change and the health effect of exposure to particulate matter. *Health Welf. Policy Forum* **2019**, *269*, 20–31.
13. Park, S.; Shin, H. Analysis of the factors influencing PM2.5 in Korea. *Korea Environ. Policy Adm. Soc.* **2017**, *25*, 227–248.
14. Kim, Y.P. Air pollution in Seoul caused by aerosols. *Korea Soc. Atmos. Environ.* **2006**, *22*, 535–553.
15. Swart, R.O.B.; Raes, F. Making integration of adaptation and mitigation work: Mainstreaming into sustainable development policies? *Clim. Policy* **2007**, *7*, 288–303. [CrossRef]

16. Klein, R.J.T.; Schipper, E.L.F.; Dessai, S. Integrating mitigation and adaptation into climate and development policy: Three research questions. *Environ. Sci. Policy* **2005**, *8*, 579–588. [CrossRef]
17. Brouwer, R.; Schaafsma, M. Modelling risk adaptation and mitigation behaviour under different climate change scenarios. *Clim. Chang.* **2013**, *117*, 11–29. [CrossRef]
18. Wilbanks, T.J.; Sathaye, J. Integrating mitigation and adaptation as responses to climate change: A synthesis. *Mitig. Adapt. Strateg. Glob. Chang.* **2007**, *12*, 957–962. [CrossRef]
19. Gifford, R. The dragons of inaction: Psychological barriers that limit climate change mitigation and adaptation. *Am. Psychol.* **2011**, *66*, 290–302. [CrossRef] [PubMed]
20. Finell, E.; Tolvanen, A.; Pekkanen, J.; Minkkinen, J.; Ståhl, T.; Rimpelä, A. Psychosocial problems, indoor air-related symptoms, and perceived indoor air quality among students in schools without indoor air problems: A longitudinal study. *Int. J. Environ. Res. Public Health* **2018**, *15*, 1497. [CrossRef] [PubMed]
21. Ferrer, R.; Klein, W.M. Risk perceptions and health behavior. *Curr. Opin. Psychol.* **2015**, *5*, 85–89. [CrossRef] [PubMed]
22. Brewer, N.T.; Chapman, G.B.; Gibbons, F.X.; Gerrard, M.; McCaul, K.D.; Weinstein, N.D. Meta-analysis of the relationship between risk perception and health behavior: The example of vaccination. *Health Psychol.* **2007**, *26*, 136–145. [CrossRef] [PubMed]
23. Menon, G.; Chandran, S. When a day means more than a year: Effects of temporal framing on judgments of health risk. *J. Consum. Res.* **2004**, *31*, 375–389.
24. Jones, C.; Hine, D.W.; Marks, A.D.G. The future is now: Reducing psychological distance to increase public engagement with climate change. *Risk Anal.* **2017**, *37*, 331–341. [CrossRef]

Article

Personal Control of the Indoor Environment in Offices: Relations with Building Characteristics, Influence on Occupant Perception and Reported Symptoms Related to the Building—The Officair Project

Ioannis Sakellaris [1,*], Dikaia Saraga [1,2], Corinne Mandin [3], Yvonne de Kluizenaar [4],
Serena Fossati [5], Andrea Spinazzè [6], Andrea Cattaneo [6], Tamas Szigeti [7], Victor Mihucz [7],
Eduardo de Oliveira Fernandes [8], Krystallia Kalimeri [1], Paolo Carrer [9] and John Bartzis [1,*]

[1] Department of Mechanical Engineering, University of Western Macedonia, Sialvera & Bakola Str.,
    50100 Kozani, Greece
[2] Environmental Research Laboratory, INRASTES, National Center for Scientific Research "DEMOKRITOS",
    Aghia Paraskevi Attikis, P.O. Box 60228, 15310 Athens, Greece
[3] Université Paris Est, CSTB-Centre Scientifique et Technique du Bâtiment, 84 avenue Jean Jaurès,
    77447 Marne-la-Vallée CEDEX 2, France
[4] The Netherlands Organization for Applied Scientific Research (TNO), P.O. Box 96800,
    2509 JE The Hague, The Netherlands
[5] ISGlobal, Institute for Global Health, 08036 Barcelona, Spain
[6] Department of Science and High Technology, University of Insubria, Via Valleggio 11, 22100 Como, Italy
[7] Cooperative Research Centre for Environmental Sciences, Eötvös Loránd University, Pázmány Péter sétány
    1/A, H-1117 Budapest, Hungary
[8] Institute of Science and Innovation in Mechanical Engineering and Industrial Management,
    Rua Dr. Roberto Frias s/n, 4200-465 Porto, Portugal
[9] Department of Biomedical and Clinical Sciences-Hospital "L. Sacco", University of Milan, via G.B. Grassi 74,
    20157 Milano, Italy
*   Correspondence: isakellaris@uowm.gr (I.S.); bartzis@uowm.gr (J.B.); Tel.: +30-246-105-6710 (J.B.)

Received: 11 July 2019; Accepted: 5 August 2019; Published: 7 August 2019

Abstract: Personal control over various indoor environment parameters, especially in the last decades, appear to have a significant role on occupants' comfort, health and productivity. To reveal this complex relationship, 7441 occupants of 167 recently built or retrofitted office buildings in eight European countries participated in an online survey about personal/health/work data as well as physical/psycho-social information. The relationship between the types of control available over indoor environments and the perceived personal control of the occupants was examined, as well as the combined effect of the control parameters on the perceived comfort using multilevel statistical models. The results indicated that most of the occupants have no or low control on noise. Half of the occupants declared no or low control on ventilation and temperature conditions. Almost one-third of them remarked that they do not have satisfactory levels of control for lighting and shading from sun conditions. The presence of operable windows was shown to influence occupants' control perception over temperature, ventilation, light and noise. General building characteristics, such as floor number and floor area, office type, etc., helped occupants associate freedom positively with control perception. Combined controlling parameters seem to have a strong relation with overall comfort, as well as with perception regarding amount of privacy, office layout and decoration satisfaction. The results also indicated that occupants with more personal control may have less building-related symptoms. Noise control parameter had the highest impact on the occupants' overall comfort.

Keywords: IEQ; perceived comfort; sick building syndrome; health effects

---

1. Introduction

Office employees spend a significant part of their time in modern office buildings that are characterized by sealed facades and complex building systems (e.g., mechanical, electrical, plumbing, controls and fire protection systems) designed to reduce energy costs through controlled indoor environment conditions. Beside central control, a wide range of degrees of personal control, such as local thermostats, windows, personal lights etc., over the indoor environment can be found in modern office buildings. Personal control has a crucial role in achieving a healthy [1–3], comfortable [4–7] and productive [8–11] environment, reducing energy consumption in buildings without sacrificing the comfort of occupants [12,13]. The effect of personal control on occupant satisfaction, especially with regards to providing well-being and comfort is an important area of study [14,15].

Several studies have focused on estimating occupants' comfort in offices and the ability to control indoor environment parameters, producing a better Indoor Environment Quality (IEQ). In 1990, Paciuk [16] studied if indoor environment comfort was affected by personal control, leading to a model using thermal control parameters. Three different types of control parameters were involved: available control (the degree and type of control made available by the environment), exercised control (the relative frequency in which occupants engage in several types of controls to obtain comfort) and perceived control (how the different degrees of available and exercised control interact to produce different levels of perceived control). The model was applied on data from ten offices (511 workstations). The results pointed out that the occupants' level of control perception at their workstation enhances their satisfaction in their working environment.

The possibility to control light satisfaction was surveyed by Collins et al. [17] in 13 office buildings. The occupants provided with a task light recorded in general a higher satisfaction with light comfort than those without a task light. In addition, both groups expressed improved light satisfaction when they had the possibility to control light conditions. In another study [18], the role of personal control in natural versus mechanical ventilated office buildings was investigated. Personal control on operable windows, electronic lightning and solar blinds in natural ventilated buildings resulted in higher levels of perceived control. It was also mentioned that control systems should be simple and in compliance with the building design, as well as with quick response to alleviate discomfort as soon as it is experienced. The level of perceived control by occupants had a small influence on the indoor environment satisfaction, as described by Haghighat and Donnini [19] through a survey in 12 office buildings. Satisfaction with respect to temperature, air quality, ventilation, air circulation and overall comfort showed a moderate correlation with perceived control. On the contrary, there was a decline in prevalence of health symptoms with an increasing amount of control over the indoor environment.

The association of temperature personal control and operable windows with reported health symptoms and complaints was indicated in the early 1990s [20]. Many years later, Toftum [21] examined occupants' comfort perception and symptoms prevalent in mechanical versus natural ventilated buildings in 24 office buildings located in Denmark. Multiple logistic regression analysis indicated that the perceived control was more important for the prevalence of symptoms and environment perception than the type of ventilation.

Zagreus et al. [22] remarked that occupants with a sense of high degree of control over environmental parameters such as temperature, air movement, air quality and noise, were more satisfied with the indoor environment. Boerstra et al. [23] examined the impact of perceived control and access to control options in occupants' health and comfort, through 64 office buildings in Europe. The link between perceived comfort and control parameters such as temperature, ventilation, shading from the sun, light, noise was investigated. The analysis showed that occupants feel more comfortable when the perceived control over temperature, ventilation and noise is high. No significant correlation was found between comfort and the different types of access to control like operable windows, type of thermostats, etc., except for solar radiation.

In a recent study, Kwon et al. [24] tried to identify the relationship between the level of personal control and users satisfaction within offices. They found that higher controllability leads to more thermal and visual satisfaction, while the results revealed the psychological impact on the users' satisfaction by indicating differences among the available control types. The psychological aspect in personal control was also raised by Luo and Cao [25] and Karjalainen [26].

The physiological and psychological aspect of IEQ satisfaction [27] and more specific thermal satisfaction makes it harder to control indoor environment conditions and provide optimal results for everyone in a given space. The ASHRAE standard 55 [27,28] or ISO 7730 [29,30] tries to give a solution in this complex relation by stating the appropriate conditions that should be met in order to establish comfort levels in offices. To achieve that, these standards consider both personal factors, such as metabolic rate and clothing level, and environmental factors, such as air temperature, mean radiant temperature, air speed and humidity. Apart from thermal comfort, other factors that can influence comfort levels are usability of a space, acoustics, ventilation, daylight and energy use in a building.

Literature review highlights the role of personal control on the IEQ in office buildings. The objective of this paper is to provide an updated overview of the personal control in office buildings and the association with occupants' perception. To the best of our knowledge, this is the first time that European employees have participated in a questionnaire survey covering simultaneously in detail records of comfort, control and health perception in office buildings. This large-scale survey was performed under the framework of the European FP7-funded project OFFICAIR [31] and included eight widely distributed across European countries (Finland, France, Greece, Hungary, Italy, Portugal, Spain, The Netherlands) with different characteristics (e.g., geographical location, climate, socio-economic status). More specifically, the aim of this study is threefold: (i) to describe the degree of personal control over indoor environments in office buildings as reported by occupants (perceived personal control) and the association with access to available controls; (ii) to investigate the associations between perceived control and building characteristics; (iii) to study the associations between perceived control and perceived comfort and health of the occupants.

2. Materials and Methods

2.1. Data Collection

This study is based on data collected between October 2011 and May 2012 in the OFFICAIR project in modern office buildings in eight European countries [32–40]. 'Modern' buildings, constructed during the last 10 years, are described by the presence of several sorts of new electronic equipment and ventilation, heating and cooling systems, making the indoor environment almost unaffected by local climate [34]. About 19 to 24 modern buildings, selected on a voluntary basis, were investigated per country, resulting in 167 office buildings.

The protocol for data collection is described elsewhere in detail [34]. After a preliminary inspection of each building, a checklist was filled in by a local investigator along with a building manager, gathering information about building characteristics (e.g., presence of solar devices, operable windows), mechanical systems (e.g., type of mechanical ventilation, heating, cooling), rooms and activities (e.g., type of work, cleaning schedules). An online survey for the building occupants was developed in the national language of the participating countries and included questions on personal control of the indoor environment as perceived by the occupant, perceived comfort and building-related symptoms. The survey was anonymous and the participants gave their consent prior to participation. The study was approved by the competent local/national ethics committees. In total, 26,735 email invitations were sent to the occupants with an average response rate of 41% across the buildings. Although the questionnaire was online and its length might have influenced the response rate, the participation rate can be considered satisfying, in line with other recent surveys [34]. The final database involved 7441 participants—52% were females and 48% were males with an average age of 41 years.

2.2. Characteristics of the Buildings and Access to Control

The available control types in each office building were obtained from the OFFICAIR checklist (details in Table 1). The checklist included information regarding the presence of solar shading devices (grouped as: not present, internal, external), type of solar shading device control, type of temperature control, presence of operable windows, type of main lights control and type of mechanical ventilation control.

Table 1. Checklist used to investigate the types of controls over the indoor environment available for occupants in the building in the OFFICAIR project.

Parameter	Items
Are there solar shading devices present? Which kind?	Not present South side only One or more other facades External vertical blinds External shutters External roller shutters External louvers External screens External window films External horizontal blinds External awnings/canopies External overhangs External vertical fins Blind between glazing Internal vertical blinds Internal louvers Atrium Double façade Other
How are the solar shading devices controlled?	No control (fixed) Individual Central down, individual up Automatic
How is the room temperature controlled?	Manual radiator valve Local thermostat at radiator/heating unit Local thermostat (e.g., on wall) Central sensor Façade sensor(s)—i.e., outside temperature Zone sensor(s) Manual control in room(s) According to occupancy Other
Are the windows operable?	Yes Yes, some Yes, but occupants are not allowed to open them No
How are main lights (e.g., ceiling or wall) controlled?	Automatic by time (building/floor/zone) Automatic with manual end control (building/floor/zone) Demand control: Daylight (photocells) Demand control: Occupants (motion sensors) Manual
What type of control system is there for mechanical ventilation?	Central—Manual (on/off) Central—Clock/Central—Demand control (temperature, CO_2, other pollutant, relative humidity) Local—Manual (on/off) Local—Clock/Local—Demand control (temperature, CO_2, other pollutant, relative humidity) Recirculation control

2.3. Perceived Control over the Indoor Environment

Five controlling parameters were set for defining the occupants' evaluation of the perceived control and IEQ: Personal control of the occupants over temperature, ventilation, shade from the sun, light and noise was investigated using the following question with a seven-point Likert-like scale answer (from "1, not at all" to "7, full control"): 'How much control do you personally have over the following aspects of your working environment?'. The combined control variables were introduced. Several combinations were used (e.g., perceived control over temperature and ventilation) and the overall combined control variable containing all the 5 parameters [23].

2.4. Personal Comfort, Reported Health Symptoms and Self-Assessed Productivity

The satisfaction of the occupants toward the following parameters was evaluated: overall comfort, temperature (overall, too hot/cold, variation), air movement, air quality (overall air quality satisfaction, humid or dry air, stuffy or fresh air, odor), light (overall light satisfaction, natural, artificial, glare), noise (overall noise satisfaction, outside noise, noise from building systems, noise within the building), vibration, amount of privacy, office layout, office decoration, and view from the windows. The following question was used: 'How would you describe the typical indoor conditions in your office environment during the past month?' or 'How would you describe the following in your office?' A seven-point Likert-like scale answer (from "1, dissatisfied" to "7, satisfied") was provided for most of the questions, except for those questions investigating two extreme conditions in contrast where a seven-point scale answer ranging from −3 to 3 was adopted (Table 2) and converted to a scale from 1 to 7 as follow: +/−3 = 1; +/−2 = 3; +/− 1 =5; 0 = 7.

Table 2. Questions used to investigate occupants' indoor environment quality perception in the OFFICAIR project.

Parameter	Sub-Parameters	Type of Answer
Overall Comfort		Seven-point Likert-like scale
Temperature	Overall temperature	Seven-point Likert-like scale
	Too hot/too cold temperature	From "−3, too hot" to "+3, too cold"
	Temperature variation	From "−3, too much" to "+3, not enough"
Air movement		From "−3, too draughty" to "+3, too still"
Air quality	Overall air quality	Seven-point Likert-like scale
	Humid/dry air	From "−3, too humid" to "+3, too dry"
	Stuffy or fresh air	Seven-point Likert-like scale
	Odor	Seven-point Likert-like scale
Light	Overall light	Seven-point Likert-like scale
	Natural	Seven-point Likert-like scale
	Artificial	Seven-point Likert-like scale
	Glare	Seven-point Likert-like scale
Noise	Overall noise	Seven-point Likert-like scale
	Outside noise	Seven-point Likert-like scale
	Noise from building systems	Seven-point Likert-like scale
	Noise within the building	Seven-point Likert-like scale
Vibration		Seven-point Likert-like scale
Amount of privacy		Seven-point Likert-like scale
Office layout		Seven-point Likert-like scale
Office decoration		Seven-point Likert-like scale
View from the windows		Seven-point Likert-like scale

The occupants were also requested to estimate their productivity at the workstation and in other locations inside the building, considering the influence of environmental conditions on a scale of 7 levels, from +30% to −30%. In addition, they were requested to record building-related health symptoms. The Personal Symptom Index-5 (PSI-5) was calculated based on the incidence of five health symptoms—dry eyes, blocked or stuffy nose, dry throat, headache, and tiredness—which are considered to be the fundamental components of sick building symptoms as mentioned by Raw et al. [41]. This indicator has a 0 to 5 score, according to the prevalence of the reported symptoms. The respective question was: 'Have you ever experienced any of the following symptoms while working in this building (or workstation) (including today)?'

2.5. Statistical Analysis

The statistical analysis of the dataset was performed in four steps (Figure 1):

i. Descriptive results were obtained.

ii. The correlation between the available control and the perceived control using the Kruskal Wallis analysis of variance test was investigated [23,42]. Groups with less than five individuals were not included in the analysis.

iii. The relationship between the occupants' overall combined perceived control and the general physical building characteristics as well as the occupant personal characteristics was examined by applying a multilevel model [43,44], to account for the three-level structure of our data (level 1-occupant, level 2-building, level 3-country). The ordered logistic regression analysis was applied using building and country as random effects and the covariates as fixed effects. Four step-by-step models were applied. The first was an empty model without any variable and with building and country variance only. In the second model, individual level variables were imported. The final version of the second model included variables, with p-value below 0.2, such as gender, age (in four groups, <35, 35–45, 46–55, >55), effort reward ratio, experience of negative events, use of air fresheners at home, type of job (managerial, professional, clerical/secretarial, other), type of job contract (full-time, part-time), and job contract duration (permanent, fixed-term). In the third model, building characteristics were imported iteratively on the second model to identify significant relations with the perceived control. Variables with a p-value below 0.2 were selected to be used in the next model. In the fourth model, both individual and building level variables were imported. The results of the fixed effects were reported in Odds Ratios (OR) and 95% Confidence Interval (CI). For the random effects, the explained variance is reported as well as the Proportional Change in Variance (PCV) between the null model and the final model with the variables.

iv. The potential relations between the occupants' perceived control and perceived comfort and reported health symptoms (PSI5) were examined by using the spearman correlation [45]. This study focused on the correlation between the various combined control scores (e.g., perceived control over temperature and ventilation) and the perceived occupants' comfort. Additionally, the dependent variable (overall comfort satisfaction) and the response-variables were expressed in values on an ordinal scale; hence, ordinal regression analysis was employed to determine the impact of the controlling parameters on overall comfort. In the regression model, the response-dependent variable was the overall comfort satisfaction and the predictor-independent variables were the satisfaction for each personal controlling parameter as evaluated by the occupants. The results are presented in the format of OR and its CI95%. The ORs were used to rank the effect of the personal controlling parameters on overall comfort. p-values <0.05 were considered as statistically significant. The statistical package IBM SPSS Statistics [46] was used for the analysis.

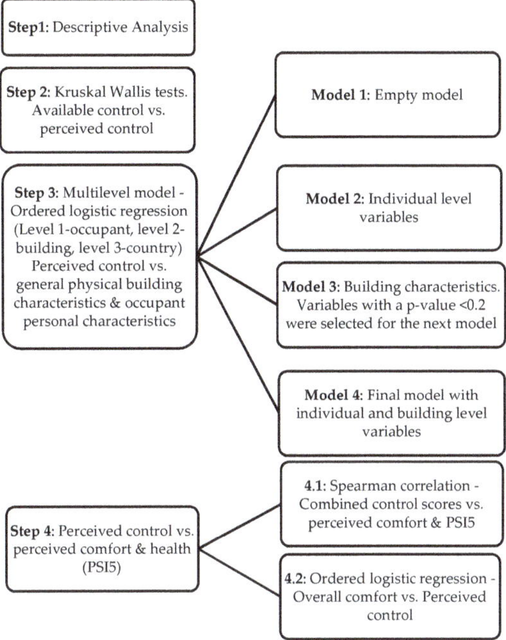

Figure 1. Schematic overview of the methodology used.

3. Results

3.1. Characteristics of Perceived Personal Control

The responses to the perceived controls (with a response rate above 99% for all control types) are presented in Figure 2. In general, noise, ventilation and temperature were perceived as poorly controlled. Regarding shading from the sun and lighting, the occupants reported the perceived control as moderate. A group of people (up to 35%) declared "no control" can be observed for all control parameters. For temperature and ventilation control, the rest of the occupants are equally distributed in the five levels of control degree. Very few occupants seem to be able to control the noise level at their workstation, where answers follow a descending rate (full control below 5%). On the contrary, regarding lighting and shading from the sun, after the distinctive "no control" group, the occupants' answers exhibited an ascending rate, indicating that they feel more able to control the light conditions (full control up to 20%).

Figure 3 presents the percentages of the office occupants who have no or low perceived control (values < 4) on their indoor working environment. The majority (63%) has no or low control on noise. Half of them declared no or low control on ventilation (53%) and temperature (47%). Almost one-third of them remarked that they do not have satisfactory levels of control for lighting and shading from sun conditions.

In Figure 4, the responses are presented based on the occupancy per room. Responses were categorized into five groups. In the first group, which is characterized by personal office rooms, occupants seem to have higher levels of control, except for ventilation. It is interesting to note that as the occupancy increases, the degree of personal control becomes lower. Offices with many occupants (30+), probably open space offices, show the lowest degrees of personal control for all parameters. Furthermore, in all groups, the occupants characterized noise control as the worst parameter, while lighting and shading from the sun control gathered higher degrees of control.

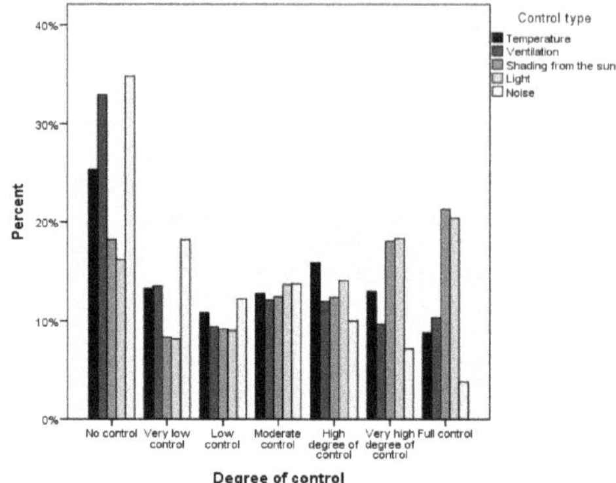

Figure 2. Percentage of occupants based on self-reported degree of control (n = 7441).

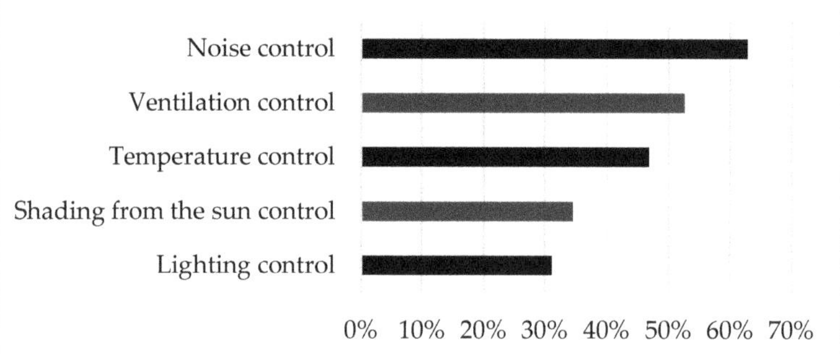

Figure 3. Percentage of occupants who reported no or low control.

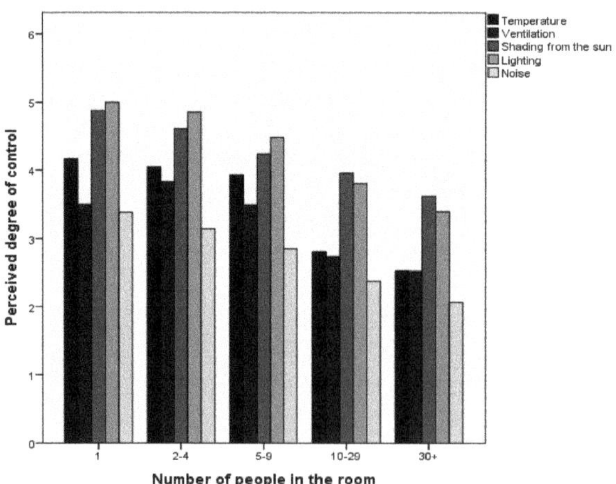

Figure 4. Mean degree of perceived control according to room occupancy (1 = no control, 7 = full control).

The satisfaction comfort towards IEQ parameters was also examined, with regard to their respective control options. The perceived satisfaction parameters were categorized in three levels of perceived control: no control (1), low control (2–3), high control (≥4) and are presented in Figure 5. For each IEQ parameter, the related personal control option was selected, e.g., for parameters about temperature satisfaction, temperature control was selected; for indoor air quality parameters, ventilation control was selected, etc. Occupants with high degrees of personal control over their working environment reported higher levels of satisfaction on average. In all cases, occupants with high perceived control reported higher satisfaction levels. Occupants without personal control evaluated satisfaction of the IEQ with lower levels. Only in some cases, 'none' and 'low personal' control options are reversed, e.g., in the case of satisfaction with odor and reflection.

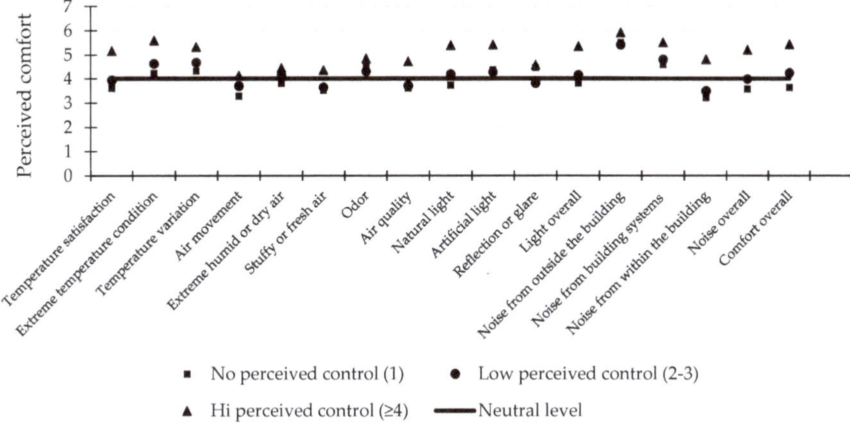

Figure 5. Perceived comfort towards indoor environment quality according to three levels of perceived control.

Figure 6 shows the degree of personal control vs. self-reported productivity at the workstation (n = 7289) and in other places in the building (n = 7154). In both cases, occupants with higher levels of personal control reported higher levels of productivity (Kruskal Wallis: $p < 0.001$). Regarding the workstations, a small increase in the low control area results in a clear increase in productivity.

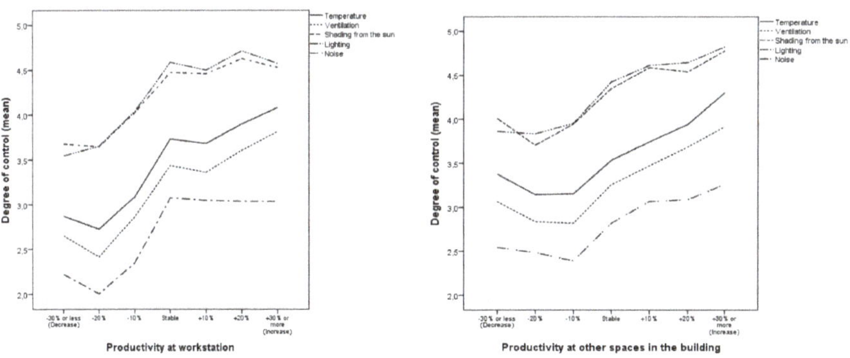

Figure 6. Self-reported productivity vs. degree of control.

3.2. Relationship between Perceived Control and Available Control

The statistically significant ($p < 0.05$) correlations between the available control and perceived control are presented in Figures 7–11.

Regarding the availability of operable windows (Figures 7–10), the scores of perceived control over temperature, ventilation, light and noise differ significantly ($p < 0.001$, $p < 0.001$, $p = 0.001$, $p < 0.001$, respectively) among buildings with operable windows, some operable windows, operable windows that people are not allowed to open, and no operable windows. The mean occupants' score over temperature in buildings with operable windows was 1.3 points higher than in buildings without operable windows ($p < 0.001$). The corresponding difference in the mean score for ventilation, light and noise was 2 ($p < 0.001$), 1 ($p < 0.001$) and 0.8 ($p = 0.001$), respectively.

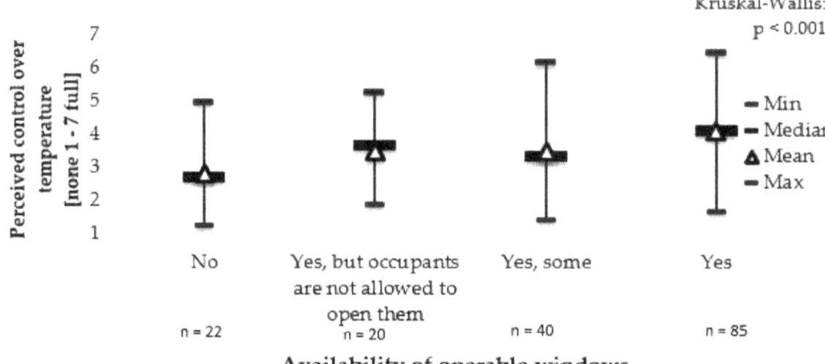

Figure 7. Perceived control over temperature vs. availability of operable windows (*n* is the number of office buildings).

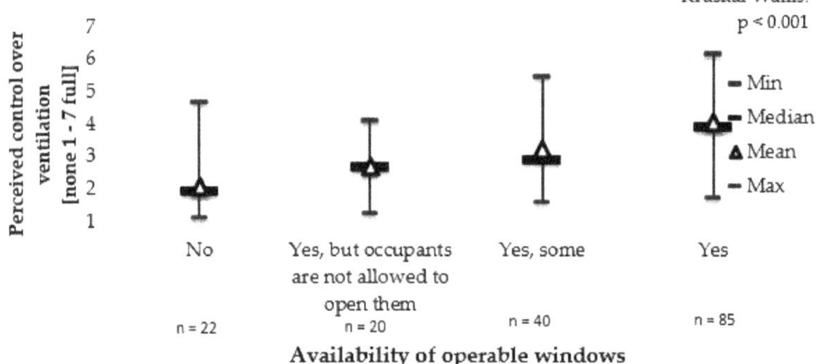

Figure 8. Perceived control over ventilation vs. availability of operable windows (*n* is the number of office buildings).

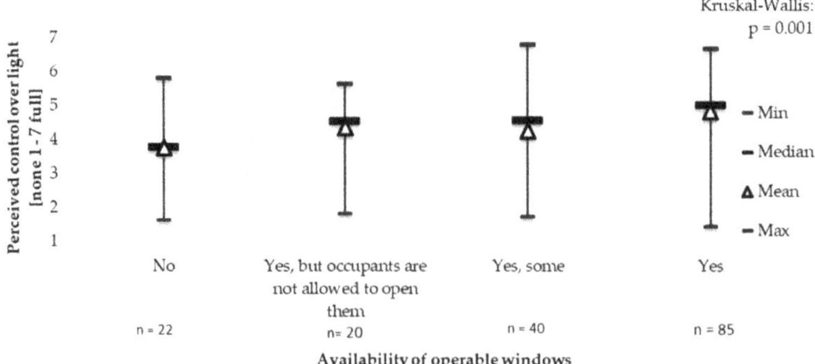

Figure 9. Perceived control over light vs. availability of operable windows (*n* is the number of office buildings).

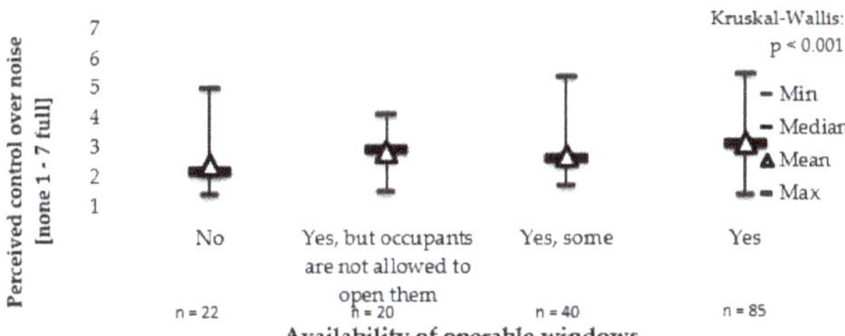

Figure 10. Perceived control over noise vs. availability of operable windows (*n* is the number of office buildings).

With regard to the presence of solar shading devices, the buildings were grouped into three categories based on the solar devices set up: internal, external, none (Figure 11). Perceived control over temperature varied significantly through the different types of the solar devices ($p = 0.028$). The highest score of perceived control over temperature was in buildings with no solar devices. This might be explained by the fact that the design of these office buildings provides adequate control over temperature without the use of solar shading devices; the multilevel regression analysis in the next step provides additional insights. No significant differences were observed between perceived control over shading and lighting in relation to the types of solar shading devices ($p = 0.635$ and $p = 0.255$, respectively, Table 3).

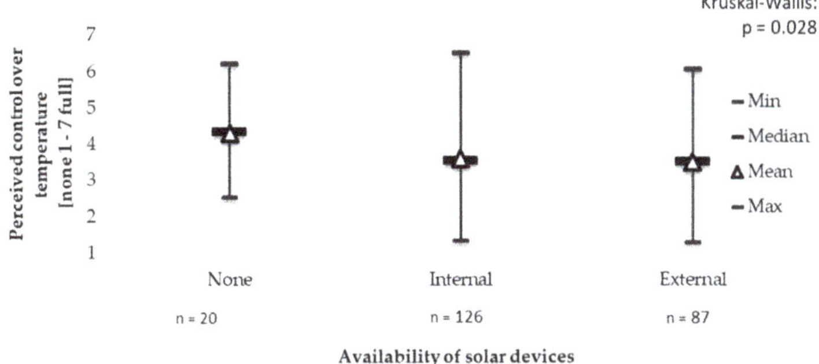

Figure 11. Perceived control over temperature vs. availability of solar shading devices (*n* is the number of office buildings).

Table 3 presents the remaining outcomes of the statistical tests performed between the available control and occupants' perceived control. Regarding temperature control, the analysis showed no significant differences ($p = 0.166$) between the perceived control over temperature and the various types of temperature controls (e.g., manual radiator valve and central sensor). The different types of operable windows did not affect the perceived control of the shading from the sun ($p = 0.100$). Subsequently, the type of control on the solar shading devices (fixed, individual, central control, automatic) was examined (Table 3). Unlike the various types of solar devices, analysis of the available different control types of the shading devices showed that there are no significant differences among the scores of the perceived control over shading, lighting and temperature ($p = 0.147$, $p = 0.710$ and $p = 0.755$, respectively). As far as the available controls of lights are concerned, no significant difference was found for perceived control over lighting ($p = 0.328$). Finally, the several types of mechanical ventilation control were examined. The occupants' perceived control over ventilation and temperature did not significantly differ ($p = 0.105$ and $p = 0.768$, respectively) from the available control types.

Table 3. Available control at building level and occupants' perceived control (from 1 = no control at all to 7 = full control) with $p > 0.05$.

	Perceived Control over Temperature	Perceived Control over Shading	Perceived Control over Light	Perceived Control over Ventilation
Type of Available Temperature Controls				
Manual radiator valve	3.38			
Local thermostat at radiator/heating unit	3.96			
Local thermostat (e.g., on wall)	3.53			
Central sensor	3.52			
Facade sensor(s)—i.e., outside temperature	4.12			
Zone sensor(s)	4.03			
Manual control in room(s)	3.70			
p-value (Kruskal-Wallis test)	0.166			
Availability of shading devices				
None		4.09	4.69	
Internal		4.43	4.43	
External		4.22	4.44	
p-value (Kruskal-Wallis test)		0.635	0.255	
Type of control of the available solar shading devices				
No control (fixed)	3.45	3.90	4.53	
Individual	3.60	4.43	4.43	
Central down, individual up	3.25	4.46	3.76	
Automatic	3.77	4.18	4.61	
p-value (Kruskal-Wallis test)	0.755	0.147	0.710	
Type of operable windows				
Yes		4.42		
Yes, some		4.15		
Yes, but occupants are not allowed to open them		4.81		
No		4.11		
p-value (Kruskal-Wallis test)		0.100		
Type of available light controls				
Manual			4.52	
Demand control: Occupants (motion sensors)			4.42	
Demand control: Daylight (photocells)			4.26	
Automatic with manual end control (building/floor/zone)			4.91	
Automatic by time (buildings/floor/zone)			4.54	
p-value (Kruskal-Wallis test)			0.328	
Type of available mechanical ventilation controls				
Central—Manual (on/off)	3.81			3.33
Central—Clock	3.62			3.23
Central—Demand control (temperature, CO_2, other pollutant, relative humidity)	3.58			3.27
Local—Manual (on/off)	4.11			4.27
Local—Clock	3.74			3.78
Local—Demand control (temperature, CO_2, other pollutant, relative humidity)	4.42			3.73
p-value (Kruskal-Wallis test)	0.768			0.105

3.3. Combined Perceived Control Versus Building Characteristics

The variances of country and building level according to Model 1 (the null model) are 0.38 and 1.17, respectively. The respective variances in Model 2, which includes the individual variables (gender, age groups, effort reward ratio, experience of negative events, use of air fresheners at home, type of job, type of job's contract, job's contract duration), were 0.52 and 0.94. The relations between the building characteristics, which were added in the model iteratively, and the combined perceived control with p-values < 0.2 are presented in Table 4 (Model 3). The strongest associations with a p-value < 0.001 were found for the building's location (suburban area), the maximum number of occupants, the documented complaints procedure, the number of people in office, as well as for the office type and availability of opening windows. Significant associations with a p-value < 0.05 were found for the number of adjacent facades with other buildings, glares from neighboring buildings, total floor area, pesticide treatment plan, smoking permission, use of portable air cleaner, floor of the workspace, partition in the offices, and noise source of occupants—distracting conversations and the location of air and exhaust devices.

Table 4. Relations between combined overall perceived control and building characteristics (Model 3).

Characteristics	n1/N1	n2/N2	OR (CI 95%)	p-Value
General building characteristics				
Location				
Mixed industrial/residential area (vs. industrial area)	696/7441	14/167	1.68 (0.76–3.69)	0.199
Commercial area (vs. industrial area)	789/7441	18/167	0.95 (0.46–1.95)	0.883
Mixed commercial/residential area (vs. industrial area)	2279/7441	50/167	1.71 (0.94–3.09)	0.076
City center, densely packed housing (vs. industrial area)	1344/7441	30/167	1.35 (0.7–2.63)	0.374
Town, with no or small gardens (vs. industrial area)	328/7441	8/167	**3.42 (1.33–8.75)**	**0.010**
Suburban, with larger gardens (vs. industrial area)	864/7441	22/167	**3.88 (1.92–7.82)**	**<0.001**
Village in a rural area (vs. industrial area)	24/7441	1/167	0.82 (0.09–7.46)	0.859
Rural area with no or few other homes nearby (vs. industrial area)	192/7441	6/167	1.27 (0.45–3.57)	0.646
Façades with adjacent buildings				
2 façades (vs. 1 façade)	1932/4664	37/102	**0.6 (0.38–0.95)**	**0.030**
3 façades (vs. 1 façade)	288/4664	9/102	1.48 (0.68–3.25)	0.322
Density of nearby obstructions				
Moderately dense (vs. very dense)	2610/7441	58/167	1.31 (0.69–2.5)	0.407
Few buildings (vs. very dense)	2992/7441	58/167	1.43 (0.74–2.76)	0.285
Free standing (vs. very dense)	1031/7441	26/167	1.95 (0.93–4.09)	0.076
Neighboring buildings with glass facades or light-colored facades causing glare				
Yes, in summer (vs. No)	215/7441	4/167	0.4 (0.13–1.27)	0.120
A little in summer (vs. No)	654/7441	12/167	**0.5 (0.25–1)**	**0.049**
Yes, in winter (vs. No)	102/7441	2/167	0.35 (0.07–1.71)	0.195
A little in winter (vs. No)	217/7441	5/167	**0.24 (0.09–0.68)**	**0.007**
Maximum number of occupants in the building				
Between 101 and 250 (vs. ≤100)	2288/7260	56/163	0.79 (0.52–1.2)	0.272
≥251 (vs. ≤100)	3544/7260	51/163	**0.39 (0.25–0.61)**	**<0.001**
Total floor area (m^2)				
Between 1441 and 3210 (≤1440)	2300/7234	53/160	0.78 (0.49–1.24)	0.295
≥3211 (≤1440)	3341/7234	54/160	**0.51 (0.31–0.83)**	**0.007**
Activities in the building besides office work				
Underground car park (vs. No)	2678/7441	44/167	0.73 (0.46–1.16)	0.178
Print shop (vs. No)	428/7441	7/167	1.92 (0.78–4.73)	0.156
Pesticide treatment plan (vs. No)	7025/7441	99/164	**0.59 (0.39–0.88)**	**0.010**
Documented complaints procedure for occupants (vs. No)	5112/7349	99/165	**0.41 (0.28–0.61)**	**<0.001**
Smoking permission				
Only outside the building (vs. No)	3916/7441	88/167	**0.55 (0.33–0.93)**	**0.026**
Only in separately ventilated rooms (vs. No)	973/7441	13/167	**0.29 (0.13–0.63)**	**0.002**
Percentage of office furniture is less than one year old and made of MDF	7025/7441	162/162	0.99 (0.99–1)	0.200
Portable air cleaner (vs. No)	225/7441	5/167	**0.33 (0.12–0.93)**	**0.035**

Table 4. *Cont.*

Characteristics	n1/N1	n2/N2	OR (CI 95%)	*p*-Value
Office characteristics				
Floor surface of the office	7410/7441	167/167	**1.05 (1.02–1.09)**	**0.001**
Number of people in the office on average				
Private (vs. 30+)	1501/7352	144/167	**7.65 (6.19–9.45)**	**<0.001**
2–4 (vs. 30+)	2134/7352	152/167	**5.15 (4.24–6.26)**	**<0.001**
5–9 (vs. 30+)	1019/7352	117/167	**2.62 (2.15–3.21)**	**<0.001**
10–29 (vs. 30+)	1492/7352	100/167	**1.48 (1.25–1.76)**	**<0.001**
Type of office				
Shared private office (vs. single person private office)	2236/7417	150/167	**0.56 (0.48–0.65)**	**<0.001**
Open space with partitions (vs. single person private office)	991/7417	115/167	**0.21 (0.17–0.26)**	**<0.001**
Open space without partitions (vs. single person private office)	2602/7441	139/167	**0.21 (0.17–0.25)**	**<0.001**
Other (vs. single person private office)	90/7417	60/167	**0.39 (0.24–0.65)**	**<0.001**
Partitions within the offices (vs. No)	4112/7441	91/167	**0.67 (0.45–0.99)**	**0.043**
Noise source of occupants–distracting conversations (vs. No)	2653/7441	64/167	**0.66 (0.45–0.98)**	**0.040**
PC or laptop monitor position				
In front of windows (vs. Not)	3018/7441	69/167	1.44 (1–2.08)	0.052
Printer/copy machines location				
In the offices (vs. on the corridor)	2976/7441	75/167	0.62 (0.39–1)	0.051
In a separate printing room (vs. on the corridor)	2501/7441	47/167	0.64 (0.39–1.05)	0.075
Operable windows				
Yes (vs. No)	3724/7441	85/163	**5.51 (3.14–9.67)**	**<0.001**
Yes, some (vs. No)	1769/7441	40/163	**3.08 (1.72–5.53)**	**<0.001**
Yes, but occupants are not allowed to open them (vs. No)	913/7441	20/163	1.54 (0.79–2.98)	0.204
Location of air supply devices inside offices—ceiling (vs. No air supply)	5783/7441	123/167	**0.63 (0.4–0.99)**	**0.046**
Location of air exhaust devices inside offices				
High (vs. None)	6189/7089	128/154	**0.46 (0.25–0.87)**	**0.017**
Low (vs. None)	382/7089	10/154	0.63 (0.24–1.63)	0.339

p-values in bold are significant at 5%. Adjusted for gender, age groups, effort reward ratio, experience of negative events, use of air fresheners at home, type of job, type of job's contract, job's contract duration. Characteristics with a p value lower than 0.20 are presented. Level 1—Occupant level, Level 2—Building level, Level 3—Country level. OR: Odd Ratio, CI: Confidence interval 95%, n1/N1: Occupants' answers/total number of occupants, n2/N2: building observations/total number of buildings.

All individual and building characteristics from Models 2 and 3 were imported to Model 4; the relevant results are presented in Table 5. The variance between buildings was equal to 0.14 in Model 4. PCV shows that 88% of the initial variance in overall perceived control was explained by the included variables. Buildings with larger total floor areas were positively associated with perceived overall control (OR 1.99, 95% CI: 1.22–3.25). The workspace floor was also positively associated with control perception (OR 1.06, 95% CI: 1.02–1.10). The existence of any kind of operable windows (OR 3.64 up to 6.53) and private and semi-private offices had the highest impact on the perceived overall control (OR 1.58 up to 3.73). On the other hand, the building's location (OR 0.24, 95% CI: 0.06–0.96) and the buildings' number of adjacent facades (OR 0.65, 95% CI: 0.44–0.97) with other buildings tended to negatively associate with the perceived control. Both high number of occupants in the building (OR 0.67, 95% CI: 0.46–0.99) and the type of office (OR from 0.23 to 0.36) had a significant negative effect on the perceived overall control. Indoor noise from distracting conversations was associated negatively with the perceived overall control (OR 0.69, 95% CI: 0.49–0.97).

Concerning individual characteristics, in addition to gender (OR 1.21, 95% CI: 1.06–1.38) and age group (OR 0.78, 95% CI: 0.61–0.99), strong association with the overall perceived control was found for the ERI (OR 0.67, 95% CI: 0.46–0.99) and the occupants' habit to use air fresheners at home (OR 1.26, 95% CI: 1.11–1.44).

Table 5. Associations between combined overall perceived control, building characteristics and individual characteristics (Model 4).

Factors	OR (CI 95%)	*p*-Value
Individual characteristics/personal activities		
Female (vs. Male)	**1.21 (1.06–1.38)**	**0.006**
Age		
<35 (vs. 55+)	0.89 (0.69–1.14)	0.345
35–45 (vs. 55+)	**0.78 (0.61–0.99)**	**0.040**
46–55 (vs. 55+)	0.95 (0.74–1.23)	0.720
Effort-reward ratio	**0.31 (0.25–0.38)**	**<0.001**
Experience of negative events (vs. No)	0.94 (0.83–1.07)	0.367
Type of job		
Managerial (vs. Other)	0.97 (0.75–1.27)	0.844
Professional (vs. Other)	0.98 (0.78–1.23)	0.873
Clerical-secretarial (vs. Other)	0.95 (0.77–1.17)	0.625
Type of job's contract–Full-time (vs. Part-time)	1.12 (0.89–1.41)	0.344
Job's contract duration–Permanent (vs. Fixed-term)	1.13 (0.91–1.39)	0.263
Air fresheners use in home (vs. No)	**1.26 (1.11–1.44)**	**<0.001**
Building characteristics		
Location		
Mixed industrial/residential area (vs. industrial area)	0.64 (0.31–1.29)	0.208
Commercial area (vs. industrial area)	1.59 (0.83–3.07)	0.164
Mixed commercial/residential area (vs. industrial area)	1.05 (0.61–1.83)	0.849
City Centre, densely packed housing (vs. industrial area)	1.15 (0.62–2.1)	0.660
Town, with no or small gardens (vs. industrial area)	2.98 (0.95–9.39)	0.062
Suburban, with larger gardens (vs. industrial area)	1.73 (0.63–4.72)	0.285
Village in a rural area (vs. industrial area)	**0.24 (0.06–0.96)**	**0.043**
Rural area with no or few other homes nearby (vs. industrial area)	1.02 (0.24–4.45)	0.974
Density of nearby obstructions		
Moderately dense (vs. Very dense)	1.11 (0.63–1.95)	0.729
Few buildings (vs. Very dense)	1.14 (0.65–2.02)	0.648
Free standing (vs. Very dense)	0.99 (0.44–2.19)	0.972
Maximum number of occupants in the building		
Between 101 and 250 (vs. ≤100)	**0.67 (0.46–0.99)**	**0.045**
≥251 (vs. ≤100)	0.92 (0.56–1.52)	0.745
Total floor area		
Between 1441 and 3210 (≤1440)	**1.99 (1.22–3.25)**	**0.006**
≥3211 (≤1440)	1.44 (0.81–2.57)	0.217
Façades with adjacent buildings		
2 façades (vs. 1 façade)	**0.65 (0.44–0.97)**	**0.033**
3 façades (vs. 1 façade)	1 (0.51–1.95)	0.992
Neighboring buildings with glass facades or light-colored facades causing glare		
Yes, in Summer (vs. No)	**3.16 (1.15–8.66)**	**0.025**
A little in Summer (vs. No)	1.52 (0.74–3.14)	0.256
Yes, in Winter (vs. No)	**0.24 (0.06–0.94)**	**0.040**
A little in Winter (vs. No)	0.49 (0.17–1.39)	0.180
Operable windows		
Yes (vs. No)	**4.81 (2.85–8.14)**	**<0.001**
Yes, some (vs. No)	**6.53 (3.76–11.34)**	**<0.001**
Yes, but occupants are not allowed to open them (vs. No)	**3.64 (1.82–7.27)**	**<0.001**
Activities in the building besides office work		
Underground car park (vs. No)	0.79 (0.52–1.19)	0.258
Print shop (vs. No)	1.4 (0.6–3.27)	0.435
Portable air cleaner (vs. No)	0.79 (0.31–2)	0.622
Pesticide treatment plan (vs. No)	1.33 (0.74–2.38)	0.336
Smoking permission		
Only outside the building (vs. No)	0.82 (0.46–1.46)	0.492
Only in separately ventilated rooms(vs. No)	0.66 (0.32–1.36)	0.260
Documented complaints procedure for occupants (vs. No)	0.77 (0.5–1.17)	0.221
Percentage of office furniture is less than one year old and made of MDF	1 (0.99–1.01)	0.800

Table 5. *Cont.*

Factors	OR (CI 95%)	*p*-Value
Office characteristics		
Floor surface of the office	1.06 (1.02–1.1)	0.002
Number of people in the office on average		
Private (vs. 30+)	2.51 (1.59–3.97)	<0.001
2–4 (vs. 30+)	3.73 (2.78–4.99)	<0.001
5–9 (vs. 30+)	2.3 (1.79–2.96)	<0.001
10–29 (vs. 30+)	1.58 (1.29–1.93)	<0.001
Type of office		
Shared private office (vs. Single person private office)	0.36 (0.23–0.55)	<0.001
Open space with partitions (vs. Single person private office)	0.23 (0.14–0.36)	<0.001
Open space without partitions (vs. Single person private office)	0.26 (0.17–0.4)	<0.001
Other (vs. Single person private office)	0.57 (0.25–1.29)	0.178
Partitions within the offices (vs. No)	1.02 (0.64–1.61)	0.935
Noise source of Occupants–distracting conversations (vs. No)	0.69 (0.49–0.97)	0.034
PC or laptop monitor position		
In front of windows (vs. Not)	1.32 (0.94–1.86)	0.103
Printer/copy machines location		
In the offices (vs. on the corridor)	0.99 (0.63–1.55)	0.956
In a separate printing room (vs. on the corridor)	1.23 (0.76–1.99)	0.396
Location of air supply devices inside offices—ceiling (vs. No air supply)	0.81 (0.51–1.28)	0.365
Location of air exhaust devices inside offices		
High (vs. None)	1.43 (0.61–3.34)	0.408
Low (vs. None)	0.93 (0.37–2.35)	0.879
County level σ2/PCV (%)	0.42/10	
Building level σ2/PCV (%)	0.14/88	

3.4. Impact of Perceived Control on Perceived Comfort and Health

3.4.1. Combined Perceived Control vs. Perceived Comfort and Health—Bivariate Analysis

The relationships between perceived control and perceived comfort are presented in Table 6. Perceived overall comfort correlated significantly and positively with perceived control over all control parameters. When the scores of the perceived control were combined, the correlation with the overall comfort increased, reaching the highest value of $r = 0.465$. The combined correlation values are stronger than the correlations between the single perceived control parameters. The strongest correlation for perceived overall temperature was indicated with control over temperature ($r = 0.420$) and not with the combined controls. The satisfaction with extreme hot or cold temperature conditions seems to have the strongest correlation only with the control over temperature ($r = 0.281$), while the combined perceived controls show equal effect. On the other hand, temperature variation was found to be more related to the combined control over temperature and shading from the sun controls ($r = 0.209$). Perceived air quality about dry or humid air and odor did not show strong correlations, either with single or with combined perceived control parameters, indicating that these parameters were more difficult to be controlled by the occupants. However, perceived overall air quality satisfaction and perceived satisfaction with fresh air were found to be more correlated with the combined control perception over temperature, ventilation and shading from the sun ($r = 0.380$ and 0.320, respectively).

Perceived comfort regarding natural light correlated positively with perceived control over shading from the sun and over light ($r = 0.370$). Moreover, artificial light perception showed correlation with the perceived control over light ($r = 0.377$). This means that occupants with higher degrees of shading and lighting controls feel more satisfied with the light levels in their offices. Perceived satisfaction with glare did not show strong correlation. In general, combined controls do not significantly affect light satisfaction.

Table 6. Correlations between combined perceived control and perceived comfort and health.

	Perceived Control over Temperature and Ventilation Combined [2: None at all-14: Full Control]	Perceived Control over Temperature and Shading from the Sun Combined [2: None at all-14: Full Control]	Perceived Control over Temperature, Ventilation and Shading from the Sun Combined [3: None at all-21: Full Control]	Perceived Control over Temperature, Ventilation, Shading from the Sun, Light and Noise Combined [5: None at all-35: Full Control]
Overall comfort [1: Unsatisfactory-7: Satisfactory]	0.367 <0.001	0.415 <0.001	0.415 <0.001	0.465 <0.001
Temperature [1: Too hot/cold- 7: Satisfactory]	0.258 <0.001	0.257 <0.001	0.255 <0.001	0.245 <0.001
Temperature [1: Varies too much/not enough variation-7: Satisfactory]	0.154 <0.001	0.209 <0.001	0.186 <0.001	0.188 <0.001
Temperature [1: Unsatisfactory-7: Satisfactory]	0.403 <0.001	0.414 <0.001	0.416 <0.001	0.43 <0.001
Air movement [1: Draughty/Still- 7: Satisfactory] Air movement	0.175 <0.001	0.164 <0.001	0.178 <0.001	0.182 <0.001
Air quality [1: Dry/humid-7: Satisfactory]	0.143 <0.001	0.142 <0.001	0.147 <0.001	0.162 <0.001
Air quality [1: Stuffy- 7: Fresh]	0.287 <0.001	0.312 <0.001	0.322 <0.001	0.342 <0.001
Air quality [1: Smelly-7: Odorless]	0.129 <0.001	0.183 <0.001	0.17 <0.001	0.195 <0.001
Air quality [1: Unsatisfactory-7: Satisfactory]	0.35 <0.001	0.363 <0.001	0.38 <0.001	0.405 <0.001
Natural light [1: Unsatisfactory-7: Satisfactory]	0.247 <0.001	0.356 <0.001	0.342 <0.001	0.361 <0.001
Artificial light [1: Unsatisfactory-7: Satisfactory]	0.245 <0.001	0.281 <0.001	0.284 <0.001	0.331 <0.001
Reflection or glare [1: Glare|-7: No glare]	0.123 <0.001	0.145 <0.001	0.142 <0.001	0.166 <0.001
Light overall [1: Unsatisfactory-7: Satisfactory]	0.295 <0.001	0.364 <0.001	0.361 <0.001	0.401 <0.001
Noise from outside the building [1: Unsatisfactory-7: Satisfactory]	−0.004 0.707	0.077 <0.001	0.048 <0.001	0.077 <0.001
Noise from building systems (e.g., heating, plumbing, ventilation, air conditioning) [1: Unsatisfactory-7: Satisfactory]	0.172 <0.001	0.191 <0.001	0.198 <0.001	0.226 <0.001
Noise from within the building other than from building systems (e.g., phone calls, colleagues chatting, photocopiers, etc.) [1: Unsatisfactory-7: Satisfactory]	0.28 <0.001	0.302 <0.001	0.304 <0.001	0.367 <0.001

Table 6. *Cont.*

	Perceived Control over Temperature and Ventilation Combined [2: None at all-14: Full Control]	Perceived Control over Temperature and Shading from the Sun Combined [2: None at all-14: Full Control]	Perceived Control over Temperature, Ventilation and Shading from the Sun Combined [3: None at all-21: Full Control]	Perceived Control over Temperature, Ventilation, Shading from the Sun, Light and Noise Combined [5: None at all-35: Full Control]
Noise overall [1: Unsatisfactory-7: Satisfactory]	0.286 **<0.001**	0.322 0.000	0.321 0.000	0.389 0.000
Vibration [1: Unsatisfactory-7: Satisfactory]	0.108 **<0.001**	0.177 **<0.001**	0.16 **<0.001**	0.193 **<0.001**
Amount of privacy [1: Unsatisfactory-7: Satisfactory]	0.333 **<0.001**	0.385 **<0.001**	0.379 **<0.001**	0.433 **<0.001**
Layout [1: Do not like at all-7: Like very much]	0.31 **<0.001**	0.367 **<0.001**	0.361 **<0.001**	0.403 **<0.001**
Decoration [1: Do not like at all-7: Like very much]	0.22 **<0.001**	0.295 **<0.001**	0.278 **<0.001**	0.315 **<0.001**
View from the windows [1: Do not like at all-7: Like very much]	0.147 **<0.001**	0.256 **<0.001**	0.233 **<0.001**	0.246 **<0.001**
PSI5	−0.249 **<0.001**	−0.251 **<0.001**	−0.263 **<0.001**	−0.289 **<0.001**

Rho and *p*-value of Spearman correlation. Significant at 5% is in bold.

Perceived comfort over outdoor noise seems to have low correlation with perceived controls. The occupants were not able to control the noise levels coming from outside probably due to inefficient available control types. Noise from building systems and noise within the building correlated positively with perceived control over noise ($r = 0.205$ and 0.377). Moreover, the occupants were more satisfied with the noise from building systems when they had higher degree of all combined controls ($r = 0.226$). The highest correlation was observed ($r = 0.408$) between the overall noise satisfaction and perceived noise control. This indicates that the occupants are more comfortable in buildings where they perceive high degree of control over noise.

The amount of privacy was found to have significant correlation with all control parameters, both single and combined, while the maximum correlation was recorded with all combined controls ($r = 0.433$). This implies that high degrees of personal control improve the privacy that the occupant needs. This positive relation was also observed for office layout and decoration perceived satisfaction with $r = 0.403$ and 0.315, respectively.

As far as perceived control and the presence of health symptoms assessed through the PSI5 are concerned, a significant negative correlation was found for all control parameters, both single and combined. The combined perceived control scores exhibited a stronger negative correlation with the PSI5.

3.4.2. Perceived Control vs. Perceived Overall Comfort—Regression Analysis

The relation between the overall comfort and five perceived control variables has been examined by applying the ordinal regression analysis. The results are presented in Table 7. The maximum OR value (1.28) corresponded to the perceived control over noise. The results showed that if the perceived control over noise increases by one unit in the 1–7 point scale, there is a 1.28 time likelihood that the overall comfort will increase by one unit. Perceived noise control was the parameter with the lowest score according to the occupants' recordings, as shown in Table 2. The impact on the overall comfort of the control over noise, lighting, temperature and shading from the sun, was almost equivalent (ORs: 1.16, 1.14 and 1.12, respectively). Lastly, the perceived control over ventilation, despite the fact that it was ranked as the second worst parameter, had the lowest impact on overall comfort (OR 1.03, 95% CI: 1.00–1.06).

Table 7. Relations between perceived control and perceived overall comfort.

Parameters	OR CI (95%)	*p*-Value
Perceived control over noise	1.28 (1.25–1.32)	$p < 0.001$
Perceived control over lighting	1.16 (1.13–1.19)	$p < 0.001$
Perceived control over temperature	1.14 (1.11–1.18)	$p < 0.001$
Perceived control over shade	1.12 (1.10–1.15)	$p < 0.001$
Perceived control over ventilation	1.03 (1.00–1.06)	$p < 0.05$

Moreover, occupants were separated in two groups: those who had no or low control and those who had high control for the sum of the five control parameters. The results are presented in Table 8. The ranking remains the same as in the previous case, but two remarks can be highlighted. First, in the "low control" group, occupants were less sensitive to controlling parameters, with lower OR values. In addition, the perceived control over ventilation was not significant ($p > 0.05$). Second, in the "high control" group, occupants were more sensitive to all controlling parameters with increased OR values. The perceived control over lighting showed an almost equal impact on overall comfort as the perceived control over noise.

Table 8. Relations between perceived control in low and high control groups and perceived overall comfort.

Parameters—Low Control Group	OR CI (95%)	*p*-Value	Parameters—High Control Group	OR CI (95%)	CI (95%)
Perceived control over noise	1.21 (1.15–1.27)	$p < 0.001$	Perceived control over noise	1.37 (1.32–1.43)	$p < 0.001$
Perceived control over lighting	1.14 (1.11–1.17)	$p < 0.001$	Perceived control over lighting	1.32 (1.25–1.39)	$p < 0.001$
Perceived control over temperature	1.11 (1.06–1.15)	$p < 0.001$	Perceived control over temperature	1.26 (1.20–1.31)	$p < 0.001$
Perceived control over shade	1.10 (1.07–1.13)	$p < 0.001$	Perceived control over shade	1.19 (1.14–1.25)	$p < 0.001$
Perceived control over ventilation	0.99 (0.95–1.04)	$p > 0.05$	Perceived control over ventilation	1.08 (1.04–1.12)	$p < 0.001$

4. Discussion

4.1. Occupants' Personal Control

The current study analyzed data from 7441 occupants in 167 European modern office buildings, sustaining and reinforcing findings from previous studies, revealing the importance of the perceived personal control over indoor environment parameters.

The overall occupants' comfort should be examined by investigating the role of every aspect of personal control. Paciuk [16] highlighted three aspects of personal control; (i) available control, (ii) exercised control; (iii) perceived control. Available control can be described by the degree and type of control made available by the environment. It can be defined by 'the degree of manipulation of thermostats and other manual controls as well as the existence of operable windows, blinds, sunshades, ventilation vanes, doors and HVAC system components'. Exercised control is defined by 'the relative frequency in which occupants engage in several types of thermal-related behaviors in order to obtain thermal comfort when needed'. The occupants' perceived control is being produced as the interaction of the different degrees of available (building controls) and exercised control (frequency of use). OFFICAIR database does not include information about the occupants' exercise control so in that study this aspect was not examined. The main idea is that personal control, both available and perceived, works as a moderator affecting the occupants' satisfaction on IEQ parameters and on the overall comfort.

Paciuk [16] examined the perceived control over temperature, ventilation, shading from the sun, lighting and noise. The results showed that occupants have moderate or low control on these parameters with a significant group of occupants, who declared that they had no access to the control of their environment. This was also remarked by Haghighat and Donnini [19], where almost 60% of the responders had no control access. Furthermore, around 30% of the participants were very dissatisfied with the level of control. An interesting relationship between the perceived productivity in offices and perceived control has emerged, where higher productivity levels were recorded when occupants perceived higher degrees of personal control. Boardass et al. [18] also pointed out the issue of productivity versus personal control, concluding in the same finding.

In this analysis, only a limited number of available control types were found to be associated with the perceived control. Perceived control over temperature, ventilation, light and noise was found to be associated only with the presence of operable windows. The effect of operable windows in the indoor environment on occupants' comfort was also raised by Brager et al. [47]. The scores of perceived temperature control differ significantly with the different types of solar shading devices. However, after importing these available control types along with various general building characteristics in the multilevel model, only operable windows remained significant. Similar findings were also reported in another study [48] where opening windows were the most frequent behavior of controlling thermal conditions. The limited effect of the available control on the perceived control was pointed out in previous studies [23,24], which found no association between available and perceived control. This could be explained by the fact that occupants could identify the available controls, but they were not allowed to use them or did not know how to set them up or the control types could not respond rapidly. The multilevel model, containing both individual and building characteristics (both general and available control types), showcased 88% variance in perceived control between buildings. A Building's location and the distance with adjacent facades from other neighboring buildings seem to be significant. The area and number of floors were also found to affect occupant control perception. The number of occupants inside the building showed strong effect, especially inside the office, as well as the type of office (private or open space with partitions). Gou et al. [49] remarked that the provision of personal control in open-plan work environments is an important means to alleviating adverse perceptions. Noise due to conversations and phone calls negatively affected the controlling perception, which was expected as people cannot control themselves the level of noise. Around one-third of the occupants reported negative perception of lighting control, which is similar to findings by Moore et al. [50]. Boardass et al. [18] and Toftum [21] also indicated that the perceived degree of controlling parameters (temperature, ventilation, lighting) decreases with the increased number of occupants inside offices. To some extent, it seems that occupants perceive controlling capabilities better when providing them with the feeling of freedom inside the office buildings. Effort-Reward Imbalance (ERI), which is a critical psychological parameter in office environments, significantly affected control perception and the occupants' perceived comfort inside offices [36]. Moreover, occupants who habitually use air fresheners in their home seem to be more sensitive to adjusting their indoor work environment. The psychological aspect of perceived control is reinforced, as observed by Luo et al. [25].

A significant positive but weak correlation was observed between perceived control and perceived comfort. This finding is in compliance with Boerstra et al. [23], Haghighat and Donnini [19] and Roulet et al. [51]. This work includes a deeper analysis of the appreciation of the perceived control. It should be noted that the analysis using the combined control parameters is not widely used in this type of studies. Combined scores of the controlling parameters were positively correlated with IEQ satisfaction, as Boerstra et al. [23] also found. However, in our study, the combined control perception affected with a stronger correlation the overall comfort, in contrast to the findings of Boerstra et al. [23]. This could be explained by the fact that occupants with higher combined control perception perceive overall comfort with more satisfaction. Combined control perception also resulted in occupants being more satisfied with the overall air quality as well as with the amount of privacy and office layout. In general, occupants who are more able to adjust their environment feel more satisfied.

Regarding building-related health symptoms, the combined scores unveil higher negative correlations with the PSI5. Office occupants reported less health symptoms related to buildings where they perceived higher degree of personal controls.

4.2. Strengths and Limitations

This study has several strengths: a relatively large sample size, a survey performed in different geographical areas across Europe, and the use of standardized procedures (questionnaire and checklist). Data on socio-demographics, psycho-social work environment, and perceived environmental quality were collected by a validated questionnaire. IEQ was assessed using both crude IEQ items (satisfaction with perceived control over temperature, ventilation, shading from the sun, light and noise, as well as satisfaction with overall thermal comfort, noise, light, and indoor air quality), and with 14 detailed indoor environmental parameters (e.g., layout, noise within the building, noise from building systems, noise outside the building, air movement). Some limitations, however, should be noted. Caution is needed when interpreting results, because data on IEQ were self-reported. Consequently, a potential (recall) bias cannot be excluded and this type of surveys do not always capture IEQ issues. A combination of objective and subjective measurements would be useful for assessing IEQ. Another limitation is the cross-sectional study design. Therefore, no causality of the identified relations can be confirmed.

5. Conclusions

The following conclusions can be drawn from the present study:

1. In general, occupants have a moderate or low access in the control of their indoor working environment. They have rated with a moderate control perception the lighting and shading from the sun parameters. Noise was the worst control parameter, while temperature and ventilation control were rated in the low control scale. Among the five control parameters, there was a significant group of occupants who were not able to control their environment at all. Nearly all occupants have no or low control on noise. Half of the occupants declared no or low control on ventilation and temperature conditions. Almost one-third of them remarked that they do not have satisfactory levels of control for lighting and shading from sun conditions. It is noteworthy that as the occupancy within the offices increases, the degree of personal control becomes lower.

2. Occupants with a higher level of personal control were reported to be more productive within their working environment. Moreover, occupants who declared high degrees of personal control reported higher levels of IEQ satisfaction.

3. Some significant correlations were found between the available controls within the building and the perceived control. The availability of operable windows had a higher impact on the occupants' control perception over temperature, ventilation, light and noise than floor area or occupancy. Perceived control over temperature differed significantly with the different types of solar devices.

4. Occupants' perceived control is related to psychological aspects. General building characteristics, such as floor number and floor area, office type, operable windows, etc., which help occupants feel freedom were positively associated with the perceived control. ERI remains a significant parameter of the controlling perception in office environments.

5. Concerning the impact of perceived control on perceived comfort, the results of the present study are in general agreement with the literature. More specifically:

   - The perceived combined control over all studied parameters is positive on the overall comfort
   - The combined control over all parameters seems to have a stronger effect on overall comfort than the single controls
   - Temperature variation seems to be more related to the combined control of temperature and shading from the sun control
   - Combined control perception over temperature, ventilation and shading positively affects the occupants and gives them the feeling of fresher air and an overall satisfaction with air quality
   - Noise from the buildings systems seems to be more affected by the combined control perception of all parameters rather than single controls
   - The combined perceived control of all parameters was found to affect the occupants' privacy, office layout and decoration satisfaction more

6. Regarding reported health symptoms, office occupants with a higher degree of personal controls reported less building-related health symptoms.

Author Contributions: Conceptualization, I.S., D.S., C.M. and J.B.; data curation, I.S.; formal analysis, I.S.; funding acquisition, J.B.; investigation, I.S., D.S., C.M., Y.d.K., S.F., A.S., A.C., T.S. and V.M.; methodology, I.S., C.M., Y.d.K. and J.B.; project administration, J.B.; resources, K.K.; supervision, D.S. and J.B.; visualization, I.S.; writing-original draft, I.S.; writing-review & editing, D.S., C.M., Y.d.K., S.F., A.S., A.C., T.S., V.M., E.d.O.F., K.K., P.C. and J.B.

Funding: This research was funded by the European Union 7th 427 Framework, grant number Agreement 265267 under Theme: ENV.2010.1.2.2-1.

Acknowledgments: This work was supported by the project "OFFICAIR" (On the Reduction of Health Effects from Combined Exposure to Indoor Air Pollutants in Modern Offices).

Conflicts of Interest: The authors declare no conflict of interest.

References

1. Huang, Y.-H.; Robertson, M.M.; Chang, K.-I. The Role of Environmental Control on Environmental Satisfaction, Communication, and Psychological Stress: Effects of Office Ergonomics Training. *Environ. Behav.* **2004**, *36*, 617–637. [CrossRef]
2. Gee, G.C.; Payne-Sturges, D.C. Environmental Health Disparities: A Framework Integrating Psychosocial and Environmental Concepts. *Environ. Health Perspect.* **2004**, *112*, 1645–1653. [CrossRef]
3. Kalimeri, K.K.; Saraga, D.E.; Lazaridis, V.D.; Legkas, N.A.; Missia, D.A.; Tolis, E.I.; Bartzis, J.G. Indoor air quality investigation of the school environment and estimated health risks: Two-season measurements in primary schools in Kozani, Greece. *Atmos. Pollut. Res.* **2016**, *7*, 1128–1142. [CrossRef]
4. Lee, S.Y.; Brand, J.L. Effects of control over office workspace on perceptions of the work environment and work outcomes. *J. Environ. Psychol.* **2005**, *25*, 323–333. [CrossRef]
5. Huizenga, C.; Abbaszadeh, S.; Zagreus, L.; Arens, E.A. Air quality and thermal comfort in office buildings: Results of a large indoor environmental quality survey. *Proceeding Healthy Build.* **2006**, *3*, 393–397.
6. Brager, G.; Baker, L. Occupant satisfaction in mixed-mode buildings. *Build. Res. Inf.* **2009**, *37*, 369–380. [CrossRef]
7. Langevin, J.; Wen, J.; Gurian, P.L. Relating occupant perceived control and thermal comfort: Statistical analysis on the ASHRAE RP-884 database. *HVAC&R Res.* **2012**, *18*, 179–194.
8. Vine, E.; Lee, E.; Clear, R.; DiBartolomeo, D.; Selkowitz, S. Office worker response to an automated Venetian blind and electric lighting system: A pilot study. *Energy Build.* **1998**, *28*, 205–218. [CrossRef]
9. Lee, S.Y.; Brand, J.L. Can personal control over the physical environment ease distractions in office workplaces? *Ergonomics* **2010**, *53*, 324–335. [CrossRef]
10. Samani, S.A. The Impact of Personal Control over Office Workspace on Environmental Satisfaction and Performance. *J. Soc. Sci. Humanit.* **2015**, *1*, 163–172.
11. Xuan, X. Study of indoor environmental quality and occupant overall comfort and productivity in LEED- and non-LEED–certified healthcare settings. *Indoor Built Environ.* **2018**, *27*, 544–560. [CrossRef]
12. Yun, G.Y. Influences of perceived control on thermal comfort and energy use in buildings. *Energy Build.* **2018**, *158*, 822–830. [CrossRef]
13. Zhao, Z.; Amasyali, K.; Chamoun, R.; El-Gohary, N. Occupants' Perceptions about Indoor Environment Comfort and Energy Related Values in Commercial and Residential Buildings. *Procedia Environ. Sci.* **2016**, *34*, 631–640. [CrossRef]
14. Al horr, Y.; Arif, M.; Katafygiotou, M.; Mazroei, A.; Kaushik, A.; Elsarrag, E. Impact of indoor environmental quality on occupant well-being and comfort: A review of the literature. *Int. J. Sustain. Built Environ.* **2016**, *5*, 1–11. [CrossRef]
15. Frontczak, M.; Schiavon, S.; Goins, J.; Arens, E.; Zhang, H.; Wargocki, P. Quantitative relationships between occupant satisfaction and satisfaction aspects of indoor environmental quality and building design. *Indoor Air* **2012**, *22*, 119–131. [CrossRef]
16. Paciuk, M. *The Role of Personal Control of the Environment in Thermal Comfort and Satisfaction at the Workplace*; University of Wisconsin-Milwaukee: Milwaukee, WI, USA, 1989.
17. Collins, B.L.; Fisher, W.; Gillette, G.; Marans, R.W. Second-Level Post-Occupancy Evaluation Analysis. *J. Illum. Eng. Soc.* **1990**, *19*, 21–44. [CrossRef]
18. Bordass, B.; Bromley, K.; Leaman, A. *User and Occupant Controls in Office Buildings*; BRE: Brussels, Belgium, 1993; p. 13.
19. Haghighat, F.; Donnini, G. Impact of psycho-social factors on perception of the indoor air environment studies in 12 office buildings. *Build. Environ.* **1999**, *34*, 479–503. [CrossRef]
20. Zweers, T.; Preller, L.; Brunekreef, B.; Boleij, J.S.M. Health and Indoor Climate Complaints of 7043 office Workers in 61 Buildings in the Netherlands. *Indoor Air* **1992**, *2*, 127–136. [CrossRef]
21. Toftum, J. Central automatic control or distributed occupant control for better indoor environment quality in the future. *Build. Environ.* **2010**, *45*, 23–28. [CrossRef]

22. Zagreus, L.; Huizenga, C.; Arens, E.; Lehrer, D. Listening to the occupants: A Web-based indoor environmental quality survey. *Indoor Air* **2004**, *14*, 65–74. [CrossRef]
23. Boerstra, A.; Beuker, T.; Loomans, M.; Hensen, J. Impact of available and perceived control on comfort and health in European offices. *Archit. Sci. Rev.* **2013**, *56*, 30–41. [CrossRef]
24. Kwon, M.; Remøy, H.; van den Dobbelsteen, A.; Knaack, U. Personal control and environmental user satisfaction in office buildings: Results of case studies in the Netherlands. *Build. Environ.* **2019**, *149*, 428–435. [CrossRef]
25. Luo, M.; Cao, B.; Ji, W.; Ouyang, Q.; Lin, B.; Zhu, Y. The underlying linkage between personal control and thermal comfort: Psychological or physical effects? *Energy Build.* **2016**, *111*, 56–63. [CrossRef]
26. Karjalainen, S. Thermal comfort and use of thermostats in Finnish homes and offices. *Build. Environ.* **2009**, *44*, 1237–1245. [CrossRef]
27. De Dear, R.; Brager, G.S. Developing an Adaptive Model of Thermal Comfort and Preference. *ASHRAE Trans.* **1998**, *104*, 1–18.
28. *ASHRAE Standard 55 Thermal Environmental Conditions for Human Occupancy*; ANSI/ASHRAE: Atlanta, GA, USA, 2013.
29. International Organization for Standardization ISO 7730:2005. Available online: http://www.iso.org/cms/render/live/en/sites/isoorg/contents/data/standard/03/91/39155.html (accessed on 26 July 2019).
30. Olesen, B.W.; Parsons, K.C. Introduction to thermal comfort standards and to the proposed new version of EN ISO 7730. *Energy Build.* **2002**, *34*, 537–548. [CrossRef]
31. OFFICAIR Project. Available online: http://www.officair-project.eu/ (accessed on 6 August 2019).
32. Nørgaard, A.W.; Kofoed-Sørensen, V.; Mandin, C.; Ventura, G.; Mabilia, R.; Perreca, E.; Cattaneo, A.; Spinazzè, A.; Mihucz, V.G.; Szigeti, T.; et al. Ozone-initiated Terpene Reaction Products in Five European Offices: Replacement of a Floor Cleaning Agent. *Environ. Sci. Technol.* **2014**, *48*, 13331–13339. [CrossRef]
33. Mihucz, V.G.; Szigeti, T.; Dunster, C.; Giannoni, M.; de Kluizenaar, Y.; Cattaneo, A.; Mandin, C.; Bartzis, J.G.; Lucarelli, F.; Kelly, F.J.; et al. An integrated approach for the chemical characterization and oxidative potential assessment of indoor PM2.5. *Microchem. J.* **2015**, *119*, 22–29. [CrossRef]
34. Bluyssen, P.M.; Roda, C.; Mandin, C.; Fossati, S.; Carrer, P.; de Kluizenaar, Y.; Mihucz, V.G.; de Oliveira Fernandes, E.; Bartzis, J. Self-reported health and comfort in 'modern' office buildings: First results from the European OFFICAIR study. *Indoor Air* **2016**, *26*, 298–317. [CrossRef]
35. De Kluizenaar, Y.; Roda, C.; Dijkstra, N.E.; Fossati, S.; Mandin, C.; Mihucz, V.G.; Hänninen, O.; de Oliveira Fernandes, E.; Silva, G.V.; Carrer, P.; et al. Office characteristics and dry eye complaints in European workers-The OFFICAIR study. *Build. Environ.* **2016**, *102*, 54–63. [CrossRef]
36. Sakellaris, I.; Saraga, D.; Mandin, C.; Roda, C.; Fossati, S.; de Kluizenaar, Y.; Carrer, P.; Dimitroulopoulou, S.; Mihucz, V.; Szigeti, T.; et al. Perceived Indoor Environment and Occupants' Comfort in European "Modern" Office Buildings: The OFFICAIR Study. *Int. J. Environ. Res. Public Health* **2016**, *13*, 444. [CrossRef]
37. Szigeti, T.; Dunster, C.; Cattaneo, A.; Cavallo, D.; Spinazzè, A.; Saraga, D.E.; Sakellaris, I.A.; de Kluizenaar, Y.; Cornelissen, E.J.M.; Hänninen, O.; et al. Oxidative potential and chemical composition of PM2.5 in office buildings across Europe—The OFFICAIR study. *Environ. Int.* **2016**, *92–93*, 324–333. [CrossRef]
38. Campagnolo, D.; Saraga, D.E.; Cattaneo, A.; Spinazzè, A.; Mandin, C.; Mabilia, R.; Perreca, E.; Sakellaris, I.; Canha, N.; Mihucz, V.G.; et al. VOCs and aldehydes source identification in European office buildings—The OFFICAIR study. *Build. Environ.* **2017**, *115*, 18–24. [CrossRef]
39. Mandin, C.; Trantallidi, M.; Cattaneo, A.; Canha, N.; Mihucz, V.G.; Szigeti, T.; Mabilia, R.; Perreca, E.; Spinazzè, A.; Fossati, S.; et al. Assessment of indoor air quality in office buildings across Europe—The OFFICAIR study. *Sci. Total Environ.* **2017**, *579*, 169–178. [CrossRef]
40. Szigeti, T.; Dunster, C.; Cattaneo, A.; Spinazzè, A.; Mandin, C.; Le Ponner, E.; de Oliveira Fernandes, E.; Ventura, G.; Saraga, D.E.; Sakellaris, I.A.; et al. Spatial and temporal variation of particulate matter characteristics within office buildings—The OFFICAIR study. *Sci. Total Environ.* **2017**, *587–588*, 59–67. [CrossRef]
41. Raw, G.J.; Roys, M.S.; Whitehead, C.; Tong, D. Questionnaire design for sick building syndrome: An empirical comparison of options. *Environ. Int.* **1996**, *22*, 61–72. [CrossRef]
42. Theodorsson-Norheim, E. Kruskal-Wallis test: BASIC computer program to perform nonparametric one-way analysis of variance and multiple comparisons on ranks of several independent samples. *Comput. Methods Programs Biomed.* **1986**, *23*, 57–62. [CrossRef]

43. Gelman, A.; Hill, J. *Data Analysis Using Regression and Multilevel/Hierarchical Models*; Cambridge University Press: Cambridge, UK, 2007; ISBN 978-0-521-68689-1.

44. Goldstein, H. *Multilevel Statistical Models*; John Wiley & Sons: West Sussex, UK, 2011; ISBN 978-1-119-95682-2.

45. Mukaka, M. A guide to appropriate use of Correlation coefficient in medical research. *Malawi Med. J.* **2012**, *24*, 69–71.

46. *IBM SPSS Statistics for Windows*; Version 22.0; IBM: Armonk, NY, USA, 2013.

47. Brager, G.; Paliaga, G.; de Dear, R.; Olesen, B.; Wen, J.; Nicol, F.; Humphreys, M. Operable Windows, Personal Control, and Occupant Comfort. *ASHRAE Trans.* **2004**, *110*, 17–35.

48. Raja, I.A.; Nicol, J.F.; McCartney, K.J.; Humphreys, M.A. Thermal comfort: Use of controls in naturally ventilated buildings. *Energy Build.* **2001**, *33*, 235–244. [CrossRef]

49. Gou, Z.; Zhang, J.; Shutter, L. The Role of Personal Control in Alleviating Negative Perceptions in the Open-Plan Workplace. *Buildings* **2018**, *8*, 110. [CrossRef]

50. Moore, T.; Carter, D.; Slater, A. User attitudes toward occupant controlled office lighting. *Light. Res. Technol.* **2002**, *34*, 207–216. [CrossRef]

51. Roulet, C.-A.; Johner, N.; Foradini, F.; Bluyssen, P.; Cox, C.; De Oliveira Fernandes, E.; Müller, B.; Aizlewood, C. Perceived health and comfort in relation to energy use and building characteristics. *Build. Res. Inf.* **2006**, *34*, 467–474. [CrossRef]

Article

An Investigation of the Effects of Changes in the Indoor Ambient Temperature on Arousal Level, Thermal Comfort, and Physiological Indices [†]

Jongseong Gwak [1,*], Motoki Shino [2], Kazutaka Ueda [3] and Minoru Kamata [2]

[1] Institute of Industrial Science, The University of Tokyo, Tokyo 153-8505, Japan
[2] Department of Human and Engineered Environment Studies, Graduate School of Frontier Sciences, The University of Tokyo, Chiba 277-8563, Japan; motoki@k.u-tokyo.ac.jp (M.S.); mkamata@k.u-tokyo.ac.jp (M.K.)
[3] Department of Mechanical Engineering, The University of Tokyo, Tokyo 113-8656, Japan; ueda@design-i.t.u-tokyo.ac.jp
* Correspondence: js-gwak@iis.u-tokyo.ac.jp; Tel.: +81-3-5452-6098
† This paper is an extended version of paper published in the 2015 IEEE International Conference on Systems, Man, and Cybernetics, held in Hong Kong, 9–12 October 2015.

Received: 30 January 2019; Accepted: 26 February 2019; Published: 3 March 2019

Abstract: Thermal factors not only affect the thermal comfort sensation of occupants, but also affect their arousal level, productivity, and health. Therefore, it is necessary to control thermal factors appropriately. In this study, we aim to design a thermal environment that improves both the arousal level and thermal comfort of the occupants. To this end, we investigated the relationships between the physiological indices, subjective evaluation values, and task performance under several conditions of changes in the indoor ambient temperature. In particular, we asked subjects to perform a mathematical task and subjective evaluation related to their thermal comfort sensation and drowsiness levels. Simultaneously, we measured their physiological parameters, such as skin temperature, respiration rate, electroencephalography, and electrocardiography, continuously. We investigated the relationship between the comfort sensation and drowsiness level of occupants, and the physiological indices. From the results, it was confirmed that changes in the indoor ambient temperature can improve both the thermal comfort and the arousal levels of occupants. Moreover, we proposed the evaluation indices of the thermal comfort and the drowsiness level of occupants using physiological indices.

Keywords: thermal comfort; arousal level; physiological indices; electroencephalography; electrocardiography

---

1. Introduction

The control of indoor environmental quality (IEQ, which consists of visual elements, olfactory elements, and thermal factors, etc.) is important for improvements of the comfort and productivity of occupants. Among the factors of IEQ, thermal factors, such as ambient temperature, radiant temperature, humidity, and air velocity, are especially related with not only the thermal sensation and thermal comfort of occupants, but also productivity and health. With the high technology of heating, ventilation, and air conditioning systems (HVAC systems), the relationship between thermal factors and thermal sensation and thermal comfort has received attention from several researchers worldwide. In previous research, indices of thermal comfort, such as the predicted mean vote (PMV), predicted percent dissatisfied (PPD), and standard new effective temperature (SET*), have been proposed based on the relationship between thermal factors and the subjective evaluation of thermal comfort sensation [1,2]. Many studies have analyzed the indoor thermal quality based on the evaluation of thermal factors using these indices [3–7].

In addition, improvements of the productivity of occupants who are working, studying, and driving in indoor spaces, such as offices, classrooms, and vehicles, are needed. It is necessary to maintain a high arousal state of the occupants to improve their productivity. In previous studies, it was shown that productivity improved when the arousal level of occupants was high, and this was related to the indoor ambient temperature that occupants felt as cool or cold [8,9]. From the results of previous studies, the conditions of the ambient temperature to improve task performance were different from the conditions of the ASHRAE (American Society of Heating, Refrigerating and Air-Conditioning Engineers)'s thermal comfort zone [10]. Therefore, the conditions of the constant ambient temperature were not appropriate to improve both the task performance and thermal comfort levels. In previous studies, it was shown that the thermal comfort of occupants was immediately changed in accordance with changes in the indoor ambient temperature [11]. The arousal level was increased by outer stimulation and was maintained at a high state for several minutes [12,13]. Based on these results of previous studies, we hypothesized that thermal stimulation due to cooling can improve the arousal level of occupants. Furthermore, thermal comfort can be improved while maintaining high arousal levels due to the removal of thermal stimulation. When considering the possibility that changes in thermal factors can improve both arousal and thermal comfort levels, the physiological effects associated with such changes in thermal factors, and how they affect the arousal level and thermal comfort of occupants are not clear. Therefore, the design requirements for changes in ambient temperature to improve both the arousal level and thermal comfort are also not clear. Therefore, a continuous and quantitative evaluation of the thermal comfort and arousal level of occupants using indices, which can be measured both continuously and quantitatively, such as physiological signals, is needed to clarify the design requirements of changes in the indoor ambient temperature to improve both the arousal level and thermal comfort.

In this study, we aimed to investigate the characteristics of changes in the arousal level and feelings of thermal comfort of occupants, and the relationship between them when thermal factors are changed. There are many factors that affect the thermal comfort and arousal level of occupants, such as the ambient temperature, indoor air velocity, mean air radiant temperature, and metabolic activity [5–7]. Especially, we focused on the changes in the indoor ambient temperature as a fundamental investigation in this study. In addition, to propose evaluation indices that can evaluate the thermal comfort and arousal level of occupants continuously and quantitatively, we investigated the relationship between the arousal level, feelings of thermal comfort, and physiological indices, which can be measured continuously and quantitatively.

2. Strategy

2.1. Investigation of the Effects of Changes in the Indoor Ambient Temperature on Arousal Levels, Thermal Comfort, and Task Performance

To verify the hypothesis that thermal stimulation, due to cooling, can improve the arousal level of occupants, and that thermal comfort can be improved while maintaining high arousal levels due to the removal of thermal stimulation, several thermal conditions were set. Subjects were asked to conduct a mathematical task and periodically evaluate their sensation values (using a subjective sensation vote) in relation to their arousal level and thermal comfort. We attempt to clarify the effect of changes in the indoor ambient temperature on the arousal level, thermal comfort, and task performance of occupants by conducting an analysis of the results of the subjective sensation vote and task performance.

2.2. Investigation of the Relation between the Subjective Evaluation Value and Physiological Parameters, and a Recommendation for Evaluation Indices

Indices that can evaluate the arousal and thermal comfort levels of occupants continuously and quantitatively are necessary to clarify the design requirements of a temperature control that can improve the arousal and thermal comfort levels of occupants. However, as it was not possible to perform a subjective sensation voting continuously, we considered clarifying the design requirements for thermal environments using changes over time. This relied on the possibility to evaluate the arousal level and feelings of thermal comfort using physiological parameters that could be measured quantitatively and continuously. Previous studies have shown that the physiological indices measured utilizing electroencephalograms (EEG), electrocardiograms (ECG), and respiration rates were effective for the evaluation of arousal levels [14,15], and the skin temperature, EEGs, and ECGs of occupants were effective for the evaluation of the thermal comfort [16,17]. Therefore, in this study, we assumed a flow of physiological responses when the occupant was stimulated by thermal factors, as shown in Figure 1, based on the above-mentioned previous studies to clarify the characteristics of the physiological parameters under the condition of changes in both the arousal level and thermal comfort. A thermal stimulation is transmitted from the sensory organ, such as warm spots and cold spots of skin, to the central nervous system, which affects thermal and comfort sensation. This change in comfort sensation then affects the autonomic nervous system, and the response is then conducted through the locomotive organ. Based on this process, we selected the skin temperature as an indicator of heat transfer between the environment and the human body, EEGs as an indicator of the reaction of the central nervous system, and ECGs as indicators of the reaction of the autonomic nervous system. Next, we attempted to search for indices that could separately evaluate the arousal level and thermal comfort by performing a multiple regression analysis utilizing the physiological parameters and subjective evaluation values of the arousal level and feelings of thermal comfort. In Figure 1, the flow of physiological responses with the thermal stimulus is shown.

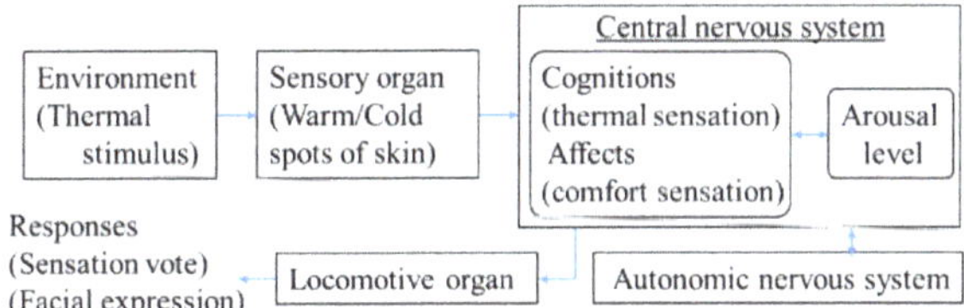

Figure 1. Flow of physiological responses with the thermal stimulus.

3. Methodology

3.1. Subjects

Ten subjects with vital statistics of a height of 173.5 ± 4.4 cm, weight of 61 ± 4.8 kg, age of 22.1 ± 1.2 years old, and who were right-handed participated in the experiment. The experimental contents and procedures, which were approved by the ethics committee of the University of Tokyo, were explained to the subjects before conducting the experiment. The subjects were then asked to avoid intense physical activity, alcohol, and caffeine for 24 h prior to the experimental session.

3.2. Experimental Task

The subjects were asked to conduct mental arithmetic tasks known as "MATH", which is based on the algorithm proposed by Tuner et al. [18]. A 1–3-digit addition or subtraction question is displayed for 2 s on the monitor, and then "equals to" is displayed for 1.5 s. Lastly, the answer is displayed for 1 s, and the next question is displayed after 0.5 s. Subjects had to determine if the answer was correct or incorrect when the answer was displayed and click the left mouse button if the answer was correct or the right mouse button if the answer was incorrect. The levels of questions in the original version of MATH consisted of 1–5 levels. In this study, the beginning level was 3 (which is 2-digit addition or subtraction), and the level of the next question was raised if the responses of the subjects were correct, and was reduced if answers were incorrect. We deduced that changing the levels affected the physiological indices. Thus, we did not change the level of the question, and fixed it to level 3. The MATH task included 50 questions for 250 seconds per set.

3.3. Experimental Conditions and Experimental Procedure

To evaluate the effects of the duration and degree of thermal stimulation, three environmental conditions (A–C) were set as follows.

- Condition A: The indoor ambient temperature was maintained at 27 °C.
- Condition B: The indoor ambient temperature was decreased from 27 °C to 20 °C, and then increased from 20 °C to 27 °C.
- Condition C: The indoor ambient temperature was decreased from 27 °C to 20 °C, and then maintained at 20 °C.

We set condition A as a thermally comfort condition [10], and condition C as a high arousal condition [8,9]. Condition B was set to verify the hypothesis that thermal stimulation due to cooling can improve the arousal levels of occupants, after which thermal comfort can be improved while maintaining high arousal levels due to the removal of thermal stimulation. There are many parameters which affect the thermal comfort and arousal levels of occupants, such as the indoor velocity, mean air radiant temperature, metabolic activity, and amount of clothing. In this study, we focused on only the indoor ambient temperature as a fundamental investigation, thus those parameters except the indoor ambient temperature were controlled in the experiment. The subjects wore short sleeves and short pants and were asked to remain in the pre-room. The room temperature was set at 27 $\pm$ 0.5 °C for approximately 1 h so that subjects could adjust to the thermal environment; they were asked to practice the "MATH" at least twice during this time. Sensors were then attached to the bodies of the subjects to measure their physiological indices. One set of tasks consisted of completing the subjective sensation vote and the "MATH" task. The subjects were asked to perform seven sets of tasks at time intervals of 10 min and to rest between each task for approximately 3 min. Physiological indices were measured from the start time (0 min) till the end time (70 min). The experimental procedure and environmental conditions are shown in Figure 2.

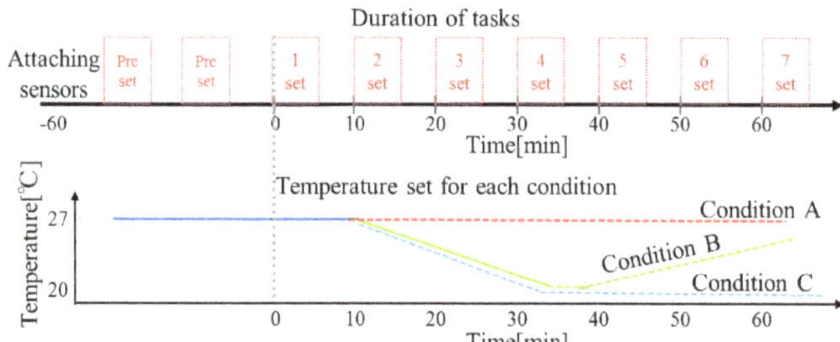

Figure 2. Experimental conditions and environmental procedure.

3.4. Measurement

3.4.1. Subjective Evaluation of the Drowsiness Level and the Thermal Comfort Sensation

Before completing the "MATH" task, the subjects were asked to complete a questionnaire related to their thermal sensation, thermal comfort sensation, and arousal level. The scale of thermal sensation was based on ASHRAE/ISO (International Organization for Standardization) [10], and was denoted using integral numbers from −3 to 3 (where −3, −2, −1, 0, 1, 2, and 3 are the meanings of cold, cool, slightly cool, neutral, slightly warm, warm, and hot, respectively.). The scale of comfort sensation was based on ISO10551 [19], and was denoted using integral numbers from −3 to 0 (where −3,

-2, -1, and 0 are the meanings of very uncomfortable, uncomfortable, slightly uncomfortable, and comfortable, respectively.). The scale of the arousal level was based on the drowsiness level of Zilberg's indicators [20], and was denoted using integral numbers from 0 to 4 (where 0, 1, 2, 3, and 4 are the meanings of alert, slightly drowsy, moderately drowsy, significantly drowsy, and extremely drowsy, respectively.).

3.4.2. Physiological Indices

(a) EEG

EEGs were recorded using an EEG-measuring instrument (EEG-1200, Nihonkohden Co., Japan) at a sampling rate of 500 Hz. EEG electrodes were attached on 16 channels (based on the internationally accepted 10–20 system, Fp1, Fp2, F7, F3, F4, F8, T7, C3, C4, T8, P7, P3, P4, P8, O1, O2). Next, raw data were processed using the Fourier transform method, and the spectral power of each frequency band, such as the content of theta wave (4–8 Hz), low-alpha wave (8–10 Hz), high-alpha wave (10–13 Hz), low-beta wave (13–20 Hz), high-beta wave (20–30 Hz), and SMR (12–15 Hz) bands, was calculated in addition to the values of the beta per alpha and alpha per high-beta for each channel.

(b) ECG

ECGs were recorded using an ECG-measuring instrument (WEB-7000 and ECG picker, Nihonkohden Co., Japan) at a sampling rate of 1000 Hz. Three electrodes were attached to the chests of subjects using the precordial leads method. The R-R interval (RRI) was calculated from the ECG waveform using MATLAB (Mathwork Co.) programs. The values of the mean of the RRI and the coefficient of variance of RRI [CVRR (100*SD/Mean of RRI)] were calculated from the RRI data. In addition, the spectral power of each frequency band, such as the very low frequency (VLF, 0.001–0.04 Hz), low frequency (LF, 0.04–0.15 Hz), and high frequency (HF, 0.15–0.45 Hz) bands, was calculated from the time series of the RRI using the fast Fourier transform method.

(c) Respiration

A thermal picker (WEB-7000, Nihonkohden Co., Japan) was used to measure the temperature of the breath of subjects. The peak values in the time series graph of the temperature were detected, and the mean and standard deviation of the respiratory cycle time were calculated.

(d) Skin Temperature

Thermocouples were attached at 7 places on the bodies of subjects based on the Hardy–Du Bois method [16] to measure the skin temperature (LT8, GRAM Co., Japan). Finally, the mean skin temperature was calculated using following Equation (1):

$$MST = 0.07\,T_1 + 0.14\,T_2 + 0.05\,T_3 + 0.35\,T_4 + 0.19\,T_5 + 0.13\,T_6 + 0.07\,T_7 \tag{1}$$

where MST is the mean skin temperature based on the Hardy-DuBois method [16], and T_1, T_2, T_3, T_4, T_5, T_6, and T_7 are the temperatures of the forehead, forearms, hands, abdomen, thighs, legs, and feet, respectively.

3.4.3. Facial Expression and Task Performance

The activities of subjects were recorded using the video camera, and the drowsiness levels were recorded by an observer at intervals of 10 s based on Zilberg's criteria [20]. After the experiment was completed, the mean of the data recorded when subjects were performing the MATH task was calculated. The task performance of MATH included the number of correct answers and the mean of the reaction time taken to solve 50 questions. Figure 3 shows a participant completing the experiment. Figure 3 shows photos of the field of experimental scene.

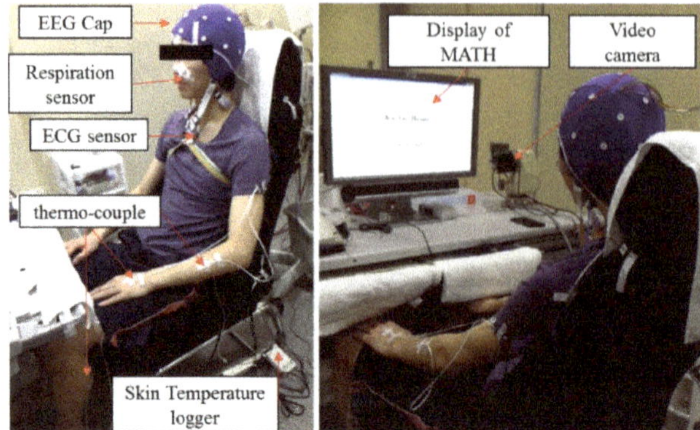

Figure 3. Field photos of the experimental scene.

4. Results and Discussion

4.1. Characteristics of the Arousal Level and Thermal Comfort Corresponding to Changes in the Indoor Ambient Temperature

4.1.1. Subjective Evaluation Value

The results of the subjective evaluation value corresponding to the drowsiness levels and the thermal comfort sensation vote of subjects under conditions A, B, and C are shown in Figures 4 and 5. We used multiple comparison based on the Bonferroni method to investigate significant differences of the subjective evaluation value due to changes in the indoor ambient temperature. There was a small change in the comfort sensation vote, but the drowsiness level increased under condition A, where the indoor temperature was maintained at 27 °C. Under condition B, the comfort sensation vote decreased to an uncomfortable state when the indoor ambient temperature dropped, and increased to a comfortable state when the indoor ambient temperature increased. Under condition C, the comfort sensation vote decreased to an uncomfortable state when the indoor temperature decreased and was then maintained in this uncomfortable state. Under conditions B and C, the drowsiness levels decreased and were maintained at an alert state, corresponding to the drop in the indoor temperature. After the completion of set 4, the comfort sensation vote increased, corresponding to the increase in the indoor temperature under condition B, but was maintained at an uncomfortable state under condition C. The drowsiness level was maintained at a low state under both conditions. According to this result, the arousal level was maintained at a high state even when the temperature increased, and subjects felt comfortable even when the temperature dropped, leading to an increased arousal level. This suggests that there is not always a dependence relationship between the arousal levels and feelings of thermal comfort. Figures 4 and 5 show that when condition B is applied in sets 6 and 7, it becomes possible to improve both the arousal level and thermal comfort of occupants by changing the indoor ambient temperature. These results show that the hypothesis that thermal stimulation due to cooling can improve the arousal level of occupants, after which thermal comfort can also be improved while maintaining a high arousal levels due to the removal of thermal stimulation was verified for condition B. Figure 4 shows the result of the thermal comfort sensation vote. Figure 5 shows the result of the drowsiness level sensation vote.

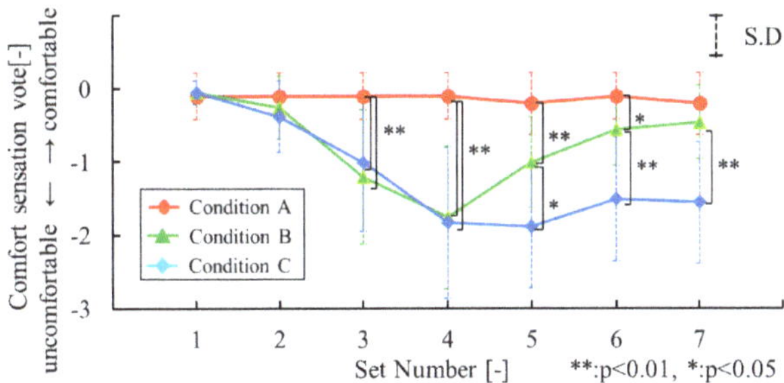

Figure 4. Results of the value of the thermal comfort sensation vote.

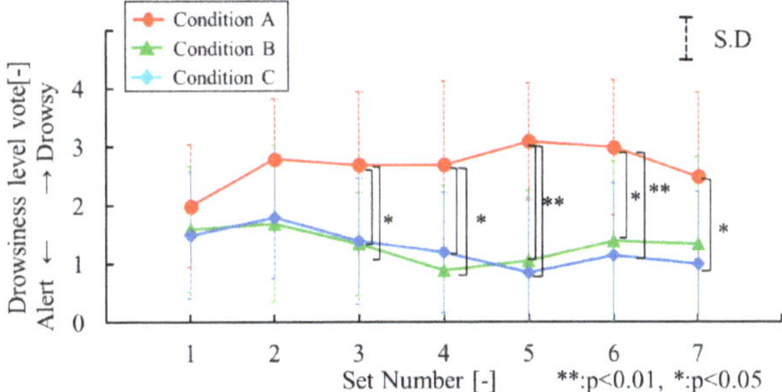

Figure 5. Results of the value of the drowsiness level sensation vote.

4.1.2. Rating Value of Zilberg's Drowsiness Level by Observer

The rating values of Zilberg's drowsiness level for each condition are shown in Figure 6, and it is evident from the figure that the results are similar to those of the subjective sensation vote. Multiple comparison based on the Bonferroni method was conducted to investigate significant differences of the drowsiness level due to the changes in the indoor ambient temperature. After the completion of set 3, the drowsiness levels under conditions B and C decreased significantly compared to that under condition A. From the result of condition B, it was confirmed that the drowsiness level was maintained at a low state after the completion of set 3.

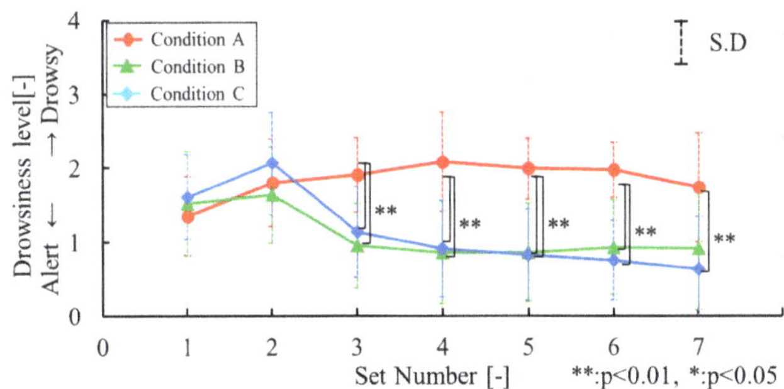

Figure 6. Drowsiness level reached when calculating MATH (analysis of facial expressions based on Zilberg's method [20]).

4.1.3. Task Performance of MATH

The results of the MATH score under each condition are shown in Figure 7, and the results of the response time taken to calculate MATH questions under each condition are shown in Figure 8. Multiple comparison based on the Bonferroni method was conducted to investigate significant differences in the task performance due to changes in the indoor ambient temperature. It can be seen from Figures 7 and 8 that the MATH score was high after set 4, and the response time taken to calculate MATH questions increases in the order of conditions C, B, and then A. Differences were observed between the conditions in the results of the subjective sensation vote and the rating value of Zilberg's drowsiness level. However, significant differences were observed only between the conditions in the result of the response time taken to calculate MATH for set 4, and the result of the MATH score for set 5. It was assumed that the MATH work was easy enough for the subjects to perform, even at a low state of arousal. Figure 6 shows the drowsiness level reached when calculating MATH (analysis of facial expressions based on Zilberg's method [20]). Figures 7 and 8 show the result of the MATH score and the response time from the display of the MATH question to the participant's click.

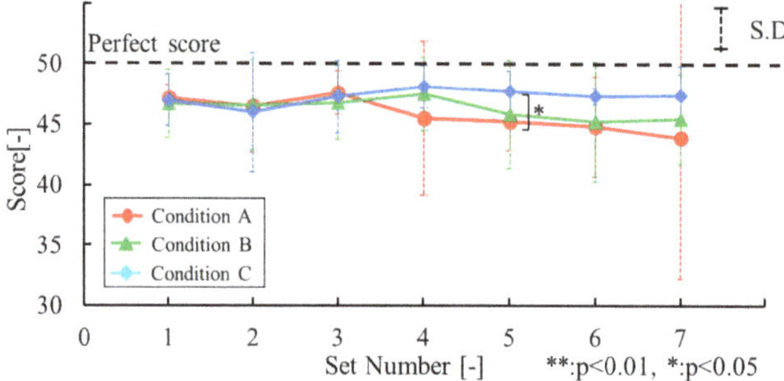

Figure 7. Result of MATH score.

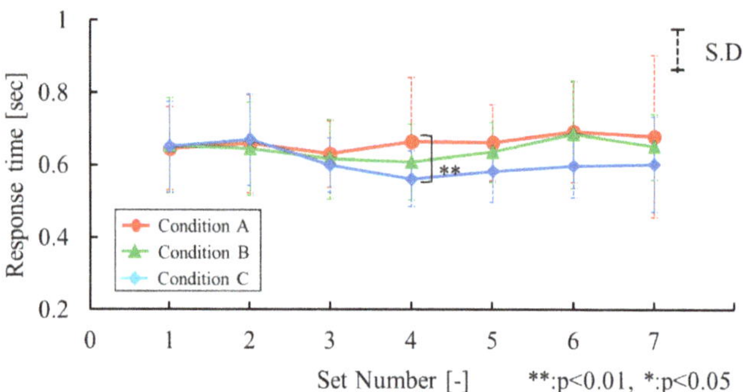

Figure 8. Result of response time.

4.1.4. Consideration of Thermal Stimulation for High Arousal Levels

In the previous studies of Yoshida et al. [12] and Mohri et al. [13], to maintain a high arousal level, the effects of several stimulation methods, such as the use of a fragrance, alarm, and shoulder oscillation, were confirmed. As a result, the sleeping rebound occurred 10 min after the stimulation in those studies. As in our results, high arousal states were maintained for over 10 min after the removal of thermal stimulation. Though further investigation of these comparisons of the time to rebound is necessary, our results showed that thermal stimulation is effective in improving the arousal levels of occupants. Especially, our method is useful in spaces where air conditioning systems are installed, because the installation of additional devices in the space to make stimulations, such as fragrances, alarms, and shoulder oscillation, is unnecessary.

4.2. Relationship between the Subjective Evaluation Value and Physiological Indices

Three hundred and twenty-nine physiological indices were calculated by using the data of the EEGs, ECGs, respiration rate, and mean skin temperature in the experiment (320 indices from EEGs, 7 indices from ECGs, 2 indices from respiration rate, and 1 index from skin temperature). By considering the result of the subjective evaluation value as the standard for the perception of state, we used correlative analysis to investigate the relationship between the subjective evaluation value and the physiological indices. According to the flow of the physiological response shown in Figure 1, changes in both the arousal level and thermal comfort affect the physiological response. To find indices (out of the 329 physiological parameters) that corresponded to the arousal level, we used data from sections when the arousal level changed, but the value of the thermal comfort sensation was maintained constantly. On the other hand, to find indices that corresponded to thermal comfort, we used data from sections when the value of the thermal comfort sensation changed and the arousal level was maintained constantly.

4.2.1. Relationship between the Drowsiness Level and Physiological Indices

As a result of the correlation analysis, the correlation coefficients between values of the drowsiness level vote and physiological indices were calculated using data from 57 sets in which the value of the drowsiness level vote varied, but the value of the comfort sensation vote remained at 0 (comfortable). There was a significant correlation between the drowsiness level and 102 physiological indices (data $N = 57$, $r > 0.339$, $p < 0.01$). The indices in relation to the ECGs were not included in these parameters; however, the indices in relation to the EEGs were included. Thus, this result showed that the relationship between the reaction of the central nervous system and changes in arousal levels is more significant than that between the reaction of the autonomic nervous system and changes in arousal levels.

4.2.2. Relationship between Thermal Comfort and Physiological Indices

As a result of the correlation analysis, the correlation coefficients between the values of the comfort sensation vote and physiological indices were calculated using the data from 98 sets in which the value of the comfort sensation vote varied, but the value of the drowsiness level vote was between 0 (alert) and 1 (slightly drowsy). There was a significant correlation between the value of the comfort sensation vote and seven physiological indices in relation to skin temperature, EEGs, and ECGs (data $N = 98$, $r > 0.259$, $p < 0.01$). The change in the mean skin temperature is related to thermal stimulation from the outer environment, which is transmitted to the central nervous system and perceived as thermal comfort, thus affecting the parameters of the EEGs. It was observed that changes in the thermal comfort affected the autonomic nervous system, which in turn affected the indices in relation to the ECGs. It appears that this process affected the result of the correlation analysis in this experiment.

4.2.3. Calculation of the Evaluation Index Using Multiple Regression Analysis

To propose an index that evaluates the changes in each arousal level and thermal comfort states, multiple regression analysis was performed using the physiological indices that have a significant correlation with the value of subjective evaluation. It was found that 102 parameters had a significant correlation with the drowsiness level vote, and seven parameters with the comfort sensation vote. These were thus considered as explanatory variables. Because an explanatory variable should have a high correlation with the subjective sensation vote, to obtain a high multiple regression coefficient, we selected an explanatory variable for use in the multiple regression analysis using the following process.

- Indices that had a significant correlation with the value of subjective evaluation were sorted according to their correlation coefficient from high to low ($x1$, $x2$, ... $x102$).
- $x1$ was selected as the explanatory variable, since it had the highest correlation coefficient with the value of subjective evaluation.

- $x2$ was selected as the explanatory variable if there was no significant correlation between $x2$ and $x1$.
- xn was selected as the explanatory variable if there was no significant correlation between xn and all the parameters selected previously as explanatory variables.

(a) Evaluation Index indicating the Arousal Level

The result of the multiple regression analysis 1 (Yd) is shown in Table 1, and the regression equation obtained from the result is expressed in Equation (2). The variables of the regression equation include the indices of the EEGs. As a result, the coefficient of determination, R^2, in relation to Yd was 0.750.

$$Yd = -3.31 + 0.134 \, Xd_1 + 2.520 \, Xd_2 + 0.088 \, Xd_3 - 0.268 \, Xd_4 + 0.271 \, Xd_5 \tag{2}$$

where Xd_1, Xd_2, Xd_3, Xd_4, and Xd_5 are the high alpha content of T7, beta per alpha content of F7, beta content of F3, low beta content of F7, and alpha content of Fp1, respectively.

Table 1. Result of multiple regression 1 (Yd).

Model Summary				
R	0.8657		Std. Error	0.7358
R^2	0.7495		data N	57
Adjusted R^2	0.7249			

Index	Coefficient	Std.Error	t	p-value
Intercept	−3.3123	0.4959	−4.6631	2.28×10^{-5}
T7_High-alpha	0.1338	0.0201	6.6478	1.96×10^{-8}
F7_Beta/Alpha	2.5196	0.3073	8.1996	7.09×10^{-11}
F3_Beta	0.0878	0.0245	3.5817	0.0008
F7_Low beta	−0.2684	0.0481	−5.5791	9.29×10^{-7}
Fp1_Alpha	0.2708	0.0315	8.5969	1.72×10^{-11}

We defined Yd as the index of the arousal level. To confirm the possibility that Yd can evaluate the arousal level even when the thermal comfort changed, we calculated Yd from the data obtained from five subjects for whom the arousal level and thermal comfort changed considerably, and then performed a correlation analysis between Yd and the evaluation value of the drowsiness level. As a result of the correlation analysis, the correlation coefficient (R) was calculated to be 0.726, and there was a significant correlation between Yd and the evaluation value of the drowsiness level. Therefore, this result suggests that Yd can be used to evaluate the drowsiness level, even when there are changes in the thermal comfort.

(b) Evaluation Index Indicating Thermal Comfort

The result of the multiple regression analysis 2 (Yc) is shown in Table 2, and the regression equation obtained from the result is expressed in Equation (3). The variable for the regression equation included the indices of the skin temperature and EEGs. As a result, the coefficient of determination, R^2, in relation to Yc was 0.528.

$$Yc = -23.372 + 0.697 \, Xc_1 + 0.172 \, Xc_2 + 0.142 \, Xc_3 \tag{3}$$

where Xc_1, Xc_2, and Xc_3 are the *MST* (Mean Skin Temperature), alpha per high-beta content of C4, and alpha per high beta content of P4, respectively.

Table 2. Result of multiple regression 2 (Yc).

Model Summary				
R	0.7267		Std. Error	0.6859
R^2	0.5276		data N	98
Adjusted R^2	0.5125			
Index	**Coefficient**	**Std.Error**	**t**	**p-value**
Intercept	-23.3721	2.5461	-9.1795	1.02×10^{-14}
MST	0.6967	0.0765	9.1119	1.42×10^{-14}
C4_Alpha/HB	0.1718	0.0586	2.9295	0.0043
P4_Alpha/HB	-0.1418	0.0296	-4.7885	6.25×10^{-6}

We defined Yc as the index of thermal comfort. To confirm the possibility that Yc can evaluate thermal comfort even when there were changes in the drowsiness level, we calculated Yc from the data of nine subjects, excluding the data of participant D whose thermal comfort vote and drowsiness level vote changed slightly. We thus confirmed the relationship between Yc and the value of thermal comfort. As a result, the correlation coefficient (R) was calculated to be 0.728, and there was a significant correlation between Yc and the value of thermal comfort ($p < 0.01$). Therefore, this result suggests that Yc can be used to evaluate thermal comfort even when there are changes in the drowsiness level.

Table 1 shows the result of the multiple regression 1 (Yd), and Table 2 shows the result of the multiple regression 2 (Yc). Figure 9 shows the relationship between the drowsiness level vote and Yd(a), and the relationship between the comfort sensation and Yc (b)

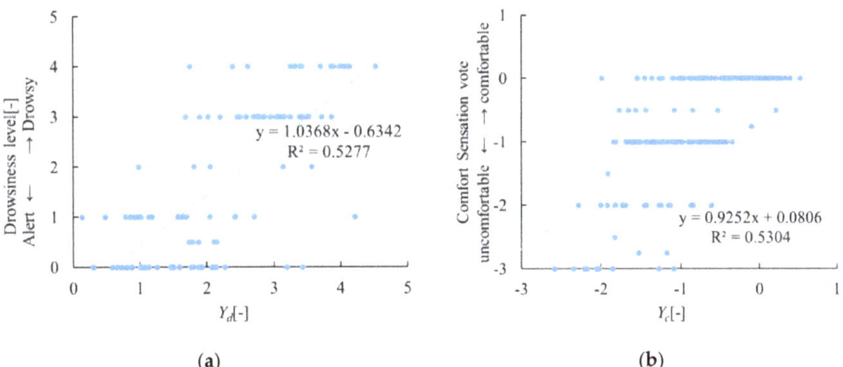

Figure 9. The relationship between the drowsiness level and Yd (**a**), the comfort sensation and Yc (**b**).

(c) Review of Continuous Evaluation using Yd and Yc

As mentioned in Section 2.2., indices that can continuously evaluate the drowsiness level and thermal comfort of occupants are needed to clarify the design requirements related to thermal factors that can improve these states of the occupants. Therefore, we conducted a time-series analysis with Yd and Yc to confirm the possibility that Yd and Yc can continuously evaluate arousal levels and thermal comfort. We calculated the time series of Yd using data from subject C, who seemed to be drowsy frequently throughout the experiment, and compared these data with those of the drowsiness level, which was recorded at intervals of 10 s by analyzing the facial expressions of the subject based on Zilberg's method [20]. The time series analyses of the normalized Yd and the drowsiness level of subject C are shown in Figure 10a,b. The figures clearly show the trend of changes in both indicators. It can be seen that Yd and values of the drowsiness level show a similar trend. For example, the red circles in Figure 10a,b indicate that values of the drowsiness level and the normalized Yd are higher than 3 (significantly drowsy) and 0, respectively, at the same time. Furthermore, it can be seen from Figure 11 that there is a similar trend between Yc and the value of the comfort sensation vote. In Figure 11a,b, the value of the comfort sensation drops to -3 (very uncomfortable) at 30 min, and at the same time, the value of Yc drops and becomes a negative number. Thereafter, the value of the comfort sensation vote remains at an uncomfortable state (-3: Very uncomfortable, -2: Uncomfortable), and similarly, the value of Yc remains at a negative value. These results suggest that Yd and Yc can be used to evaluate the drowsiness level and thermal comfort of occupants continuously. Figures 10 and 11 show, respectively, changes in the normalized Yd (a) and drowsiness level (b), (subject C) and changes in Yc (a) and the comfort sensation vote (b) (subject B).

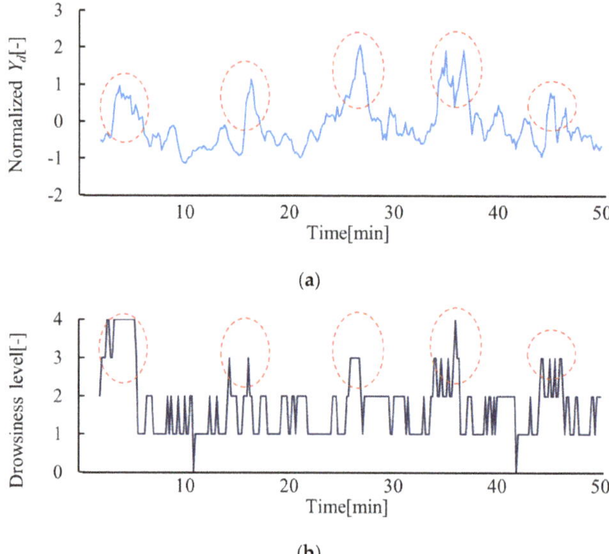

Figure 10. Changes in the normalized Yd (**a**) and drowsiness level (**b**), (subject C): (**a**) Time series of normalized Yd, (**b**) Time series of the value of the drowsiness level.

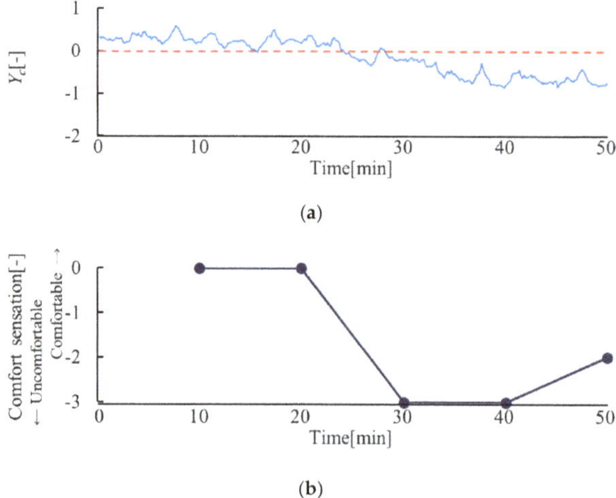

Figure 11. Changes in Yc (**a**) and the value of the comfort sensation vote (**b**): Time series of the value of the comfort sensation vote, (subject B).

(d) Review of the Utility and Validity of the Evaluation Index

To review the utility and validity of the evaluation index, as an example, we set the threshold value in relation to Yd and the drowsiness level. When the threshold value of Yd is sets as 2.3, as shown in Figure 12, Yd can be used as the classifier, which can classify the drowsiness level between level 0, 1, 2, or more with an 86.7% accuracy. This result suggests that there is a significant difference in the physiological reaction between the drowsiness level of 1 (slightly drowsy) and 2 (moderately drowsy), and is a similar trend to the results of a previous study about the relation between the arousal levels and physiological indices of a driver [21], though the task of the subjects is different from that of the previous study (arithmetic task vs driving). This result also suggests that Yd is a valid indicator of the drowsiness level of occupants to classify between the drowsiness level of 1 (slightly drowsy) and 2 (moderately drowsy). Figure 12 shows the setting of the threshold value in relation to Yd.

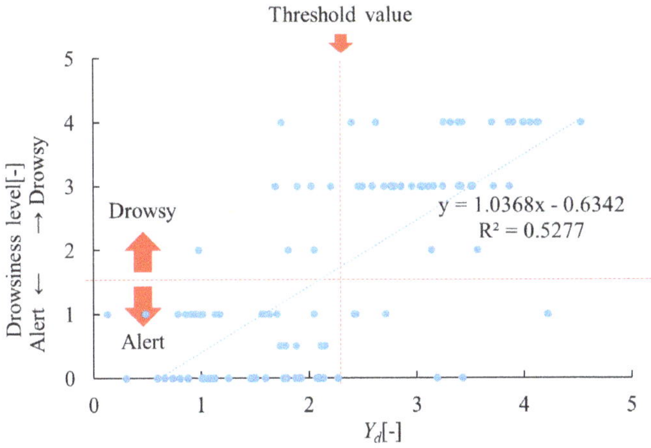

Figure 12. Setting of the threshold value in relation to Yd.

5. Conclusions

In this study, we aimed to design a thermal environment that can improve the indicators of both the arousal level and feelings of thermal comfort. We hypothesized that thermal stimulation due to cooling can improve the arousal level of occupants, after which thermal comfort can be improved while maintaining high arousal levels due to the removal of thermal stimulation. To verify the hypothesis, we measured physiological indices, values of subjective evaluation, and performances of arithmetic tasks throughout several thermal conditions in which the indoor ambient temperature was changed. In addition, we investigated the relationships between them to identify the indices that can be used to evaluate the arousal levels and thermal comfort of occupants. As a result, the following findings were noted:

- When the indoor ambient temperature decreased and then increased, both the arousal level and thermal comfort of occupants remained at high levels. This result suggests that the hypothesis of this study was verified and changes in the indoor ambient temperature can be used to improve both thermal comfort and the arousal level of occupants.
- We proposed the evaluation indices of thermal comfort and the drowsiness level of occupants. It was observed that the drowsiness level and thermal comfort of occupants can be evaluated quantitatively and continuously using Yd and Yc, which were obtained from the equation consisting of physiological indices in relation to EEGs, ECGs, and skin temperature.

In future work, we will investigate the relationship between the change patterns of the indoor ambient temperature, thermal comfort, and arousal levels of occupants. After that, we aim to design a novel thermal environment, considering all comfort parameters based on these findings, with the aim of improving both the arousal levels and feelings of thermal comfort of occupants, and then carry out an evaluation of the validity of the designed thermal environment.

Author Contributions: Conceptualization, J.G., M.S. and M.K.; methodology, J.G., M.S., K.U. and M.K.; software, J.G.; validation, J.G.; formal analysis, J.G.; investigation, J.G.; resources, M.S., K.U. and M.K.; data curation, J.G.; writing—original draft preparation, J.G.; writing—review and editing, J.G., M.S., K.U. and M.K.; visualization, J.G.; supervision, M.S. and M.K.; project administration, M.S. and M.K.; funding acquisition, M.S. and M.K.

Funding: This research received no external funding.

Conflicts of Interest: The authors declare no conflict of interest.

Formula Symbols

Index	Meaning	Unit
MST	Mean skin temperature	°C
T_1	Temperature of the forehead	°C
T_2	Temperature of the forearms	°C
T_3	Temperature of the hands	°C
T_4	Temperature of the abdomen	°C
T_5	Temperature of the thighs	°C
T_6	Temperature of the legs	°C
T_7	Temperature of the feet	°C
Yc	Index of thermal comfort	
Yd	Index of drowsiness level	
Xd_1	High alpha content of T7	
Xd_2	Beta content per alpha content of F7	
Xd_3	Beta content of F3	
Xd_4	Low beta content of F7	
Xd_5	Alpha content of Fp1	
Xc_1	Mean skin temperature	°C
Xc_2	Alpha content per high beta content of C4	
Xc_3	Alpha content per high beta content of P4	

References

1. ISO7730:2005. Ergonomics of the thermal environment-Analytical determination and interpretation of thermal comfort using calculation of the PMV and PPD indices and local thermal comfort criteria
2. De Dear, R.; Brager, G.S. Developing an adaptive model of thermal comfort and preference. *ASHRAE Trans.* **1998**, *104*, 145–167.
3. Yang, Y.; Li, B.; Liu, H.; Tan, M.; Yao, R. A study of adaptive thermal comfort in a well-controlled climate chamber. *Appl. Therm. Eng.* **2015**, *76*, 283–291. [CrossRef]
4. Cannistraro, M.; Cannistraro, G.; Restivo, R. Some Observations on the Influence on Exchanges Radiative. *Int. J. Heat Techonol.* **2015**, *33*, 115–122. [CrossRef]
5. Cannistraro, M.; Lorenzini, E. The Applications of the New Techonologies E-Sensing in Hospitals. *Int. J. Heat Techonol.* **2016**, *34*, 551–557. [CrossRef]
6. Cannistraro, M.; Cannistraro, G.; Restivo, R. Smart Controll of Air Climatization System in Function on the Values of the Mean Local Radiant Temperature. *Smart Sci.* **2015**, *3*, 157–163. [CrossRef]
7. Liu, J.; Yao, R.; McCloy, R. A method to weight three categories of adaptive thermal comfort. *Energy Build.* **2012**, *47*, 312–320. [CrossRef]
8. McIntyre, D.A. *Indoor Climate*; Applied science publishers: London, UK, 1980.

9. Gohara, T.; Iwashita, G. Disscussion of relationship between indoor environmental comfort and performance in simple task. *J. Environ. Eng.* **2003**, *572*, 75–80. [CrossRef]

10. ASHRAE-55. *Thermal Environmental Conditions for Human Occupancy*; American Society of Heating, Refrigerating and Air-conditioning Engineers: Atlanta, GA, USA, 2004.

11. Kondo, E.; Kurazumi, Y.; Horikoshi, T. Human physiological and psychological reactions to thermal transients with air temperature step changes: A Case of Climacteric Aged Females in Summer. *J. Hum. Living Environ. Jpn.* **2014**, *21*, 75–84.

12. Yoshida, M.; Kato, C.; Kakamu, Y.; Kawasumi, M.; Yamasaki, H.; Yamamoto, S.; Nakano, T.; Yamada, M. Study on Stimulation Effects for Driver Based on Fragrance Presentation. In Proceedings of the MVA2011 IAPR Conference on Machine Vision Applications, Nara, Japan, 13–15 June 2011; pp. 332–335.

13. Mohri, Y.; Kawaguchi, M.; Kojima, S.; Yamada, M.; Nakano, T.; Mohri, K. Arousal retention effect of magnetic stimulation to car drivers preventing drowsy driving without sleep rebound. *Trans. Inst. Electr. Eng. Jpn.* **2016**, *136*, 383–389. [CrossRef]

14. Mardi, Z.; Naghmeh, S.; Ashtiani, M.; Mikaili, M. EEG-based drowsiness detection for safe driving using chaotic features and statistical tests. *J. Med. Signals Sens.* **2011**, *1*, 130–137. [PubMed]

15. Furman, G.D.; Baharav, A.; Cahan, C.; Akselrod, S. Early detection of falling asleep at the wheel: A heart rate variability approach. *Comput. Cardiol.* **2008**, *35*, 1109–1112.

16. Hardy, J.D.; Dubois, E.F. The technic of measuring radiation and convection. *Nutrition* **1938**, *5*, 461–475. [CrossRef]

17. Yao, Y.; Lian, Z.; Liu, W.; Jiang, C.; Liu, Y.; Lu, H. Heart rate variation and electroencephalograph-the potential physiological factors for thermal comfort study. *Indoor Air* **2009**, *19*, 93–101. [CrossRef] [PubMed]

18. Turner, J.R.; Hewitt, J.K.; Morgan, R.K.; Sims, J.; Carroll, D.; Kelly, K.A. Graded mental arithmetic as an active psychological challenge. *Int. J. Psychophysiol.* **1986**, *3*, 307–309. [CrossRef]

19. SO10551. First edition 1995-05-15, Ergonomics of the thermal environment–Assessment of the influence of the thermal environment using subjective judgement scales.

20. Zilberg, E.; Burton, D.; Xu, Z.M.; Karrar, M. Methodology and initial analysis for development of non-invasive and hybrid driver drowsiness detection systems. In Proceedings of the 2nd IEEE International Conference Wireless Broadband and Ultra Wideband Communications, AusWireless, Sydney, Australia, 27–30 August 2007; p. 16.

21. Gwak, J.S.; Shino, M.; Hirao, A. Eary detection of driver drowsiness using machine learning based on physiological signals, behavioral measures and driving performance. In Proceedings of the 21th IEEE International Conference on Intelligent Transportation Systems, Maui, Hawaii, USA, 4–7 November 2018; pp. 1794–1800.

 *applied sciences*

Article

Probability of Abnormal Indoor Air Exposure Categories Compared with Occupants' Symptoms, Health Information, and Psychosocial Work Environment

Katja Tähtinen [1,2,*], **Sanna Lappalainen** [2], **Kirsi Karvala** [2], **Marjaana Lahtinen** [2] and **Heidi Salonen** [1]

1   Department of Civil Engineering, Aalto University, 02150 Espoo, Finland; heidi.salonen@aalto.fi
2   Finnish Institute of Occupational Health, Healthy Workspaces, P.O. Box 18, 00032 Työterveyslaitos, Finland; sanna.lappalainen@ttl.fi (S.L.); kirsi.karvala@ttl.fi (K.K.); marjaana.lahtinen@ttl.fi (M.L.)
*   Correspondence: katja.tahtinen@ttl.fi; Tel.: +358-40-584-3360

Received: 29 November 2018; Accepted: 21 December 2018; Published: 28 December 2018

Abstract: Indoor air problems are complicated and need to be approached from many perspectives. In this research, we studied the association of four-level categorisation of the probability of abnormal indoor air (IA) exposure with the work environment-related symptoms, group-level health information and psychosocial work environment of employees. We also evaluated the multiprofessional IA group assessment of the current indoor air quality (IAQ) of the hospital premises. We found no statistical association between the four-level categorisation of the probability of abnormal IA exposure and the employees' perceived symptoms, health information, and perceived psychosocial work environment. However, the results showed a statistical association between perceived symptoms and man-made vitreous fibre sources in ventilation. Furthermore, extensive impurity sources in the premises increased the employees' contact with health services and their perceived symptoms. The employees perceived stress and symptoms in all categories of abnormal IA exposure, which may be related to IAQ or other factors affecting human experience. Prolonged process management may influence users' experiences of IAQ. The results suggest that an extensive impurity source in premises does not always associate with the prevalence of perceived symptoms. We conclude that indoor air questionnaires alone cannot determine the urgency of the measures required.

Keywords: perceived indoor air quality; building research; indoor air questionnaires; psychosocial work environment; categorisation; ventilation; mould; moisture; man-made mineral fibres

---

1. Introduction

To assess the health significance, urgency, and extent of required indoor air quality (IAQ) measures, property owners and occupational health and safety professionals need reliable information on the buildings' conditions and impurity sources. Information is also needed regarding the experiences and health of the users of the premises, and on the cooperation in indoor air (IA) solution processes. When all the factors affecting the IAQ problem have been properly assessed, the degree, timing, and possible prioritisation of measures can be decided on. Properly timed and targeted measures have important implications for the economy, health, and well-being. IAQ problems can be controlled and good IAQ achieved if (i) the factors affecting the indoor environment are under control, (ii) the indoor environment is perceived as good and healthy [1–3], and (iii) good practices are in place for maintaining the indoor environment and solving indoor air (IA) problems [3,4].

IAQ problems are often the result of many different factors and their interaction or complex combination. In addition to moisture and mould damage, several other factors and their interactions, such as material emissions [5], ventilation deficiencies [6], system impurities [7], outdoor and soil impurities [8], human activities in the premises [9], IA temperature, and dry air [10] can cause IAQ problems. Office-type buildings have no established measurement methods for all IA pollutants, and no health-based limit values for most of them [2,11].

Several IA pollutant sources can cause symptoms and harm to the users of premises. A recent review concluded that the greater the presence of moisture and mould damage in the premises, the greater the risk of respiratory health effects [11]. Man-made vitreous fibres (MMVF) in the ventilation system may cause upper respiratory irritation and skin symptoms among users of the premises [5,12–14]. Volatile organic compounds (VOC) from building materials may cause sensory irritation [5,7]. IAQ problems may also affect sick leaves and work efficiency [15].

It has been estimated that predictive property management reduces the IAQ-related symptoms of premises users [16], and some evidence shows that repairing moisture and mould damage and removing contaminants from buildings can reduce respiratory symptoms [17,18] and improve work efficiency [19].

In addition to IAQ factors, several other factors at workplaces, such as stress, poor cooperation, heavy workload, and individual factors may also affect perceived IAQ and play a role in IAQ problems [20,21]. These problems should be examined from a wider perspective, and experience of the users of the premises and the psychosocial environment should be considered [20–23]. It has been suggested that good practices for solving complex or prolonged IAQ problems are well organised and involve long-term multiprofessional cooperation between experts [23,24].

The aim of this study was to test the use of the holistic approach in determining the urgency of the measures required from the perspective of building health. It can be divided into the following sub-aims: (i) to evaluate the relation between the four-level categorised probability of abnormal IA exposure and employees' work environment-related symptoms, group-level health information, and psychosocial work environment, (ii) to assess the relation between ventilation system deficiencies and employees' work environment-related symptoms and (iii) to evaluate the impact of prolonged IAQ problem solution processes on perceived IAQ.

This paper uses the term probability of abnormal IA exposure, which means a comprehensive method of categorising the results of building and ventilation system research. The method used to assess the probability of abnormal IA exposure is presented in our earlier study [25].

2. Materials and Methods

2.1. Materials

This study is based on two research and development projects conducted at the Finnish Institute of Occupational Health (FIOH). These projects were carried out between February 2013 and April 2014, in 27 hospital buildings (studied area altogether about 130,000 m^2), which form a unified building complex located in two Finnish hospital districts. Background information on the buildings' earlier history and documents revealed that parts of the buildings had IAQ problems. We investigated altogether 111 building floors or sections and selected forty building floors or sections on which to focus in more detail, from the premises in which both the IA questionnaire and the assessment of probability of abnormal IA exposure were carried out. We also conducted the building investigations and abnormal IA exposure assessments were still carried out in building premises that were not workplaces or were not in use. The oldest building was built around 1902 and the newest in 2010. One half of the buildings (48%) were built between the 1940s and 1950s. All of them were of stone or different combinations of stone materials and were mostly multistorey and divided between many hospital department areas and hospital functions. Some of the buildings had been repaired in several different stages and these renovations varied greatly. The ventilation system of the buildings was mostly mechanical extract and supply ventilation. However, several different ventilation systems and machines served different parts of the building. The maintenance, repair, reliability, and age of the ventilation systems varied considerably across the floors or sections of even one building.

The same IA researcher group conducted all the building research and ventilation system assessments. All the data were analysed by the same multi-professional group of experts, which comprised IA researchers, a civil engineer, an occupational health physician, a microbiologist, and a ventilation and building health specialist.

2.2. Assessment of Probability of Abnormal IA Exposure

We carried out systematic building examinations that covered (i) structural and architectural plan surveys, (ii) maintenance staff interviews, (iii) examinations and openings of high-risk building structures, (iv) moisture- and mould-damaged range and severity authentications, (v) assessments of ventilation systems, (vi) assessments of air leaks from or through damaged structures, (vii) assessments of air pressure differences, and (viii) assessments of other IA pollutants or pollutant sources in the buildings [25] (Figure 1). We collected building investigation and IAQ measurement results and used a four-level categorisation method to assess the probability of abnormal IA exposure [25] (Figure 1).

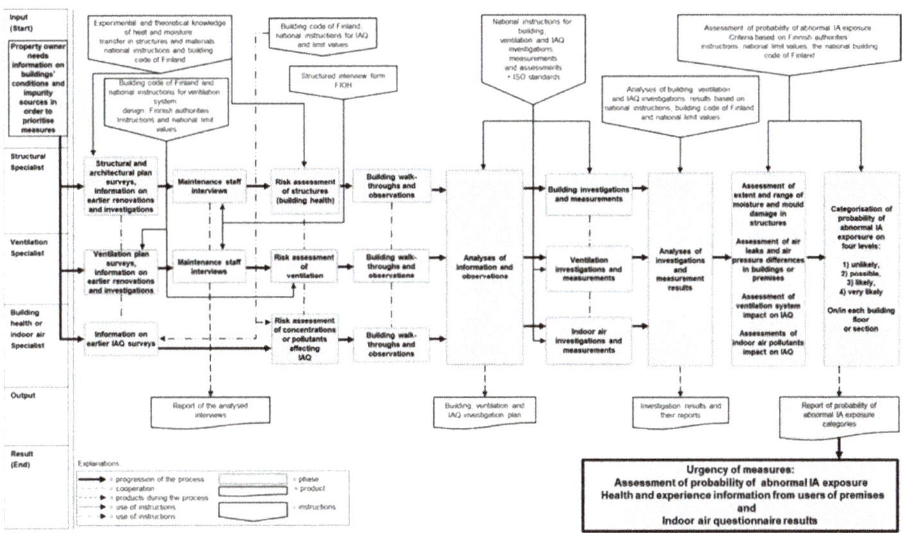

Figure 1. Process for assessing the probability of abnormal indoor air (IA) exposure [25].

We collected categorised parameters for the final assessment of the probability of abnormal IA exposure. This probability was categorised as: (1) probability of abnormal IA exposure unlikely, (2) probability of abnormal IA exposure possible, (3) probability of abnormal IA exposure likely, and (4) probability of abnormal IA exposure very likely [25] (Table 1). In cases of moisture and mould damage, air leaks from or through damaged structures to IA must be examined simultaneously with indoor negative pressure. In the main criteria for assessing the probability of abnormal IA exposure, the predominant IA impurity source is a determining criterion. The probability of abnormal IA exposure arises when the national limit values (IAQ, material samples, ventilation) are exceeded, structures or systems have found damaged, or IAQ pollutant sources that are known to affect indoor air quality and building health are found [25]. The national maximum limit values for IA concentrations, microbial growth on building material, MMVF and asbestos in dust, and specific detailed methods for evaluating building and ventilation conditions and IAQ are presented in our earlier research [25].

Table 1. Main criteria and categories for assessing probability of abnormal IA exposure in buildings.

Categories	Main Criteria for Assessing Probability of Abnormal IA Exposure in Buildings.
Unlikely	No moisture or mould damage in structures. No air leaks from or through damaged structures. Ventilation system can be controlled by indoor pressure difference from the building envelope. Room acoustic materials and ventilation system have no man-made vitreous fibres (MMVF) sources. Indoor air quality corresponds to national reference values and guidelines set for the premises.
Possible	Mould-damaged structure type is not widespread in building and repairs are easily definable (less than 1 m^2). A few or single air leaks from or through damaged structures or from surrounding premises. Room acoustic materials or ventilation system have MMVF sources and fibres may end up in the indoor air or on surfaces.[1] Concrete floor has extensive moisture, which can cause water vapour damage to permeable floor coating (emissions).[1] Indoor air quality does not correspond to national reference values or the guidelines set for the premises, and an indoor air impurity source has been identified.[1]
Likely	Building or premises have widespread mould-damaged structure. Repairs are significant and affect a large part of the (one) structure of the building or premises, e.g., whole base floor structure. There is recurrent damage in the type of structure. Air leaks from or through damaged structures or from surrounding premises and moisture or mould-damaged materials are regular and recurrent in the structure, occasionally there is negative pressure in the premises and/or air-tightness is risky. Indoor air quality does not correspond to national reference values or the guidelines set for the premises, and an indoor air impurity source has been identified.[1] Creosote has been used in the structure and air leaks into the indoor air from the structure. There is also a notable smell of creosote (e.g., naphthalene) in the indoor air.[1]
Very likely	The building or premises has a great deal of extensive mould damage in several structures. The extent of repairs is significant and affects several structures in the building or premises e.g., whole façade and whole base floor. There is recurrent damage in the type of the structures. Air leaks from or through damaged structures are regular and recurrent, negative pressure is significant in the premises and/or air-tightness is very risky. Indoor air quality does not correspond to national reference values or the guidelines set for the premises, and an indoor air impurity source has been identified.[1] Creosote has been used in the structures and air leaks into the indoor air from the structures. In addition, concentrations of polycyclic aromatic hydrocarbons (PAH) or separate components exceed the set national values and guidelines.[1] Dust sample tests have found asbestos fibres in the premises, and the pollution source has been defined.[1] Indoor radon concentrations exceed the set national values and guidelines (400 Bq/m^3 [26]).[1]

[1] The assessment must take into account the extent and impact of the problem and impurity source.

2.3. Employees' Experiences of Indoor Air Quality and Psychosocial Work Environment

To study the users', i.e., employees' experiences of the work premises, work environment-related symptoms and psychosocial work environment, we used FIOH's validated and frequently used IA questionnaire, which is based on Örebro's [27] indoor climate questionnaire [28,29]. To study perceived stress, we used a validated single-item measure of stress symptoms [30]: "Stress means a situation in which a person feels tense, restless, nervous or anxious or is unable to sleep at night because his/her mind is troubled all the time. Do you feel this kind of stress these days?". The response options were: (1) not at all, (2) just a little, (3) some, (4) quite a lot, and (5) very much. In the analyses, we combined the levels (1) not at all and (2) just a little into one level, and levels (4) quite a lot and (5) very much into one level.

We sent the questionnaire to 3608 hospital employees, of whom 2669 responded. The total response rate was 74%, with a range of 51% to 93%. The surveys were conducted in the spring from February to April and in the autumn in November, in 2013. We selected 40 IA questionnaire groups for the study, totaling 1558 respondents. The selected IA questionnaire groups were in premises in which the probability of abnormal IA exposure assessment had already been performed. The employees did not know the results of the assessment prior to responding to the IA questionnaires. This is a questionnaire-based study, in which participation was voluntary and performed no intervention on individuals, according to Finnish legislation it did not require ethics committee handling.

2.4. Group-Level Information from Occupational Health Services and Multiprofessional Indoor Air Group

We obtained information on the assessment of the group-level health of employees from occupational health services (OHS). The information covered employees' health from 43 building sections or floors. The group-level information from the OHS did not contain information on how many employees had work environment-related health symptoms in the building sections or on the floors. In the survey, we used short forms to ask about the following issues in relation to employees' health: (i) case of new onset asthma or aggravation of existing asthma, (ii) having to change workroom because of IAQ and work environment-related symptoms, (iii) increased amount of employee visits to OHS due to IAQ-related issues, and (iv) increased sickness absences due to respiratory symptoms. The hospital's multiprofessional IA group also provided information on the estimated duration of the IAQ problem solution on every building floor or section.

2.5. Statistical Analyses

Statistical analyses were carried out using IBM SPSS Statistics program 25.0 with a statistically significant level of $p < 0.05$. The statistical analysis used weighted averages of group response rates. The Mann–Whitney U test studied the differences between the probability of abnormal IA exposure categories (*unlikely, possible, likely,* and *very likely*) and the employees' complaints about their work environment-related symptoms and psychosocial work environment. This test also compared the difference between the two groups' (yes/no) ventilation adequacy, ventilation MMVF sources, ventilation moisture problems, and expired ventilation lifespan and the employees' complaints about their work environment-related symptoms. Fisher's exact test studied the relation between the weekly work environment-related symptoms experienced by the employees and the categorised group-level information on employee health. The group-level health information was categorised as 'yes' and 'no' as follows: case of new onset asthma or aggravation of existing asthma, having to change workroom or workplace because of work environment-related symptoms, increased amount of employee visits to OHS, and increased sickness absences due to respiratory symptoms.

3. Results

Probability of Abnormal IA Exposure and Employees' Experience

All building floors or sections (total 111) were investigated and we were able to assess the probability of abnormal IA exposure on or in 95 building floors or sections. In the case of forty building floors or sections, the assessment of the probability of abnormal IA exposure and IA questionnaire could both be conducted in the same areas (the IA questionnaire group was located in the area that was assessed as belonging to an abnormal IA exposure category). The probability of abnormal IA exposure was assessed as *unlikely* for 5% (n = 2), *possible* for 40% (n = 16), *likely* for 45% (n = 18), and *very likely* for 10% (n = 4) of the selected forty floors or sections of the buildings. In the *likely* and *very likely* categories, these floors or sections had wide moisture and mould damage in their structures together with air leaks from damaged materials to the IA and often had a detected MMVF source in the ventilation system as well as ventilation deficiencies (Table 2 and criteria in Table 1). These categories also had other impurity sources (Table 1), but moisture and mould damage in the building structures were dominant. The higher (more abnormal) the assessed category of probability of abnormal IA exposure, the more insufficient the ventilation was, the more often the lifespan of the ventilation system was exceeded and the more often MMVF sources were detected in the ventilation system from the categories likely and very likely (and not from unlikely category) (Table 2).

Table 2. Ventilation survey findings are included in all the sections or floors in which probability of abnormal IA exposure was assessed (95 floors or sections). (All building floors or sections (111) were investigated and we were able to assess the probability of abnormal IA exposure on or in 95 building floors or sections.).

Assessed Probability of Abnormal IA Exposure on/in Building Floors or Sections (n = 95)	Lifespan of Ventilation System Had Been Exceeded n (%)	Insufficient Ventilation n (%)	MMVF (Man-Made Vitreous Fibres) Source in Ventilation System n (%)	Moisture Problem in Ventilation System n (%)
Unlikely (n = 7)	7 (100)	3 (43)	3 (43)	0 (0)
Possible (n = 39)	23 (59)	22 (56)	12 (31)	6 (15)
Likely (n = 37)	26 (70)	24 (65)	26 (70)	15 (41)
Very likely (n = 12)	10 (83)	9 (75)	9 (75)	1 (8)

The perceived weekly work environment-related symptoms among the employees exceeded the corresponding number of weekly work environment-related symptoms in FIOH's reference data [28,29] (Figure 2). Even in premises in which the building research showed no source of contamination, some employees perceived weekly symptoms more than those in FIOH's reference data [28,29] (Figure 3). An analysis of the differences between the probability of abnormal IA exposure categories (*unlikely, possible, likely, very likely*) and the employees' weekly perceived work environment-related symptoms (headaches, concentration difficulties, irritation of the eyes and nose, irritation of the skin on the face and hands, hoarse throat, coughing, coughing at night, shortness of breath, wheezing, fever or chills, joint pain, and muscular pain revealed no statistically significant differences (Table 3).

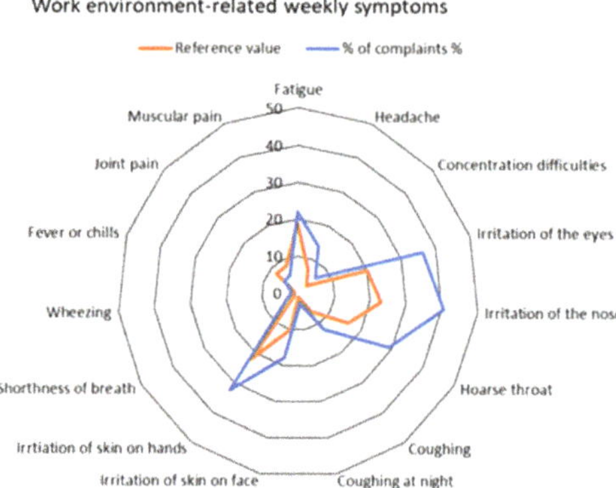

Figure 2. In the buildings assessed for probability of abnormal IA exposure, weekly work environment-related symptoms (n = 1558) were perceived more often than in the Finnish Institute of Occupational Health's (FIOH) comparable reference data [28].

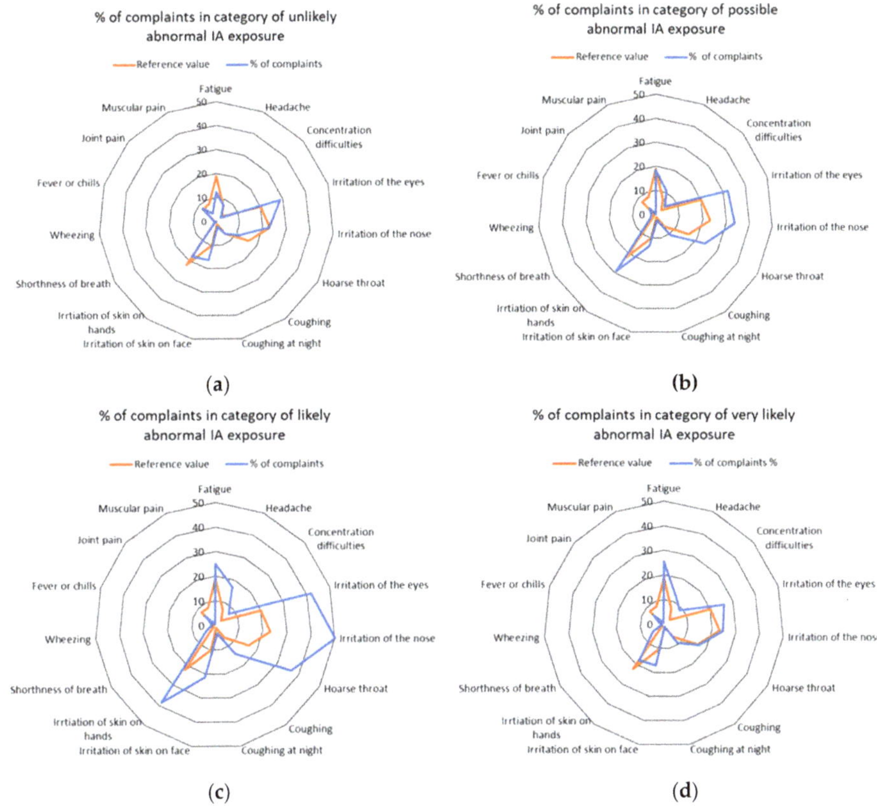

Figure 3. Perceived work environment-related symptoms of users of premises in assessment of probability of abnormal IA exposure categories, compared to FIOH reference data [28] (**a**) unlikely (n = 61), (**b**) possible (n = 618), (**c**) likely (n = 760), (**d**) very likely (n = 118).

Table 3. An analysis of the differences between the probability of abnormal IA exposure categories (unlikely, possible, likely, very likely) and the employees' weekly perceived work environment-related symptoms.

Variable (Weekly Symptoms)	Unlikely 1 n = 2 (*p*-Value)	Possible 2 n = 16 (*p*-Value)	Likely 3 n = 18 (*p*-Value)	Very Likely 4 n = 4
Fatigue	1 vs. 2 (0.673)	2 vs. 3 (0.220)	3 vs. 4 (0.609)	
	1 vs. 3 (0.257)	2 vs. 4 (0.570)		
	1 vs. 4 (0.355)			
Headache	1 vs. 2 (0.778)	2 vs. 3 (0.157)	3 vs. 4 (0.481)	
	1 vs. 3 (0.208)	2 vs. 4 (0.777)		
	1 vs. 4 (0.643)			
Concentration difficulties	1 vs. 2 (0.672)	2 vs. 3 (0.233)	3 vs. 4 (1.000)	
	1 vs. 3 (0.165)	2 vs. 4 (0.507)		
	1 vs. 4 (0.064)			
Irritation of the eyes	1 vs. 2 (0.888)	2 vs. 3 (0.262)	3 vs. 4 (0.125)	
	1 vs. 3 (0.378)	2 vs. 4 (0.777)		
	1 vs. 4 (0.814)			
Irritation of the nose	1 vs. 2 (0.779)	2 vs. 3 (0.073)	3 vs. 4 (0.061)	
	1 vs. 3 (0.130)	2 vs. 4 (0.925)		
	1 vs. 4 (0.643)			

Table 3. *Cont.*

Variable (Weekly Symptoms)	Unlikely 1 n = 2 (*p*-Value)	Possible 2 n = 16 (*p*-Value)	Likely 3 n = 18 (*p*-Value)	Very Likely 4 n = 4
Hoarse, dry throat	1 vs. 2 (0.399) 1 vs. 3 (0.130) 1 vs. 4 (1.000)	2 vs. 3 (0.101) 2 vs. 4 (0.508)	3 vs. 4 (0.061)	
Coughing	1 vs. 2 (0.481) 1 vs. 3 (0.378) 1 vs. 4 (0.355)	2 vs. 3 (0.147) 2 vs. 4 (0.636)	3 vs. 4 (0.287)	
Coughing at night	1 vs. 2 (0.941) 1 vs. 3 (0.792) 1 vs. 4 (0.411)	2 vs. 3 (0.652) 2 vs. 4 (0.289)	3 vs. 4 (0.237)	
Irritation of skin on face	1 vs. 2 (0.260) 1 vs. 3 (1.000) 1 vs. 4 (1.000)	2 vs. 3 (0.133) 2 vs. 4 (0.508)	3 vs. 4 (0.551)	
Irritation of skin on hands	1 vs. 2 (0.888) 1 vs. 3 (0.378) 1 vs. 4 (1.000)	2 vs. 3 (0.152) 2 vs. 4 (0.705)	3 vs. 4 (0.349)	
Shortness of breath	1 vs. 2 (0.562) 1 vs. 3 (0.509) 1 vs. 4 (0.355)	2 vs. 3 (0.508) 2 vs. 4 (0.502)	3 vs. 4 (0.663)	
Wheezing	1 vs. 2 (0.374) 1 vs. 3 (0.178) 1 vs. 4 (0.480)	2 vs. 3 (0.342) 2 vs. 4 (0.600)	3 vs. 4 (0.275)	
Fever or chills	1 vs. 2 (0.352) 1 vs. 3 (0.736) 1 vs. 4 (0.623)	2 vs. 3 (0.308) 2 vs. 4 (0.562)	3 vs. 4 (1.000)	
Muscular pain	1 vs. 2 (1.000) 1 vs. 3 (0.44) 1 vs. 4 (1.000)	2 vs. 3 (0.161) 2 vs. 4 (0.298)	3 vs. 4 (0.932)	
Joint pain	1 vs. 2 (0.324) 1 vs. 3 (0.900) 1 vs. 4 (0.643)	2 vs. 3 (0.283) 2 vs. 4 (0.570)	3 vs. 4 (0.898)	
Other work environment-related symptoms	1 vs. 2 (0.176) 1 vs. 3 (0.074) 1 vs. 4 (0.060)	2 vs. 3 (0.099) 2 vs. 4 (0.288)	3 vs. 4 (0.831)	

Statistically significant level of $p < 0.05$.

As regards the IA questionnaire results, most of the employees (88%) felt that their work was often stimulating and interesting, 74% believed they would receive help from their colleagues if needed, 21% often had the opportunity to influence their own work and working conditions, and 53% had no feelings of stress. A heavy workload was reported by 14% of the employees, which is below FIOH's reference value [28]. Stress was reported by 16% of the employees and 4% believed that they would not get help from colleagues if needed. Stress and lack of help from colleagues were perceived more often than in FIOH's comparable reference data [28]. Stress was perceived more often than in FIOH's reference data in every category of abnormal IA exposure (Table 4). Most often, stress was perceived in premises in which the probability of abnormal IA exposure was estimated as being *unlikely* (Table 4). An analysis of the differences between the probability of abnormal IA exposure categories (*unlikely, possible, likely, very likely*) and the employees' perceived psychosocial work environment and stress revealed no statistically significant differences.

Table 4. Perceived psychosocial work environment and stress according to probability of abnormal IA exposure categories (*unlikely, possible, likely, very likely*).

Question on Psychosocial Work Environment	Unlikely n = 61 (%)	Possible n = 619 (%)	Likely n = 763 (%)	Very Likely n = 118 (%)	FIOH's Reference Value %
Do you regard your work as interesting and stimulating?					
Yes, often	52 (88.1)	540 (87.5)	684 (89.9)	102 (86.4)	82
Yes, sometimes	7 (11.9)	66 (10.7)	61 (8.0)	11 (9.3)	16
No, seldom or rarely	0 (0)	11 (1.8)	16 (2.0)	5 (4.2)	0
Do you have too much work?					
Yes, often	12 (20.7)	95 (15.5)	98 (12.9)	15 (12.7)	20
Yes, sometimes	36 (62.1)	360 (58.7)	489 (64.4)	71 (60.2)	64
No, seldom or rarely	10 (17.2)	158 (25.8)	172 (22.7)	32 (27.1)	16
Do you have opportunities to influence your working conditions?					
Yes, often	38 (32.8)	120 (19.5)	161 (21.2)	38 (32.8)	21
Yes, sometimes	61 (52.6)	335 (54.3)	379 (49.9)	61 (52.6)	51
No, seldom or rarely	17 (14.7)	162 (26.3)	220 (29.0)	17 (14.7)	28
Do your fellow workers help you with work-related problems?					
Yes, often	33 (55.9)	463 (75.2)	569 (74.8)	83 (70.3)	79
Yes, sometimes	24 (40.7)	129 (20.9)	171 (22.5)	28 (23.7)	19
No, seldom or rarely	2 (3)	24 (3.9)	21 (2.8)	7 (5.9)	2
Do you feel stress?					
Quite a lot or very much	15 (26.3)	92 (15.1)	105 (15.0)	16 (14.0)	10
Some	12 (21.1)	181 (29.7)	243 (32.4)	40 (35.1)	28
Not at all or just a little	30 (52.6)	337 (55.2)	403 (53.7)	58 (50.9)	63

In addition, an analysis of the differences between the probability of abnormal IA exposure categories' (*unlikely, possible, likely, very likely*) and the employees' group-level health information (obtained from OHS) revealed no statistically significant differences. However, the more abnormal the probability of the IA exposure category, the more employees contacted OHS due to IAQ-related issues from the categories possible, likely and very likely (Table 5).

Table 5. IAQ-related group-level health information on employees (obtained from OHS), according to building floors or sections in which probability of abnormal IA exposure categories (*unlikely, possible, likely, very likely*) were assessed.

IAQ-Related Health Information	Unlikely n = 2 (%)	Possible n = 16 (%)	Likely n = 20 (%)	Very likely n = 5 (%)
Some employees have new asthma or aggravation of previous asthma [1]				
Yes	2 (100)	5 (31)	5 (25)	3 (60)
No	0 (0)	10 (63)	14 (70)	2 (40)
No information	0 (0)	1 (6)	1 (5)	0 (0)
Some employees have changed work premises or work places due to IAQ-related symptoms [1]				
Yes	1 (50)	4 (25)	4 (20)	2 (40)
No	1 (50)	11 (69)	16 (80)	3 (60)
No information	0 (0)	1 (6)	0 (0)	0 (0)
The amount of employee contacts with OHS due to IAQ-related issues has increased [1]				
Yes	2 (100)	5 (31)	10 (50)	3 (60)
No	0 (0)	10 (63)	10 (50)	2 (40)
No information	0 (0)	1 (6)	0 (0)	0 (0)
The amount of sickness absence due to respiratory symptoms has increased [1]				
Yes	1 (50)	4 (25)	8 (40)	1 (20)
No	0 (0)	7 (44)	11 (55)	3 (60)
No information	1 (50)	5 (31)	1 (5)	1 (20)

[1] Group-level health information does not contain information on how many employees have IAQ-related health symptoms on building floors or in sections or the assessed categories. IAQ: indoor air quality; OHS: occupational health services.

The results show a statistical association between detected MMVF sources in ventilation systems and perceived work environment-related symptoms and a statistical association between ventilation system age and perceived work environment-related symptoms (Table 6). The hospital's multiprofessional IA group estimated the duration of the IAQ problem solution process on every building floor or in each section, and the attempts to solve the indoor air problems in the *unlikely*, *possible*, *likely*, and *very likely* categories had lasted a year or more (Table 7). The hospital's multiprofessional IA group also estimated 'No IAQ problems' in premises in which the research group had assessed the probability of abnormal IA exposure as being *possible* and *likely* (Table 7).

Table 6. Statistical significance of differences (*p*-value) between weekly reported work environment-related symptoms and ventilation factors (yes/no) studied.

Perceived Work Environment-Related Symptoms Weekly	Respondents N = 1558 n	Technical Lifespan of Ventilation System Had Been Exceeded	Moisture Problem in Ventilation System	MMVF Source in Ventilation System	Insufficient Ventilation or Ventilation System Did Not Match Purposes of Facilities
Fatigue	532	NS	NS	0.005	NS
Headache	266	0.035	NS	0.002	NS
Concentration difficulties	164	0.049	NS	0.005	NS
Irritation of the eyes	559	0.014	NS	0.006	NS
Irritation of the nose	656	0.003	NS	0.001	NS
Hoarse, dry throat	475	0.001	NS	0.001	0.025
Coughing	225	0.016	NS	0.022	NS
Coughing at night	50	0.008	NS	NS	NS
Irritation of skin on face	373	NS	NS	NS	NS
Irritation of skin on hands	508	0.035	NS	0.003	NS
Shortness of breath	64	0.045	NS	NS	NS
Wheezing	32	0.001	NS	NS	0.044
Fever or chills	42	NS	NS	NS	NS
Muscular pain	160	0.027	NS	0.003	NS
Joint pain	193	0.039	NS	NS	NS
Other work environment-related symptoms	75	0.005	NS	0.001	0.027

NS: not significant. Statistically significant level of $p < 0.05$.

Table 7. Estimated duration of IAQ problems on each building floor or in each section studied.

Estimated Duration of IAQ Problems: Number of Cases	Unlikely (n = 2)	Possible (n = 16)	Likely (n = 18)	Very Likely (n = 4)
No IAQ problems	1	7	6	0
Duration of IAQ problems less than a year	0	2	4	0
Duration of IAQ problems one year or more	1	7	8	4

For each building floor or section, we looked at all the collected data (results of assessed probability of abnormal IA exposure, IA questionnaires results, group-level health information and IA group information) at the same time. The most urgent measures were required for floors or sections on/in which the probability of abnormal IA exposure was *likely* or *very likely*, the health information (Table 5) and IA questionnaire results (Figure 3a–d) indicated health problems, and solutions were delayed (Table 7).

4. Discussion

The detailed examination of the buildings and the four-level categorisation of the probability of abnormal IA exposure made it easier to organise the outcomes and obtain a clear picture of the many factors affecting the IAQ of the buildings and premises. The strength of our research was its multiprofessional approach, which took into account the employees' perceived work environment-related symptoms and health information, multiprofessional IA group information and the results of the technical building investigations.

According to previous studies, observed indoor mould and moisture damage indicates an increased health risk, and the greater the mould and moisture damage, the more prevalent respiratory symptoms in adults [11]. This study found no statistically significant differences between the four-level categories and employees' perceived work environment-related symptoms, perceived psychosocial environment, or OHS group-level information on employees' health. However, extensive impurity sources in the premises increased some employees' perceived work-related symptoms. Our earlier study [31] shows that all the symptoms perceived by the employees were very similar to the work-related symptoms examined in this study. The results showed a statistical association between MMVF sources in ventilation systems and perceived IAQ and work environment-related symptoms. Other studies have achieved similar results [5,14]. In addition, the age of the ventilation system was associated with perceived symptoms. The more the probability of abnormal IA exposure was estimated to differ, the more prevalent were ventilation system deficiencies and the MMVF sources in the ventilation system. Therefore, IAQ problems were usually affected by impurities from both the building and ventilation, which shows that many IAQ factors can affect perceived symptoms.

Employees perceived more work environment-related symptoms and stress and lack of social support from colleagues than those reported in FIOH's comparative data from (damaged and nondamaged) hospital buildings, a result which may be related to the poor condition of the buildings or other factors affecting the human experience. On the other hand, the amount of employee contact with OHS increased due to IAQ and work environment-related issues in all the categories of the probability of abnormal IA exposure. An earlier study has shown an association between symptoms and work strain [21], the psychosocial work environment [4,32], and individual factors (e.g., gender, age) [21,33]. The risk of experiencing the workplace as harmful has shown to be higher among employees who report mould problems than those who report ventilation problems in workplaces [34].

In our research, symptoms were common, stress was high, and the amount of contact with OHS in IAQ and work environment-related issues was great on floors or in sections in which the category of abnormal IA exposure was assessed as being *unlikely* and building floors or sections that were undamaged. Workers in non-damaged buildings have also shown to have IAQ and work environment-related symptoms [29,33]. Overall, the employees perceived stress in every category of abnormal IA exposure more than the amount reported in FIOH's reference data [28]. On the other hand, the employees often perceived their work as stimulating and interesting in all categories of the probability of abnormal IA exposure. An earlier study has also shown that hospital employees find their work more interesting and stimulating than office workers [28,32]. As the buildings and the factors affecting IAQ have been carefully studied, the building technology and IAQ alone may not explain the prevalence of the perceived work environment-related symptoms and the stress and the amount of IAQ- and work environment-related contact with OHS. These issues may also be partly affected by other factors influencing human experience, such as work-related and organizational factors not investigated in this study, or individual human factors.

Based on the IA group's information, attempts to solve the indoor air problems in several buildings had lasted more than a year. The IA group's evaluation of no IAQ problems in the premises often contradicted our building investigation results. One reason for these prolonged problem-solving processes may be that building-related problems or impurity sources have been unclear or building investigations incomplete. An earlier study also had similar findings [4]. Prolonged or unclear problem-solving processes may have increased health concerns and distrust of the problem-solving process and influenced the IAQ-related experience of employees. Thus, careful decision-making procedures are important, especially when people feel threatened by IAQ-related risks [34].

The employees did not know the results of the assessment of the probability of abnormal IA exposure prior to responding to the IA questionnaires, which contributes to the reliability of the study. The differences between previous studies and our research results may result from differences in the building research methods. Many studies are based on observations of visible damage or indications of damage, and hidden damage has remained unclear [11]. Our study was very detailed, and we also investigated hidden damage, in addition other pollutant sources. Possible limitations of our research may be that our data concerned only a small number of buildings with no IAQ problems or impurities affecting IAQ, which may have affected data distribution. Due to this, we had no reliable comparison survey of damaged and nondamaged buildings or premises. The method for assessing the probability of abnormal IA exposure is very pragmatic and is always based on strong technical expertise in building technology and IAQ. In addition, the criteria for the probability of abnormal IA exposure recommend taking into account many impurity sources in a building, based on national instructions, regulations and limit values in the field of the built environment and IAQ. However, the assessment also involves a researcher's subjective view. OHS' information was collected at group-level, and as the results were analysed categorically, they may not have provided sufficiently accurate results. In addition, the questions concerning the psychosocial work environment were quite limited in the IA questionnaire. They covered qualitative and quantitative workload, opportunities to influence one's working conditions, and social support at work. Although these are essential factors in the light of the stress theory, they fail to provide a comprehensive picture of the psychosocial work environment. The questionnaire did not include, for instance, factors such as organizational changes and questions concerning leadership.

The probability of abnormal IA exposure provides us with a holistic picture of the many factors affecting the IAQ of buildings and premises. In this case, the most urgent measures could have been identified more easily and holistically. Moreover, in the premises that needed the most urgent repairs, employee contact with OHS was increased, and the employees' perceived work environment-related symptoms indicated poor IAQ. On the other hand, in the premises in which no technical problems were found, the employees still perceived more work environment-related symptoms and stress than those reported in FIOH's reference data. Thus, IAQ problems should always be analysed from many perspectives; (i) the building's technical condition, (ii) perceived IAQ and psychosocial work environment, (iii) OHS information, and (iv) measures for solving IAQ problems, which may all affect the experience of IAQ problems. Indoor air questionnaires can serve as a parallel method with technical investigations in the building.

5. Conclusions

The four-level categorisation of the probability of abnormal IA exposure provides a comprehensive and systematic way of ranking building sectors from the perspective of building health. The method is based on national instructions for building and ventilation investigations, building codes, and limit values, and is therefore systematic and partially established. Thus, it can be applied in different environments. The method may also be used in other countries (with similar environments to that in Finland), if the national instructions, limit values, and building codes are taken into account and applied. The method enabled the holistic identification of the most urgent measures. This may help property owners allocate resources for proper repairs and also help OHS identify employees' IAQ and work environment-related symptoms. The results suggest that the extensive impurity source in premises does not always associate with the prevalence of perceived IAQ and work environment-related symptoms. Therefore, the solution to the IAQ problem is more specific when technical survey results, the health and experience information of the users of the premises, as well as the problem solution process are taken into account. The results also suggest that IA questionnaires alone cannot determine the urgency of the measures required. Possible limitations of our research are that the study was only conducted in hospital buildings and premises. Limitations may be that our data concerned only a small number of buildings with no IAQ problems or impurities affecting IAQ: this may have affected data distribution. The method for assessing the probability of abnormal IA exposure is, however, very pragmatic and always based on strong technical expertise in building technology and IAQ. Further studies should assess the probability of abnormal IA exposure in different work environments (e.g., offices), and the associations between the probability of abnormal IA exposure categories with perceived IAQ and the health of employees. They should also assess the impact of the IAQ solution process perceived IAQ. The results of further studies may possibly validate the method.

Author Contributions: K.T. wrote the paper and analysed the data; S.L. and K.K. contributed to the study design and data analysis and participated in writing the paper; M.L. contributed to the data analyses and participated in writing the paper; H.S. was the principal supervisor of the work and participated in writing the paper.

Funding: This research received no external funding.

Acknowledgments: The authors acknowledge the contribution of Leena Aalto, Veli-Matti Pietarinen; Ulla-Maija Hellgren, Jouko Remes and Pauliina Toivio and all co-workers in building research.

Conflicts of Interest: The authors declare no conflict of interest.

Abbreviations

The following abbreviations are used in this manuscript:

FIOH	Finnish Institute of Occupational Health
IA	Indoor air
IAQ	Indoor air quality
OHS	Occupational health service
PAH	Polycyclic aromatic hydrocarbon
VOC	Volatile organic compound

References

1. Frontczak, M.; Wargocki, P. Literature survey on how different factors influence human comfort in indoor environments. *Build. Environ.* **2011**, *46*, 922–937. [CrossRef]
2. WHO. *Dampness and Mould—Who Guidelines for Indoor Air Quality*; World Health Organisation: Geneva, Switzerland, 2009.
3. Lappalainen, S.; Lahtinen, M.; Palomaki, E.; Korhonen, P.; Niemelä, R.; Reijula, K. Comprehensive procedure for solving indoor environment problems. In Proceedings of the NAM Nordic Work Environment Meeting, Espoo, Finland, 31 August–2 September 2009; p. 93.
4. Lahtinen, M.; Huuhtanen, P.; Kahkonen, E.; Reijula, K. Psychosocial dimensions of solving an indoor air problem. *Indoor Air* **2002**, *12*, 33–46. [CrossRef] [PubMed]
5. Salonen, H.; Lappalainen, S.; Riuttala, H.; Tossavainen, A.; Pasanen, P.; Reijula, K. Man-made vitreous fibers in office buildings in the helsinki area. *J. Occup. Environ. Hyg.* **2009**, *6*, 624–631. [CrossRef] [PubMed]
6. Sundell, J.; Levin, H.; Nazaroff, W.W.; Cain, W.S.; Fisk, W.J.; Grimsrud, D.T.; Gyntelberg, F.; Li, Y.; Persily, A.K.; Pickering, A.C.; et al. Ventilation rates and health: Multidisciplinary review of the scientific literature. *Indoor Air* **2011**, *21*, 191–204. [CrossRef] [PubMed]
7. Wolkoff, P.; Wilkins, C.K.; Clausen, P.A.; Nielsen, G.D. Organic compounds in office environments—Sensory irritation, odor, measurements and the role of reactive chemistry. *Indoor Air* **2006**, *16*, 7–19. [CrossRef] [PubMed]
8. Airaksinen, M.; Pasanen, P.; Kurnitski, J.; Seppänen, O. Microbial contamination of indoor air due to leakages from crawl space: A field study. *Indoor Air* **2004**, *14*, 55–64. [CrossRef] [PubMed]
9. Nevalainen, A.; Täubel, M.; Hyvärinen, A. Indoor fungi: Companions and contaminants. *Indoor Air* **2015**, *25*, 125–156. [CrossRef]
10. Nordström, K.; Norbäck, D.; Akselsson, R. Effect of air humidification on the sick building syndrome and perceived indoor air quality in hospitals: A four month longitudinal study. *Occup. Environ. Med.* **1994**, *51*, 683–688. [CrossRef]

11. Mendell, M.J.; Kumagai, K. Observation-based metrics for residential dampness and mold with dose–response relationships to health: A review. *Indoor Air* **2017**, *27*, 506–517. [CrossRef]
12. Hellgren, U.-M.; Hyvärinen, M.; Holopainen, R.; Reijula, K. Perceived indoor air quality, air-related symptoms and ventilation in finnish hospitals. *Int. J. Occup. Med. Environ. Health* **2011**, *24*, 48–56. [CrossRef]
13. Redlich, C.; Sparer, J.; Cullen, M.R. Sick-building syndrome. *Lancet* **1997**, *349*, 1013–1016. [CrossRef]
14. Schneider, T. Dust and fibers as a cause of indoor environment problems. *Scand. J. Work. Environ. Health* **2008**, *33*, 10–17.
15. Milton, D.K.; Glencross, P.M.; Walters, M.D. Risk of sick leave associated with outdoor air supply rate, humidification, and occupant complaints. *Indoor Air* **2000**, *10*, 212. [CrossRef] [PubMed]
16. Cox-Ganser, J.; Rao, C.; Park, J.; Schumpert, J.; Kreiss, K. Asthma and respiratory symptoms in hospital workers related to dampness and biological contaminants. *Indoor Air* **2009**, *19*, 280–290. [CrossRef] [PubMed]
17. Mendell, M.; Brennan, T.; Lee, H.; Odom, J.D.; Offerman, F.J.; Turk, B.H.; Wallingford, K.M.; Diamond, R.C.; Fisk, W.J. Causes and prevention of symptom complaints in office buildings. *Facilities* **2006**, *24*, 436–444. [CrossRef]
18. Sauni, R.; Verbeek, J.H.; Uitti, J.; Jauhiainen, M.; Kreiss, K.; Sigsgaard, T. Remediating buildings damaged by dampness and mould for preventing or reducing respiratory tract symptoms, infections and asthma. *Cochrane Database Syst. Rev.* **2015**, *76*, 1. [CrossRef] [PubMed]
19. Wargocki, P.; Lagercrantz, L.; Witterseh, T.; Sundell, J.; Wyon, D.P.; Fanger, P.O. Subjective perceptions, symptom intensity and performance: A comparison of two independent studies, both changing similarly the pollution load of an office. *Indoor Air* **2002**, *12*, 74. [CrossRef]
20. Brauer, C.; Mikkelsen, S. The influence of individual and contextual psychosocial work factors on the perception of the indoor environment at work: A multilevel analysis. *Int. Arch. Occup. Environ. Health* **2010**, *83*, 639–651. [CrossRef]
21. Magnavita, N. Work-related symptoms in indoor environments: A puzzling problem for the occupational physician. *Int. Arch. Occup. Environ. Health* **2015**, *88*, 185–196. [CrossRef]
22. Finell, E.; Haverinen-Shaughnessy, U.; Tolvanen, A.; Laaksonen, S.; Karvonen, S.; Sund, R.; Saaristo, V.; Luopa, P.; Ståhl, T.; Putus, T.; et al. The associations of indoor environment and psychosocial factors on the subjective evaluation of indoor air quality among lower secondary school students: A multilevel analysis. *Indoor Air* **2017**, *27*, 329–337. [CrossRef]
23. Lahtinen, M.; Lappalainen, S.; Reijula, K. Multiprofessional teams resolving indoor-air problems-emphasis on the psychosocial perspective. *Scand. J. Work. Environ. Health* **2008**, *34*, 30–34.
24. Carrer, P.; Wolkoff, P. Assessment of indoor air quality problems in office-like environments: Role of occupational health services. *Int. J. Environ. Res. Public Health* **2018**, *15*, 741. [CrossRef] [PubMed]
25. Tähtinen, K.; Lappalainen, S.; Karvala, K.; Remes, J.; Salonen, H. Association between four-level categorisation of indoor exposure and perceived indoor air quality. *Int. J. Environ. Res. Public Health* **2018**, *15*, 679. [CrossRef] [PubMed]
26. Finnish Ministry of the Environment, Department of Built Environment. *D2 the National Building Code of Finland, Health. Indoor Climate and Ventilation of Buildings, Regulations and Guidelines*; Department of Built Environment: Helsinki, Finland, 2012.
27. Andersson, K. Epidemiological approach to indoor air problems. *Indoor Air* **1998**, *8*, 32–39. [CrossRef]
28. Hellgren, U.-M.; Palomaki, E.; Lahtinen, M.; Riuttala, H.; Reijula, K. Complaints and symptoms among hospital staff in relation to indoor air and the condition and need for repairs in hospital buildings. *Scand. J. Work Environ. Health* **2008**, *34*, 58–63.
29. Reijula, K.; Sundman-Digert, C. Assessment of indoor air problems at work with a questionnaire. *Occup. Environ. Med.* **2004**, *61*, 33–38. [PubMed]
30. Elo, A.-L.; Leppänen, A.; Jahkola, A. Validity of a single-item measure of stress symptoms. *Scand. J. Work Environ. Health* **2003**, *29*, 444–451. [CrossRef]
31. Tähtinen, K.; Lappalainen, S.; Karvala, K.; Salonen, H. A comprehensive approach to evaluating the urgency of iaq measures. In Proceedings of the 15th Conference of the International Society of Indoor Air Quality & Climate (ISIAQ), Philadelphia, PA, USA, 22–27 July 2018.
32. Lahtinen, M.; Sundman-Digert, C.; Reijula, K. Psychosocial work environment and indoor air problems: A questionnaire as a means of problem diagnosis. *Occup. Environ. Med.* **2004**, *61*, 143. [CrossRef]

33. Bakke, J.; Moen, B.; Wieslander, G.; Norbäck, D. Gender and the physical and psychosocial work environments are related to indoor air symptoms. *J. Occup. Environ. Med.* **2007**, *49*, 641–650. [CrossRef]
34. Finell, E.; Seppälä, T. Indoor air problems and experiences of injustice in the workplace: A quantitative and a qualitative study. *Indoor Air* **2018**, *28*, 125–134. [CrossRef]

 *applied sciences*

Article

Combined Model for IAQ Assessment: Part 1—Morphology of the Model and Selection of Substantial Air Quality Impact Sub-Models

Michał Piasecki * and Krystyna Barbara Kostyrko

Department of Thermal Physic, Acoustic and Environment, Building Research Institute, Filtrowa 1, 00-611 Warsaw, Poland; k.kostyrko@itb.pl
* Correspondence: m.piasecki@itb.pl; Tel.: +48-22-5796-149

Received: 17 July 2019; Accepted: 9 September 2019; Published: 18 September 2019

Abstract: Indoor air quality (IAQ) is one of the most important elements affecting a building user's comfort and satisfaction. Currently, many methods of assessing the quality of indoor air have been described in the literature. In the authors' opinion, the methods presented have not been collected, systematized, and organized into one multi-component model. The application purpose of the assessment is extremely important when choosing IAQ model. This article provides the state-of-the-art overview on IAQ methodology and attempts to systematize approach. Sub-models of the processes that impact indoor air quality, which can be distinguished as components of the IAQ model, are selected and presented based on sensory satisfaction functions. Subcomponents of three potential IAQ models were classified according to their application potential: IAQ quality index, IAQ comfort index, and an overall health and comfort index. The authors provide a method for using the combined IAQ index to determine the indoor environmental quality index, IEQ. In addition, the article presents a method for adjusting the weights of particular subcomponents and a practical case study which provides IAQ and IEQ model implementation for a large office building assessment (with a BREEAM rating of excellent).

Keywords: indoor environment quality; IEQ; PPD; IAQ; TVOC; BREEAM assessment; occupant satisfaction

---

1. Introduction

1.1. State-of-the-Art Indoor Air Quality Measurement Systems

Approximately 30 years ago, people began to realize that buildings not only provide them with a sense of security, but can also significantly affect their health and well-being. This is particularly important due to the fact that people spend an increasing amount of time in closed indoor environment. Air quality and ventilation approaches were initially based on the users' dissatisfaction with the scent of the human body and, as such, the understanding of indoor air quality (IAQ) had serious limitations. Large quantities of pollutants and their sources clearly influence the indoor comfort of building inhabitants, as well as their health. In 1998, Fanger [1] presented an approach to the quantitative determination of perceived IAQ based on the level of dissatisfaction of residents caused by bad odors and irritants, smoke, and other sources of pollution. This approach provided two new measures of IAQ: the olf, which quantifies the pollution generated from a strong source of human bio-pollutant in the range of the impact of emitted odors on perceived air quality, and the decipol, measuring the perceived air quality in an indoor space with a source of pollution of one olf at a ventilation rate of 10 l/s. The number of emitted olfs per floor unit in different types of buildings and the amounts of pollutants from tobacco smoking (in olfs) can then be determined. Consideration of only the odour of the human body, without taking into account the influence of pollutants from various other sources (for example emissions from construction products), was very limited. At this stage, the determination of IAQ did not take into account the significant differences among contaminants and did not distinguish their specific impacts on health or comfort. The study of emissions undetectable by the senses (such as carbon monoxide and other pollutants that affect health at concentrations below their odor threshold), together with health-effects thresholds, has become particularly important. The range of air pollutants that should be considered as IAQ components is very difficult to determine, because the composition of pollutants constantly changes due to the fact of their dynamic nature, secondary reactions, sorption processes, and other physical and chemical phenomena occurring in indoor environments, thus it cannot be inherently determined, as illustrated in Figure 1.

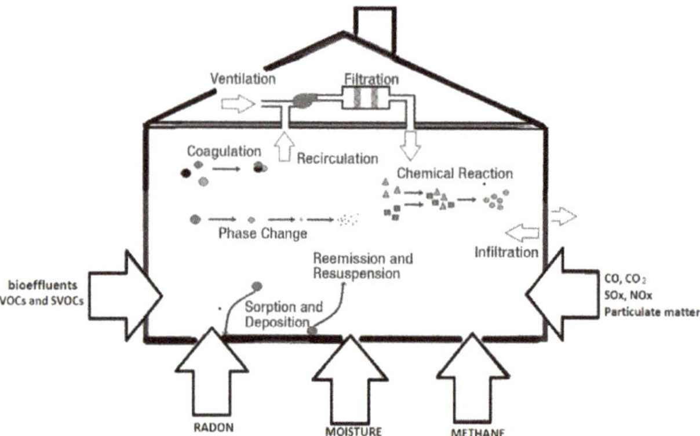

Figure 1. Overview of the physical and chemical processes of potential pollution sources in the indoor environment of a building. (VOCs—Volatile Organic Compounds, SVOCs—semi Volatile Organic Compounds).

Smoking alone emits more than 7000 different compounds, many of which are harmful [2] for humans and animals and may transport biological pollutants that can act as allergens. People and household animals emit gases which are unpleasant, transfer pathogens, and cause diseases. These examples show that there are many paths for penetration of and exposure to the sources of pollution in indoor environments.

In connection with the growing need to determine levels of indoor air pollution, new centers performing tests and new methods have been created considering the ability to analyze an increasing number of harmful substances. Measurement of pollutant concentrations in the air is generally a task performed by experts mainly in accredited laboratories and the results are published in scientific journals, technical reports, and, eventually, in guidelines, e.g., those of American Society of Heating Refrigerating and Air Conditioning Engineers ASHRAE [3]. The presence and concentrations of pollutants are often detected and measured without careful consideration of the significance of these measurements, and the pollutants measured may not be the most widespread or the most harmful. Some emissions are incorrectly grouped together; for example, more than one million volatile organic compounds (VOCs) are known and their toxicities are generally unknown, but they are often reported as a single value and referred to as the total VOCs (TVOCs) component. Frequently, carbon dioxide is used as an indicator of IAQ, although it does not have such a negative effect on the health of residents in the concentrations in which it is usually found in buildings. In our opinion, CO_2 is rather a marker of human bioeffluents. Examples of different understandings of the set of typical pollutants in an indoor environment are shown in Table 1 [4]. This table provides recommended values from the results of the European project HealthVent [5], which aimed to develop health-based ventilation guidelines. Table 1 also includes recommendations provided by the Word Health Organization (WHO) on the acceptable levels of pollutant concentrations [6,7], as well as recommendations from other organizations, such as China's IAQ standard values [8]. Different approaches to the IAQ issue mean that the exposure limits assumed in the various source materials differ.

Table 1. Acceptable levels of pollutant concentrations occurring in indoor air, according to World Health Organization (WHO) recommendations supported by the results of the EU research program HealthVent, values recommended by the standard EN 16798-1:2019 (replacing EN 15251:2007), values recommended by the Chinese indoor air quality (IAQ) standard GB/T 18883-2002 and for acceptable levels of pollution for the certification of office buildings in Hong Kong.

Pollutant	WHO Guidelines for IAQ with Updates [6,7]	HealthVent Project [5,9]	EN 16798-1:2019 [10]	IAQ Standards for China [8,11,12]	IAQ Certification Hong Kong [13]
CO_2			<500 ppm beyond outdoor level	485 ppm	<1000 ppmv 1800 mg/m^3 (8 h)
CO	100 mg/m^3 (15′) 35 mg/m^3 (1 h) 10 mg/m^3 (8 h) 7 mg/m^3 (24 h)	19 mg/m^3 (8 h)	100 mg/m^3 (15′) 35 mg/m^3 (1 h) 7 mg/m^3 (24 h)	1 mg/m^3	10 mg/m^3 7000 μg/m^3 (8 h)
Formaldehyde HCHO	0.1 mg/m^3 (30′)	0.03 mg/m^3 (30′)	0.1 mg/m^3 (30′)	10 μg/m^3	<0.1 mg/m^3 (8 h)
Benzene	>0.17 mg/m^3	<outdoor concentration	No safe level can be determined	0.11 mg/m^3	17 μg/m^3
NO_2	40 μg/m^3 (1 year) 200 μg/m^3 (1 h)	40 μg/m^3 (1 week)	40 μg/m^3 (1 year) 200 μg/m^3 (1 h)	10 μg/m^3	150 μg/m^3 (8 h)
SO_2	20 μg/m^3 (24 h)		20 μg/m^3 (24 h)	20 μg/m^3 (24 h)	
Naphthalene	0.01 mg/m^3 0.02 (1 year)	0.01 mg/m^3 (1 year)	0.01 mg/m^3 (1 year)		0.01 mg/m^3 (8 h)
Trichloroethene	>2.3 μg/m^3				230 μg/m^3 (8 h)
Tetrachloroethene	0.25 mg/m^3 (1 year)	0.25 mg/m^3 (1 year)	0.25 mg/m^3 (1 year)		0.25 mg/m^3 (8 h)
Respirable particulate matter $PM_{2.5}$	10.0 μg/m^3 (1 year)		10 μg/m^3 (1 year)	15 μg/m^3 (1 year) 35 μg/m^3 (24 h)	100 μg/m^3 (8 h)
PAH *	>0.012 ng/m^3		No safe level can be determined		1.2 ng/m^3 (8 h)
TVOC **			1000 μg/m^3	600 μg/m^3	600 μg/m^3 (8 h)

* PAH Para-Aminohippuric Acid—cyclic aromatic hydrocarbons; ** TVOC Total Volatile Organic Compounds; Reference [13] provides additional TVOC certification tests for new office buildings by determining (at the ppbv level) the content of carbon tetrachloride, chloroform, 1,2- and 1,4-dichlorobenzene, ethylbenzene, toluene, and o-, m- and, p-xylene.

Theoretical work on a combined IAQ model allowing aggregation of the results of the assessment of components affecting humans [14] is not yet well recognized in the literature. However, studies on IAQ indicators, which aim to provide a quantitative description of indoor air pollution, have been conducted since the nineties. In 2003, a significant study by Sekhar et al. [15] was published related to the standard indoor pollutant index (IPSI), the disease symptom index in the building symptom index (BSI), and to the often-cited works by Moschandres and Sofuoglu [16,17] on the indoor environmental index (IEI), indoor air pollution index (IAPI), and the indoor pollutant standard index (IPSI). The IAPI characterizes air pollution in an office with a single number: the index. The index value ranges between zero (lowest pollution level, i.e., best indoor air quality) and 10 (highest pollution level i.e., worst indoor air quality). The IAPI is a composite index; sub-indices ed are aggregated using the arithmetic mean in conjunction with a tree-structured calculation scheme. This scheme gives rise to some reservations, because at the top of the tree-structured calculation scheme is the IEI (calculated as the arithmetic mean of the IAPI and the IDI (indoor air discomfort index), and the combination of IAQ sensation and thermal conditions does not appear until later.

While considering the indicators for the quantitative description of pollution, the proposal of the IEA Working Group named "Defining the Metrics of IAQ" should also be mentioned. This group prepared, in 2017, the document entitled "In the Search of Indices to Evaluate the Indoor Air Quality of Low-Energy Residential Buildings" [18]. The group made the following assumption for the categorization of various indicators: there should be one index per individual pollutant and a dimensionless coefficient should be specified to evaluate the IAQ, provided that the current (observed) concentrations of a given pollutant c_j are related to the ELVs (exposure limit values) concentration $c_{j,ELV}$.

$$IAQ_{index}(j) = I_j = \frac{c_j}{c_{j,ELV}} \tag{1}$$

The index is calculated for each individual pollutant [18], which is specific only for this exact pollutant. The report showed that aggregation can be performed by addition, by taking the maximum value or by other methods, in an attempt to define metrics that can be used to evaluate IAQ. The assumption was that the reference value usually refers to health risks (accounting for chronic or acute effects), but other metrics can also be used, (e.g., odor or irritation threshold). There are two important properties to be considered when aggregating sub-indices: ambiguity and eclipsing. As a result of the analysis, the authors concluded "that there are problems with model aggregation methods. In the aggregation model $I_{agg} = I_1 + I_2$, ambiguity creates a false alarm and in the aggregation model $I_{agg} = 1/2(I_1 + I_2)$, eclipsing underestimates the effect" [18]. Therefore, the discussion remains open [18]. The report also showed how there are large spreads of concentrations of individual pollutants (up to seven rows), even in the group of pollutants for which sub-indices were built. It determined the difficulties of building a weighted scheme based on the simplest percentage adjustment of the concentration shares and, thus, the share of the mass of pollutants to be removed by ventilation.

The current state of knowledge does not provide information authorizing the omission of certain pollutants. Hence, taking into account the lack of data on the characteristics of each chemical compound and consideration of the "removal efficiency" [19] requires us to abandon thinking about the adjustment of many individual pollutants, and to focus only on the creation of a model based on the representative and target components. In this state of knowledge, there are hopeful studies and proposals with a grey combined $\sum IAQ_{index}$ model and the grey clustering model for IAQ indicators proposed by Zhu and Li in 2017 [20] is particularly interesting, especially when the relationships between system factors and the system's IAQ behavior and the interrelationships among the factors are uncertain. At first, all specific indoor air pollutants and related parameters should be measured. However, this is a very complex and time-consuming process. On the basis of the characteristics and correlations of the pollutants, the indoor air quality can be characterized by representative indicators. Studies [20] have pointed out that respirable particulates, CO_2 and TVOCs, were the three most representative and independent environmental parameters which can be used as an evaluation index of indoor air quality in office buildings. Since each indicator represents a class of pollutants with similar sources and dissemination characteristics, this index group avoids unreliability due to the fact that these indicators are "too small" because of critical concentration depression. A data pretreatment method must be used in the calculation procedure, reflecting the differences in concentration levels among different pollutants, but also expressing their influence on the comfort and health of the indoor occupants. Moreover, the measured pollutant concentrations can be used to predict the probable levels of other parameters, and good agreement was found between the predictions and measured values.

1.2. The Research Questions

The main research question contained in the paper concerned whether it was possible with the current state of knowledge to create and use in practice an IAQ model that was based on a unified and coherent approach for input indoor air parameters (such as pollutant concentrations, odor levels, and moisture content) and provided one output parameter (we proposed occupant satisfaction, IAQ_{index} (in %)). The authors looked for physical equations for the IAQ_{index}'s subcomponents and dependencies for their predicted occupant satisfaction functions with a pollutant concentration c_j ($PD = f(c_j)$ in %) which could be used as a model for subcomponents.

This paper's intention was to provide an IAQ model with a step-by-step process which can be used to determine the value of the overall indoor environmental quality index (in %) including another three components: thermal comfort, acoustic comfort, and lighting quality. The innovative approach and added value of this article is in the use of the proposed IAQ model in practice and the relatively simple calculation of the overall IEQ value (with an uncertainty estimation) using the actual results of measurements in the Building Research Establishment Environmental Assessment Method BREEAM certified case study office building. The authors also provided occupant satisfaction functions for CO_2, TVOCs, and formaldehyde HCHO in two variants: with experimental *%PD* values taken from the literature and for these pollutants' *%PD* values converted from an Air Quality Index system (see Section 2.2.).

2. Methods

2.1. Research Content and Strategy

The proposed IAQ model is presented in Sections 2.2–2.7. The model is later used to analyze the case study of an office building described in Section 2.9. Figure 2 presents the subsequent research steps from theory to practical application. Section 2.8 shows the method for determining indoor environmental quality IEQ_{index} where IAQ index is a subcomponent/part of the IEQ_{index} model. In order to determine the IAQ and IEQ, physical measurements of the indoor environment in the building were conducted using the experimental approach provided in Section 2.10. Based on these physical indoor measurements the IAQ and IEQ indexes (number of occupants satisfied with the indoor air and overall indoor quality, respectively) were assessed (see Section 3) and discussed (see Section 4).

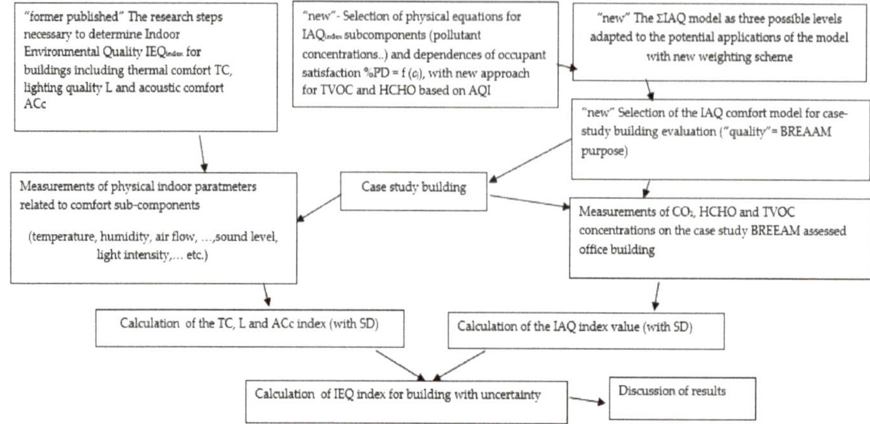

Figure 2. To determine the indoor air quality and indoor environmental quality (IEQ) indexes for the case study. (TC—thermal comfort, L—light quality, ACc—acoustic comfort).

2.2. The IAQ Model Proposal—Basic Assumptions

In the IAQ model construction process (our proposal), the commonly accepted approach is to transform individual concentrations of pollutants into subcomponents before they are aggregated into a single index (occupant satisfaction in %). However, summation of sub-indices can lead to situations in which all are under individual health thresholds, but the final indicator shows when the threshold has been exceeded. Conversely, the averaging of partial sub-indices can lead to an overall indicator showing an acceptable IAQ, even though one or more partial indicators are larger than their individual thresholds. One solution is to use the maximum value of all sub-indices to create the final form of the $\sum IAQ_{index}$. Taking these issues into consideration, the authors created the $\sum IAQ$ model with three complication levels adapted to the purposes of potential applications of the model, as presented in Figure 3;

i. Certification of a building, e.g., via the BREEAM system using (three sub-indices), called "quality";

ii. Design, including perceptible contaminants affecting comfort and using the IAQ index when calculating the IEQ (five sub-indices), called "comfort";

iii. Complex design, with the $\sum IAQ_{index}$ representing both comfort and health (seven sub-indices) called "comfort/health".

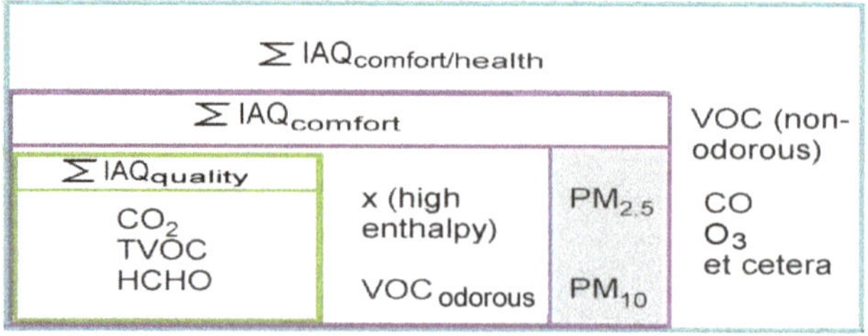

Figure 3. $\sum IAQ$ model has three possible levels (i.e., similar to a Russian doll structure) adapted to the potential applications of the model.

The simplest one, "quality", is an inner part of the $\sum IAQ_{comfort}$ model and can be used separately for simple applications with the main purpose of supporting a green building certification, e.g., via the

BREEAM system using three components, i.e., CO_2, HCHO, and TVOC. This model is used later on the case study of a BREEAM building.

Figure 1 shows the processes influencing the morphology of the $\sum IAQ$ model and Figure 3 provides a list of the pollutants for which three IAQ submodels were built, containing human-perceived contaminants ($IAQ_{quality}$ and $IAQ_{comfort}$ models), but also the $IAQ_{comfort/health}$ model for both perceptible and imperceptible pollutants, i.e., those that are not perceptible by humans but affect health and require additional energy for intensive ventilation for health reasons. There are potential sub-indices, such as $IAQ(VOC_{non\text{-}odorous})$ or $IAQ(CO)$ [5,7]. Dust pollutants may have their sub-index both in the comfort model (if reliable curves of human sensory perception of PM concentrations are known) or in the $IAQ_{comfort/health}$ model if their health impact is considered to be the dominant feature. Considering the types of pollutants harmful to health assigned to the sub-indices of IAQ, we only consider the most important air pollutants (i.e., target emissions) that were given in the WHO guide in 2010 [5,7]. Submodels of processes that impact on air quality in indoor environments, which can be distinguished as components of the IAQ model, were based on sensory satisfaction functions (index of occupant dissatisfaction (PD) with the level of air pollution). Subcomponents of the three potential IAQ models were classified according to their future potential applications: in the assessment of environmental quality index IEQ (models $IAQ_{quality}$ and $IAQ_{comfort}$) or in the design of ventilation taking into account all possible harmful-to-health pollutants (model $IAQ_{comfort/health}$). In our opinion, such systematization creates order and has a practical dimension as presented later on in the case study. The following are the target pollutant groups:

i. In the air quality model $\sum(IAQ)_{quality}$, the IAQ index subcomponents were assigned to the selected three pollutants. The submodels for the IAQ were CO_2, TVOC, and formaldehyde HCHO, as recommended by References [5,10,21,22];

ii. In the $\sum(IAQ)_{comfort}$ model, thepreviously provided simplified $IAQ_{quality}$ subcomponents for the three main pollutants were extended with a set of selected compounds $VOC_{odorous}$, related to the collection of IAQ sub-indices ($VOC_{odorous}$) with an unknown cardinality, increased appropriately for the number of dominant pollutants. In addition, we provided a conditional deluge of two more components: (1) calculated using the enthalpy of hot and humid air (high enthalpy h > 55 kJ/kg [23]), the percentage of persons dissatisfied with respiratory cooling with humid air at relatively high temperatures and (2) the percentage of persons dissatisfied with indoor pollution with respect to dust pollution (PM_{10} and $PM_{2.5}$), measured via panel tests. The introduction of a dust-pollution subcomponent to the IAQ model may be debatable, because some experimenters [24,25] underline the unique results of sensory tests of discomfort from dust, and the influence of "emissions" of respiratory dust particles on satisfaction is still under-researched. Considering the above, we expected two variants of the comfort model: with PD (PM_{10}, $PM_{2.5}$) or without this factor;

iii. In the overall $\sum IAQ$ model, comfort $(IAQ)_{comfort}$ and health risk $(IAQ)_{health}$ indicators were used, and, hence, this model was called $(IAQ)_{comfort/health}$. Models for subcomponents of IAQ not perceived by humans but influencing health, can be borrowed from the index set in the AQI (air quality index) system [26–28], which was adapted to assess the quality of indoor air based on, and in accordance with, the concepts of the air quality assessment system used globally by the American EPA.

Values of AQI indices published on active EPA websites using the air quality index system were introduced for application in US federal regulations in 1999 [28]. Currently, the AQI system for outdoor air includes the following pollutants: ozone, particulate pollutants (PM_{10} and $PM_{2.5}$), carbon monoxide CO, sulfur dioxide SO_2, and nitrogen dioxide NO_2. To convert a specific air pollutant concentration to an AQI, the EPA developed a tool called the AQI Calculator, which is an open resource [29]. This system (referring to the index from 2004 [16] and the indoor pollutant standard index (IPSI)) was further developed, and the proposed IAQI for indoor air presented by Wang et al. [30] in 2008 and a newer proposal [27] from 2017 for a similar but narrower set of indices, also for indoor air, were both modeled on it. The AQI and IAQI indicators showed an increase in the level of impact on human health with increasing concentrations of air pollution. There are some detected difficulties here, since "AQI is a piecewise linear function of the pollutant concentration" [27]. The calculated values of the AQI [31] or IAQI [30] indices, over the entire 0–500 scale calculated from the measured concentrations of selected contaminants or in the part of the scale corresponding to the IAQ rating, ranged from "good" to "unhealthy", and can be converted to *PD%* for use in the model equation, (IAQ)comfort/health. Concentrations will be significant when the uncertainties of scale conversions are estimated. Authors believe that their way of converting the AQI scale to *PD%* (which is similar to the method of conversion of the IEQ components' ordinal scales from the OFFICAIR EC project [32] to *PD%* scale; for example, the occupant percentage dissatisfied with noise [33,34] should be accepted in light of the expected results of a metrological analysis of the reliability of the combined $\sum$IAQ model.

Subcomponent models (physical functions, *PD%*) of the IAQ model for all individual air pollutants are presented later in this section.

2.3. $\sum$IAQ Model Weighting Scheme Considering Air Pollution Ventilating

To obtain a comprehensive picture of IAQ in a building, it is necessary to measure the number of pollutants with different individual concentrations. There are methods that weigh sub-indices [21] but the problem is finding an effective weighting scheme and understanding how to adjust them in the overall model of all the pollutants in $\sum$IAQ. For this reason, we proposed an adjustment method for the weights. In our opinion, provided in detail in References [33,34], and also according to Reference [22], the best weighting scheme, which would lead to a credibly aggregated model of IAQ composed of many extractable components (sub-indices), would be a system based on concentration values (the "excess masses" of pollutants to be regarded as loadings for the ventilation system). Therefore, the we aimed to determine the individual pollutants assigned to the IAQ model, their concentrations, c_j, as the inputs of the IAQ submodels, and their "excess concentrations" originating from emissions or determined within indoor environments. Thus, it was possible to determine directly the energy requirements for ventilation purposes and the required minimum global ventilation rate. Determining the input concentration value, c_j, for each IAQ sub-index enables the determination of the total mass of pollutants in the air, which is the basis for determining the air change rate $N_{1,\dots 7}$ (overall air change rate), assuming that the model includes all significant IAQ pollutants.

Currently, according to References [35–37], the most common assumption made is that pollution from VOC$_j$ compounds arises only from emissions due to the presence of construction or finishing materials (for $j = 1, 2, \dots n$) (it can be assumed that the source i of an emission is the entire indoor environment and then $i = 1$) from the zero state. The physical model for determining the ventilation rate in indoor environments polluted with VOC-type pollutants from building materials is given by EN 16798-1:2019 [10], assuming that design parameters for indoor air quality are derived using limit values for substance concentrations. In accordance with ECA Report Number 11 [36], the design ventilation rate required to dilute an individual substance emitted from building materials is calculated as:

$$Q_h = \frac{G_h}{C_{h,j} - C_{h,0}} \times \frac{1}{\varepsilon_v} \tag{2}$$

where Q_h is the ventilation rate required for dilution in m^3 per second, G_h is the emission rate of the substance in micrograms per second, $C_{h,j}$ is the guideline value of the substance in micrograms per m^3, $C_{h,0}$ is the concentration of the substance in the supply air in micrograms per m^3, and ε_v is the ventilation effectiveness.

In fact, in a building with active "indoor chemistry" (see Figure 1), the use of this formula seems to be increasing. Taking into account the dynamic nature of the processes of generating various pollutants, the approach to IAQ and its components should be changed and subcomponents should be treated as pollution load processes, increasing in number not only due to the emission processes but also due to the generation of bio-pollution, water evaporation, and even dust infiltration from outside. It is also possible to set steady-state (initial) concentrations of pollutants and to determine the expected time courses of removal of these pollutants by means of ventilation (curves $\sum c_j = f(\tau)$), at constant values of air change rate per hour ACH (h^{-1}). Such ventilation rate calculations for CO_2 were developed in 1997 by Persily [38] and similar ones were provided in 2017 by Gyot [39] at the Berkeley National Laboratory. These calculations are not very accurate, as shown in the general demonstration graph in Figure 4.

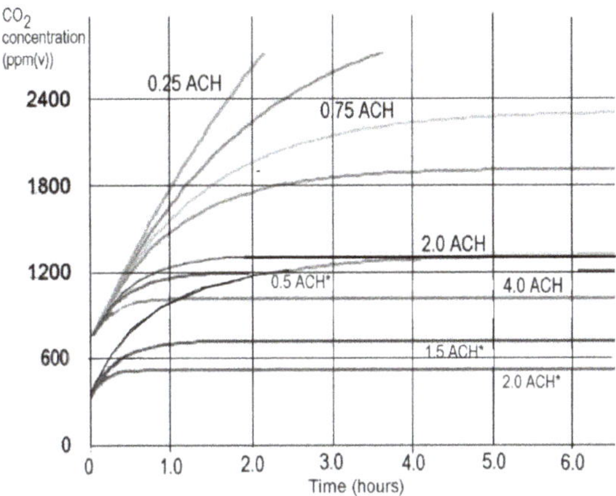

Figure 4. Of CO_2 above outdoors levels with two people in the contaminated building (ACH—air change rate) and in a typical clean office room (ACH*).

Less accurate time-dependent curves of the total minimum ventilation rate ACH needed for "contaminant exhaustion" can be determined using programs [40] based on generic engineering equations for the sum of pollutants $\sum c_j$. The generic equation for pollution concentration (the ratio of the amount of polluting product to the amount of fluid in the space (such as air in a room) can be calculated from the following equation:

$$c = q/(n \cdot V \cdot (1 - e^{-Nt}))\tag{3}$$

where c is the pollution concentration in the space (or in the room) with perfect mixing (m^3/m^3) or (kg/kg), q is the amount of pollution added to the space (m^3/h) (kg/h), N is the air change rate per hour (h^{-1}), V is the volume or mass of the space (m^3) or (kg), e is the number 2.72, and t is time (h). If the initial concentration (at $t = 0$) in the space and the concentration in the supply fluid is zero, after some time the concentration in the room will stabilize. The ventilation rate graph for an amount of pollution $q = 1$ and a volume of space $V = 1$, shows the values of $\sum c_j$, similar to Figure 4. In order to obtain more exact values of the VOC concentrations remaining in the room for the air change rate function, it is possible to use the published dependencies or for the assumed volumes V of ventilated spaces with determinate concentrations, as they can be determined experimentally. A simplified method for determining the ventilation rate N from a simple formula for the time, t, course of a trial ventilation was provided by the Japanese researchers Noguchi et al. [41] in 2016. A description of this method is worth reading. Based on the temporal changes of the TVOC concentration measured using a PID_{TVOC} meter [42], the air change rate N or the ventilation rate F was estimated using the following method. Assuming perfect mixing of the air in the room and a constant TVOC emission rate E, the concentration change of TVOC in the room can be expressed by the following equation:

$$C(t) = C_j + \left(\frac{E}{F}\right)\left(1 - e^{-\frac{F}{V}t}\right) = C_j + \left(\frac{E}{F}\right)\left(1 - e^{-Nt}\right)\tag{4}$$

Finally, Equation (5) can be expressed with one unknown parameter, the air exchange rate N as:

$$\log\left(\frac{C_{st} - C_j}{C_{st} - C(t)}\right) = Nt \tag{5}$$

The initial concentration c_j can be determined from the experimental results. After a long time, when the exponential term in Equation (5) can be assumed to be zero, the concentration $C(t)$ becomes constant. The steady-state concentration C_{st} can be determined from the temporal change in the experimental results where the concentration levels off.

The rule that the IAQ model should include a weighting scheme, referring to the variation in the share of pollutants in the IAQ, has been noted in References [35,37]. According to the first proposal, the weighting system is based on the differentiation of coefficients R_j, which are the ratios of the real concentration values (or mass of pollutants) to the values of reference concentrations, representing the so-called relative masses of non-eliminated pollutants. This can also be represented by the desirable reduction in the level of pollution by means of ventilation and, thus, also by the energy requirements. For one emission source, the proposed system with a coefficient R_j for a given pollutant C_j has the formula:

$$R_j = \frac{y_j}{I_j}, \; j = 1, 2, \ldots m \tag{6}$$

where I_j is the ratio of the gas-phase concentration to the reference concentration value. For example, for the LCI (Lowest Concentration of Interest) value for the jth compound emitted from the building material, the factor R_j is dimensionless, since y_j is the gas-phase concentration for the jth compound in $\mu g \cdot m^{-3}$, I_j is the lowest concentration of interest (LCI) [41] for the jth compound in $\mu g \cdot m^{-3}$ and m is the number of all elected compounds. The weight coefficients W_j are used as the weights of the equations in the IAQ index, according to:

$$W_j = \frac{R_j}{\sum\limits_{j=1}^{m} R_j} \qquad j = 1, 2, \ldots \ldots m \tag{7}$$

The authors of Reference [35] justified adjusting the coefficients where $\sum R_j \leq 1$, but they did not explain the physical meaning of this condition. We believe that further discussion should include the issue of whether the "relative mass" of contamination expressed by Equation (6) has a proper place in the weighting scheme for the equations. The dimensionless quantity (7) does not have a sound physical meaning [37]. In our opinion, this type of calculation method is debatable.

One should strive to cover all the sub-indices of the combined IAQ_{ij} model with a weighting scheme that would give VOCs a share in the total energy requirements for ventilation. The term "relative mass" should correspond to weights rationally proportional to the energy expenditure for ventilation of individual pollutants (IAQ sub-indices). Therefore, our future work will focus on introducing weights for all expressions in the overall IAQ model equation. However, since this is currently not possible without an adjustment method adapted to weight determination for very small concentrations, it was decided to present our model as an interim solution. We proposed the use of weights based on the "excess concentration" values within the pollutant categories only with similar and comparable orders of concentration values, for example the $VOC_{odorous}$ and $VOC_{non-odorous}$ categories. The rules of adjustment with boundary conditions will be provided and justified in the follow-up article to this report.

2.4. $\sum IAQ$ Model Scheme Morphology

According to the new proposal, for models with air quality sub-indices $IAQ(P_j)$ (developed with the standard EN 16798-1:2019 as a reference for IEQ model creation [22,33] and with assumptions described in Reference [34]), in the case where indoor air has many pollutants, $P_1, \ldots j$, the combined ΣIAQ_{index} equation is:

$$\Sigma IAQ_{index} = W_{P1} \cdot IAQ(P_1)_{index} + W_{P2} IAQ(P_2)_{index}, \ldots, W_{Pj} IAQ(P_j)_{index} \qquad (8)$$

where the $W_{P1}, \ldots _{Pj}$ weighting system for IAQ components is created on the basis of the arithmetic mean and the concept of "excess concentration" is introduced only for groups of pollutants with similar concentration values. There is a difference in concentration Δc_j between the observed concentration of pollutant c_j and the reference concentration c_{ref} (c_{ELV} or c_{LCI}), which is below the current concentration in contaminated rooms. Thus, the excess concentration is:

$$\Delta c_j = c_j - c_{ref} \qquad (9)$$

The weights $W_{1, \ldots j}$ for all three IAQ models are determined on the basis of arithmetic means or by adjusting all the values of Δc_j in a given model using Equation (10):

$$W_j(IAQ_{comfort/health}) = \frac{\Delta c_j}{\sum\limits_{j=1\ldots7} \Delta c_{1\ldots7}} \qquad (10)$$

where the sum of the adjusted weights W_j of all ventilated pollutants described with sub-indices should be unity. The weight values for a given IAQ model, (e.g., $IAQ_{comfort}$) may be different, but the sum of the sub-index weights must be ≤ 1.0.

The values of the reference concentrations are the concentration levels that are acceptable or recommended as limit values for various pollutants P_j. In the case of the $\Sigma IAQ_{quality}$ submodel as part of the IEQ_{index} model, weights should be used for the $VOC_{odorous}$ (HCHO and TVOC) reference threshold concentrations of odors. The weights in the weighting system should be adjusted to unity according to the Equation (11):

$$W_{HCHO} = \frac{\Delta c_j}{\sum\limits_{j=2,3} \Delta c_{2,3}} = \frac{(c_j - c_{ref})}{(c_{TVOC} - c_{ref}) + (c_{HCHO} - c_{ref})} \qquad (11)$$

There are, however, non-typical cases in which the scales have different values. This is the case for formaldehyde, the concentration of which is many times lower in the building than the threshold level c_{th}. According to the WHO [7], the admissible value c_{ref} is also higher than the concentration in the building. In this case, the authors recommend taking the reference value as zero. Then, the weight W_{HCHO} described by Equation (11) (for two pollutants), would not be negative ($c_{HCHO} - c_{ref}$). The ASHRAE Guideline 10 (2011) [3] recommends that the IEQ model (and appropriate weights, W_i) should contain synergy effects of environmental parameters included in the subcomponents and their sensory perceptions.

Figure 5 shows the extended IAQ_{index} model with its sub-indices treated as components of the IEQ_{index}, but also with sub-indices of the $IAQ_{comfort/health}$ type, i.e., pollutants that do not belong to the IEQ model but are important to health and the energy balance of a building with a mechanical ventilation system. The experimental dependencies of the percentage of persons dissatisfied, $\%PD$, and the values of the concentrations of pollutants, c_j, sensed in indoor air in the appropriate ranges are of fundamental significance in the sub-indices relevant to the IEQ model [43].

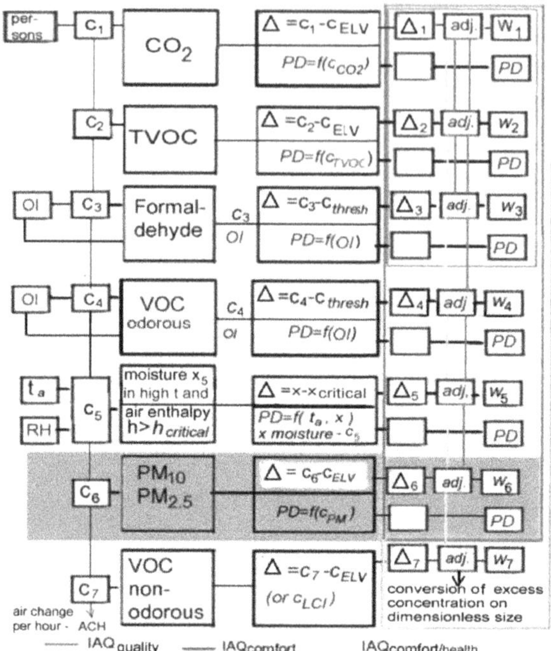

Figure 5. ΣIAQ_{index} model with weighting scheme.

From the dependencies, expressed as the curves for $PD(CO_2)$ or $PD(VOC_{odorous})$, the equations of the models are derived Equations (12)–(14):

$$\Sigma IAQ_{quality} = W_1 \cdot IAQ(CO_2) + W_2 \cdot IAQ(TVOC) + W_3 \cdot IAQ(HCHO) \tag{12}$$

$$\Sigma IAQ_{comfort} = W_1 \cdot IAQ(CO_2) + W_2 \cdot IAQ(TVOC) + W_3 \cdot IAQ(HCHO) \\ + W_4 \cdot IAQ(VOC_{odorous}) + W_5 \cdot IAQ(h) \tag{13}$$

$$\Sigma IAQ_{comfort/health} = W_1 \cdot IAQ(CO_2) + W_2 \cdot IAQ(TVOC) + W_3 \cdot IAQ(HCHO) + W_4 \cdot IAQ(VOC_{odorous}) \\ + W_5 \cdot IAQ(h) + W_6 \cdot IAQ(PM_{2.5}, PM_{10}) + W_{7a} \cdot IAQ(VOC_{non-odorous}) + W_{7b} \cdot IAQ(CO) + W_{7c} \cdot IAQ(NO_2) \tag{14}$$

The scheme of the $\Sigma IAQ_{comfort/health}$ model consists of seven (or more) components or IAQ submodels and these are models for the various types of pollutants: $IAQ(CO_2)$, $IAQ(TVOC)$, $IAQ(HCHO)$, $IAQ(VOC_{odorous})$, $IAQ(h)$, $IAQ (PM_{2.5}, PM_{10})$, and the selected $IAQ(VOC_{non-odorous})$. The $IAQ(VOC_{odorous})$ and $IAQ(VOC_{non-odorous})$ models should be multiplied, depending on the number of dominant VOC pollutants, and, hence, the $\Sigma IAQ_{comfort/health}$ model will, in practice, have more than seven components.

The inputs of each IAQ submodel are unit concentrations in air of a given pollutant, c_j (in the case of $IAQ(h)$. This is the moisture content x in well-known units "g of water vapor (g_w) per kg of dry air (kg_a)", converted to a concentration of c_j in $\mu g_{water}/m^3$ (or H—absolute humidity in g_w/m^3 which is a measure of water-vapor density). In some cases, it is necessary to convert the pollution-derived parameter to VOC concentration (conversion of the odor intensity OI to VOC concentration is described later in this section).

From the concentration values, total air pollution can be calculated, and, subsequently, also the energy needed to ventilate the indoor air pollution. When the concentration levels of pollutants are variable and are increasing due to the presence of emissions, then the formulas given in Reference [36] and the amended standard EN 16798-1:2019 are used to calculate the required ACH ventilation rate. When the level of contamination is set (or quasi-fixed) and the volume and other parameters of the ventilated room are known, it is possible to calculate ventilation-time curves, i.e., maximum ventilation curves for ACH ventilation rate to reach concentration levels ELV, LCI or the olfactory threshold level, according to Reference [40] or another adequate equation.

There are two outputs of each IAQ submodel as described below:

1. The weights of the weighting system for the model $\sum\text{IAQ}_{\text{quality}}$ and hypothetically $W_{1,\ldots\,5}$ for the model $\sum\text{IAQ}_{\text{comfort}}$ or $W_{1,\ldots,7}$ for the model $\sum\text{IAQ}_{\text{comfort/health}}$ (in a hypothetical model with a set adjustment method). These should reflect the energy load of the IAQ expressed by the theoretically assumed increase in the current concentration of pollutant c_j relative to the reference concentration of $c_{j,ref}$, which determines the level of this concentration intended to be obtained by ventilation.

2. The *PD%* with the IAQ as a function of air pollution concentration. These values, determined in panel tests, reflect the impact of the interaction of air with a given pollutant at the actual level of concentration, estimated via panelists' sensations/perceptions ($PD = f(c_j)$ in %).

Examples of measurable physical parameters for the purpose of IAQ and IEQ calculations (see case study) are given in the following section.

For the construction of the combined model $\sum\text{IAQindex}$ with a weighting scheme useful for aggregating sub-indices, we proposed the model presented in Figure 5. In this scheme, the combined IAQ model is shown as the basic assumption for the aggregation of all sub-indices. First, the model was cut by a cross-connected vertical connection regarding the inputs of submodels—the calculation of the sum of the masses of all pollutants $\sum c_j$ in the ventilated space, using the values for the inputs of all submodels of IAQ concentration values of contaminants. The sum of the concentrations of all air pollutants expressed as mass units of pollution per m^3 of volume (which can be read after multiplication by V (m^3) as the mass to be displaced by ventilation), is the basis for calculating the "air change rate per hour" (ACH), the minimum air exchange rate needed to reduce the observed mass level of air pollution in a ventilated room (see Reference [40]). The second connection concerns the submodel outputs—the conversion of excess concentration to a dimensionless value, which allows to for the weighting scheme of the $\sum\text{IAQ}_{\text{index}}$ combined model and the weights of individual IAQ submodels to be determined.

Additional assumptions were as follows:

i. The sum $\sum\Delta c_{1,\ldots\,7}$, which admittedly constitutes an excess mass increase of the sum of pollutants described by the submodels, was treated only as a "virtual energy load of the building" for ventilation conducted for the elimination of pollutants, and therefore, when adjusting the weights, the possibility of dividing the excess concentration by the sum of concentrations should be considered.

ii. The percentage of persons dissatisfied *PD(IAQ_{component})* determined experimentally in sensory studies using panelists' sense of air quality during their exposure to an internal environment deteriorated by a given contamination component ($PD = f(c_j)$), was derived from the literature or direct experiments.

Values of weights $W_{1,\ldots,j}$ in sets of three, five or more components of the three $\sum IAQ_{index}$ models (see Equations (12)–(14), were adjusted to a value of unity by dividing $\Delta c_{1,\ldots,j}$ of each component by the sum of excess concentrations $\sum \Delta c_{1,\ldots,j}$ in μ/m^3. We proposed the use of c_{ref} values, apart from the values of c_{LCI} [18], c_{ELV}, and the threshold values c_{th} for odorous compounds, were as follows.

i.   For IAQ(CO_2) and IAQ(HCHO), the c_{ELV} concentrations were derived from EN 16798-1:2019 [10];

ii.   For IAQ(TVOC) and IAQ($VOC_{odorous}$), the threshold concentrations, c_{th}, for identified odorous compounds or mixtures are from Reference [44];

iii.   For the IAQ(h), the water-vapor concentration H (g_w/m^3), recalculated from the moisture content x ($g_w/kg_{dry\ air}$) using the gas constant for the water vapor and the actual temperature, the value of h up to the critical value for "high enthalpy of humid air" was evaluated using the formula:

$$h = 1.006t_a + x\cdot(2501 + 1.805t_a) \tag{15}$$

where h is the specific enthalpy of humid air (kJ/kg), which must be >55 kJ/kg. The EN 16798-1:2019 standard [10] recommends a limit for the dehumidification of air of $12g_w/kg_{dry\ air}$ (this value must be converted to a c value in g_w/m^3).

i.   For IAQ($VOC_{non\text{-}odorous}$), the c_{ELV} value in cases where no established LCI values were derived from the EN 16798-1:2019 standard.

ii.   For IAQ($PM_{2.5}$, PM_{10}), the c_{ELV} values were derived from the WHO [7] or other organizations (Tables 1 and 2).

The proposed reference values of pollutants forming the sub-indices of the IAQ model are given in Table 2. With reference to the concentration values of c_{LCI}, it should be noted that according to References [18], this value is typically acquired by dividing occupational exposure limits by a safety factor (100 or 1000). Concentration c_{LCI} is taken from Lowest Concentration of Interest (EU-LCI) from European Commission lists.

However, the model values for exposure limit values (ELVs) of indoor air pollution, in accordance with the recommendations of the health-based ventilation guidelines [5], should be adopted in accordance with the current WHO guidelines given in the periodically issued *WHO Air Quality Guidelines* [7].

2.5. Selection of Submodels for Pollutant Components

Our overall selection of physical subcomponent equations and dependences for $\%PD = f(c_j)$ is presented in Table 3. The models presented were used to determine the IEQ for the sample building. The highlighted pollutants were taken into account in the case study building assessment.

Table 2. Exemplary pollution concentration reference values for low-pollution building.

Component of Pollution P_j	Reference Concentration	Reference Value	Reference/Recommendation List
CO_2 (outdoor 350 ppm)	c_{ELV}	380 ppm	EN 16798-1:2019
TVOC	c_{ELV} c_{th}	<300 µg/m^3 50 µg/m^3	EN 16798-1:2019 MV. Jokl [45]
HCHO	c_{ELV}	30 µg/m^3 (30 min) 100 µg/m^3 (30 min) 9 µg/m^3 (1 year)	EN 16798-1:2019 WHO Guidelines for IAQ (2010) IEA-AIVC Report. Annex 68 (2017)
	or c_{th}	300 µg/m^3 60 µg/m^3	AIHA Odor Thresholds for Chemicals with Established Health Standards (2013)
VOC$_{odorous}$ (as in EN 16516) Naphthalene Nitrogen dioxide Ammonia Ozone Carcinogenic VOCs	c_{th} or c_{ELV}	10 µg/m^3 (1 year) 20 µg/m^3 (1 year) 70 µg/m^3 (1 year) 100 µg/m^3 (8 h) 5 µg/m^3	AIHA Odor Thresholds for Chemicals (2013) EN 16798-1:2019
Moisture content x in indoor air (enthalpy >55 kJ/kg, t_a = 23 °C and 60% RH) x [g$_{water}$/kg$_{dry\ air}$] $\Rightarrow$ H [g$_{water}$/m^3]	> unacceptable value of moisture content x at 23 °C.	x > 12 g/kg RH > 60%	EN 16798-1:2019
PM$_{10}$	c_{ELV}	20 µg/m^3	WHO Air Quality Guidelines (2010) EU Directive 2008/50/EC introduced additional PM$_{2.5}$ objectives targeting the exposure of the population to fine particles.
PM$_{2.5}$	c_{ELV} $c_{DIRECTIVE}$	10 µg/m^3 2 25 µg/m^3 (1 year)	
VOC$_{non\text{-}odorous}$ (EN 16516 includes pollutants with limit values on concentration that have been identified): Carbon monoxide Carcinogenic [1] VOCs	c_{ELV} or c_{LCI}	 7 mg/m^3 (24 h) 5 mg/m^3	EN 16798-1:2019 EN 16516; WHO Guidelines for IAQ (2010)

[1] Limit values and carcinogenic effect: the level of PAHs, particles, benzene, and trichloroethylene should always be kept as low as possible.

Table 3. Selection of physical equations for IAQ_{index} components and dependences for $PD = f(c_j)$.

Sub Component	Input Parameters	Sensory Equations to Calculate the %PD Value	References
CO_2	$c(CO_2)$ (in ppm)	$PD_{IAQ(CO_2)} = 395 \cdot \exp(-15.15 \cdot C_{CO_2}{}^{-0.25})$ $c_{IAQ(CO_2)} = 55\,833 \cdot (\ln(PD) - 5.98)^{-4}$ $IAQ = 6 \times 10^{-0.07} c_{CO_2}^2 - 0.0025 \times c_{CO_2} + 1.9416$ $PMV_{CO_2} = 6.364 \times log \frac{c_{co_2}}{485}$ $PD = 100 - 95 \cdot \exp(-0.03353 \cdot PMV^4 - 0.2179 \cdot PMV^2)$	[36] [45] [46] [47] [48]
TVOC [1]	$c(TVOC)$ $\Rightarrow$ PD $IAQI \Rightarrow$ PD^*	$c_{TVOC} = 16{,}000(\ln(PD) - 5.988)^{-4};$ $c_{TVOC}(\mu g/m^3)$ $PD_{(IAQ)TVOC} = 405 \cdot \exp(-11{,}3 \cdot c_{TVOC}{}^{-0.25})$ In Taiwan EPA Indoor Air Quality Index System (IAQI). Nine major indoor air pollutants are included: PM10, PM2.5, CO_2, CO, O_3, HCHO, TVOC, bacteria, and fungi. We proposed the use of a calibration curve IAQI = f(c_{TVOC}) for conversion to PD^* %	[49] [22,45] [30]
HCHO	$c(HCHO)$ $(pi)^2 \Rightarrow$ $OI^3 \Rightarrow$ PD $c(HCHO)$ $\Rightarrow$ $IAQI \Rightarrow$ PD^*	$PMV_{HCHO} = 2\log \frac{c_{HCHO}}{0.01}, \qquad c_{HCHO}(mg/m^3)$ $PD = 100 - 95 \cdot \exp(-0.03353 \cdot PMV^4 - 0.2179 \cdot PMV^2)$ $PD_{IAQ(OI)} = \dfrac{1}{1 + \frac{1}{\exp(2.14 \cdot OI - 3.81)}}$ In the Taiwan EPA Indoor Air Quality Index[6] System (IAQI). Nine major indoor air pollutants are included: PM10, PM2.5, CO_2, CO, O_3, HCHO, TVOC, bacteria and fungi. We proposed the use of the calibration curve IAQI = f(c_{HCHO}) for conversion [7] to PD^*%	[47] [48] [50,51] [30]

Table 3. *Cont.*

Sub Component	Input Parameters	Sensory Equations to Calculate the %PD Value	References
			[52]
	$c_{VOC} \Rightarrow OI^4$	Formulae for the conversion of odorant concentration c_j to odor intensity OI for recalculation [4] of the conversion of odor intensity to odorant concentration.	
	$c_{VOC} \Rightarrow OI^3$	Conversion of the chemical concentrations c_j (mg m^{-3}) into odor concentrations c_{OD} (ou$_E$ m^{-3}) and odor intensities OI; $c_{OD,0}$ – unity of odor concentration ($c_{OD,0}$ = 1ou$_E$ m^{-3}) $$OI_j = k_j \log c_{OD,j} + 0.5$$	[53]
SVOC and VOC$_{odoros}$	$OI^3 \Rightarrow ACC_{VOC} \Rightarrow PD$	The mean of acceptability votes as a function of the mean of intensity votes. $$ACC = -0.45\,OI + 0.93,\ R^2 = 0.979$$ ACC in scale: from -1 to $+1$ OI odor intensity in scale: from 0 to 5 $$PD_{IAQ(OI)} = \left(\frac{\exp(-0.18-5.283ACC}{1+\exp(-0.18-5.283ACC} \right)3100$$	[54] [55]
	OI^2 in scale (pi) recalculation to $\Rightarrow OI^3$	The perceived intensity in pi units is determined by comparing the intensity of the sample with different specified intensities of the reference substance, (e.g., acetone). Concentrations for 1 to n (pi) follow a linear gradation of the acetone concentration. Confidence intervals should be within ± 2 pi. Recalculation (pi) to OI	[51]
			[34]

Table 3. *Cont.*

Sub Component	Input Parameters	Sensory Equations to Calculate the %PD Value	References
			[10]
Moisture x in high enthalpy h of moist air [5] (in room temperature t_a when air enthalpy $h > 55$ kJ/kg)	x (g_w/kg_a) $\Rightarrow$H (g_w/m^3) (H-absolute humidity)	The recommended criteria for dimensioning of humidification and dehumidification. It is recommended to limit the absolute humidity value to $x = 12 g_w/kg_a$ or H(g_w/m^3) $$h = 1.006t + 0.622(2501 + 1.84t) \cdot \frac{0.01 \cdot RH \cdot \exp(23.58 - 4043/(t+273.15-37.58))}{Patm + 0.01 \cdot RH \cdot \exp(23.58 - 4043/(t+273.15-37.58))}$$ IAQ acceptability equation $ACC = ah + b$ where a and b are different for different pollutants $$PD = \frac{100}{1+\exp(-3.58+0.18(30-t_a)+0.14(42.5-0.01p_v))}\%$$ $ACC = 5.63 + 0.46 \ln RH - 1.32 \ln h$ uncertainty of acceptability ACC is ± 0.12. $$PD_{IAQ(h)} = \left(\frac{\exp(-0.18-5.283ACC)}{1+\exp(-0.18-5.283ACC)}\right)3100$$	[56] [23] [57] [55]
Particulate matter [6] PM$_{2.5}$ PM$_{10}$	cPM2.5 cPM10	IAQI (Indoor Air Quality index) [6] is the proposal for a system comparable to the World AQI EPA system cPM$_{2.5} \Rightarrow$IAQI$\Rightarrow PD*\%$ cPM$_{10} \Rightarrow$IAQI$\Rightarrow PD*\%$	[30]
Air pollutants [7] VOC$_{non\text{-}odorous}$ CO, NO$_2$	c_{VOC}	IAQI [6] value for c_{VOC} was calculated with System IAQI [27] by interpolation and after which the index IAQI must be converted [7] to $PD*\%$	[30]

[1] TVOC, according to Reference [45], represents a narrow chromatographic picture that excludes, for example, the lower aldehydes, e.g., formaldehyde. [2] The measurement of the intensity of the odors in the building in which emissions from construction materials occur can be performed by a panel of participants, where the room is treated as a "test room for background odor" according to Section 6.8.1 of ISO 16000-28:2013, "Indoor Air—Part 28: Determination of odor emissions from building products using test chambers" (2013) [51]. The assessment of the 90% confidence level is possible through the use of a 15-pi odor intensity scale with a reading uncertainty of ±2 pi. [3] *OI* is perceived odor intensity on a six-level scale from 0 to 5 (no odor = 0, slight odor = 1, moderate odor = 2, strong odor = 3, very strong odor = 4, and overpowering odor = 5). [4] Based on the study Kim and Kim (2014) [52] selected 22 odorants with a similar chemical structure (structural formula) and determined an equation to convert the concentration x in ppm to intensify y, with odor OI on a scale of zero to five.

The study included odors from the distribution of food. These 22 odorants can be divided into five chemical groups: (1) reduced sulfur compounds, (2) carbonyls, (3) nitrogenous compounds, (4) VOCs, and (5) volatile fatty acids. For example, for the group of reduced sulfur compounds for compound number 1, the conversion equation for H_2S is $Y = 0.950\log X + 4.14$, for the carbonyl compounds group for compound number 10, ammonium NH_3, it is $Y = 670\log X + 2.38$, and for the VOC group for styrene it is $Y = 1.420\log X + 3.10$. [5] If one considers a cooling system that removes heat from a space but does not remove moisture unless condensation occurs, such as radiant cooling without dehumidification in a ventilation system, the importance of humidity is very clear. The sensible cooling of air in a room (no change in absolute humidity) from 25 °C and 60% RH to 20 °C (process a–b for $x = 0.012$ kg_w/kg_a).This value, coupled with a high temperature, $t_a = 25$ °C, is accepted as the critical limit value for dehumidification by EN16798-1:2019 [10] and can be converted into an absolute humidity H (g_w/m^3). These processes are expected to significantly increase thermal comfort and air quality. Nevertheless, the same change in enthalpy of the air h can be achieved by simply reducing the humidity by 10% RH and keeping the temperature constant (processes a–c). Since IAQ is a function of enthalpy, these are expected to be the same. Here, a change in humidity of 10% RH at constant temperature is equivalent to a change in temperature of 5 or 6 °C at constant moisture content in air x ($kg_w/kg_{dry\ air}$). [6] The IAQI system [30] adopts the methodology of AQI to set up the range of IAQI values from zero to 500, including 50, 100, 150, 200, 300, and 500. The IAQI values of 100 and 150 correspond to the concentrations in the Taiwan IAQG standard. Other IAQI values between 50 and 200 correspond to the concentration rankings given by several reference resources including the US EPA AQI system [31]. All on-site concentrations of indoor air pollutants (HCHO, TVOC, PM10, PM2.5, CO, CO_2, O_3, bacteria, fungi, SO_2, and NO_2) are combined using the IAQI system. The IAQI values are calculated on the basis of the concentration value c using the interpolation method for each air pollutant. In the IAQI system, the index range 0–50 is "good" with a significance level of "little or no risk", 51–100 is "moderate" where "sensitive persons or those with respiratory symptoms are concerned", 101–150 is "unhealthy for sensitive groups", 151–200 is "unhealthy for all individuals", 201–300 is "very unhealthy—more serious health effects for everyone for short-term exposure", and 301–500 is "hazardous" with "health warning of emergency conditions for everyone". Therefore, the comfort scale for IAQI is adequate for index values from zero to 200 and our proposal is to use the "hypothetical" scale of PD^* of 0–100% in this range, converted from IAQI. [7] The method of conversion of the IAQI scale to the PD^* scale is based on experiences gained during research [27]. The IAQI values in a health-risk scale can be given in % for persons giving a verbal answer of "no risk", "moderate", and "unhealthy" as their health risk evaluation. Therefore, when determining, for instance, the concentration function values of the $\%PD_{IAQI}$(TVOC) and $\%PD_{IAQI}$(HCHO) components, it is necessary to recalculate the scale IAQI = f(c) in the range from zero to 200 to the scale PD^* from 0 to 100%. There are break points in the new PD^*% scale: for IAQI values 0–50, PD^*% is 0–25, for IAQI values 51–100, PD^*% is 26–50, for IAQI values 101–150, PD^*% is 51–75, and for IAQI values 151–200, PD^*% is 76–100.

The air quality indexes (i.e., AQI [31] and IAQI [30]) are piecewise linear functions of the pollutant concentrations. At the boundary between AQI categories, there is a discontinuous jump of one AQI unit. To convert from concentration c_j to I_j (in the converted scale index I_j will be PD_j, Equation (16) is used.

$$I_j = ((I_{high} - I_{low})/(c_{high} - c_{low})) \cdot (c_p - c_{low}) + I_{low} \tag{16}$$

where I_j is the air quality index I in the $PD^*\%$ scale, c_{low} is the pollutant concentration break point, which is $\leq c_j$, c_{high} is the pollutant concentration break point, which is $\geq c_j$, I_{low} is the index break point corresponding to c_{low}, I_{high} is the index break point corresponding to c_{high} and c_p is the truncated (to an integer) actual concentration for the pollutant. Little data exist on the AQI's metrological reliability for AQI and IAQI. Only in the EPA Air Program undertaken at Cornell University [26] is there a previous review of the quality assurance requirements for AQI.

2.6. The Representative VOCs for Indoor Environment

The time when IAQ studies focused on a class of contaminants referred to as volatile organic compounds (VOCs) is bygone. The analytical methodology available was the primary basis for this focus, but the recent broadening of analytical methods has led to growing realization that other compounds (i.e., SVOCs) beyond traditional VOCs are implicated in IAQ problems. The choice of VOCs remains a challenge in IAQ assessment. Moreover, VOCs is somewhat vague term, the definition of which is not universally agreed upon. It has been defined in terms of vapor pressures and boiling points, as well as molecular chain lengths detectable by chromatographic techniques. Due to the complexity of VOC emission profiles, it is tempting to simplify the analysis and reporting of emissions by grouping all detected compounds together. The first problem with this approach is that individual compounds have highly variable health and/or comfort effects, the result being that concentration alone is not predictive of IAQ effects. Levels of concern vary by orders of magnitude, so a collective concentration will not correlate with IAQ. Second, VOC detection and quantification are highly method dependent. A given sampling and analysis system cannot capture or respond to all the VOCs present in any indoor environment or in the test chamber for a given emitting material. Thus, the term "total" is misleading. The important aspect of IAQ submodel selection is the strategy defined by the US EPA as "VOCs—Total versus Target: Irritancy, Odor and Health Impact". The representative 90 target VOCs were presented by Canada's National Research Council's Institute for Research in Construction (NRG-IRC) in collaboration with several academic and governmental partners, including Health Canada. The compounds were selected based on health impact, occurrence in indoor air, known emission from building materials, as well as suitability for detection and quantification by gas chromatography-mass spectrometry (GC-MS) or high-performance liquid chromatography (HPLC). Our list of target VOCs was actually representative for indoor environment and recommended by the HealthVent project and is provided in Table 1.

2.7. Steps of ΣIAQ_{index} Calculation

After selection of the IAQ model type ($\Sigma IAQ_{quality}$ or $\Sigma IAQ_{comfort}$), the IAQ_{index} evaluation was carried out using the complex model ΣIAQ from Figure 5, which should contain the following stages.

(a)   Calculation of the total concentration of pollutants in the ventilated space or the total mass of air pollutants per m^3, the level of which is to be reduced by the ventilation process (taking into account the ventilated volume of the room and the emissions present).

(b)   Selection of the IAQ_{index} model shape from the models defined by Equations (12)–(14), with the provision that due to the multiplication of the submodels for $IAQ(VOC_{odorous})$, the number of subcomponents of the $IAQ_{comfort}$ model will be more than five.

(c)   Processing the input data of the submodels to obtain the concentration value c_j, e.g., converting the measured *OI* value into a concentration value for a given pollutant c_j in $\mu g/m^3$.

(d) Calculation of the excess concentration values for each identified contaminant (Table 2) $\Delta c_j = c_j - c_{ref}$.

(e) Calculation of the sum of excess concentrations (see Table 2), $\sum \Delta c_j$.

(f) Calculation of adjusted weights Wj for the selected model equations. $IAQ_{quality}$ and $IAQ_{comfort}$, are determined on the basis of arithmetic means or by adjusting all the values of Δc_j in a given model using Equation (10), only for groups of pollutants with similar concentration values.

(g) Calculation of the value of the ventilating air flow for the environment described in the IAQ_{index} model in accordance with the requirements of the standard EN 16798-1:2019 (a method using the criteria for the ventilation required for the individual substance emitted) [10].

(h) Calculation of given IAQ environmental input parameters, including concentrations of pollutants c_j assigned to submodels. The PD values from their sensory equations (Table 3) are presented as the dependence of the percentage of persons dissatisfied $PD = f(c_j, \ldots)$, from one of the formulas from Table 3, in order to determine this function.

(i) Selection of the $\sum IAQ_{quality}$ model equation (with weights W_1, W_2, and W_3) or the $IAQ_{comfort}$ model equation (with weights $W_1, \ldots W_5$ or more) and calculation from Equation (13) of the value with adjusted weights, followed by multiplication of IAQ submodels (Equation (8)) and insertion as a term of the IEQ_{index} in Equation (18) [34].

When selecting the $\sum IAQ_{comfort/health}$ model type, an IAQ_{index} evaluation is carried out using the combined model $\sum IAQ$ from Figure 5, which should contain the following steps.

(a) Calculation of the total concentration of pollutants $\sum c_j$ in the ventilated space or the total mass of air pollutants per m³, the level of which is to be reduced by the ventilation process (taking into account the ventilated volume of the room and the emissions present).

(b) Choosing the IAQ_{index} model (12) from among the models defined by Equations (12)–(14), with the provision that by multiplying the $IAQ(VOC_{non\text{-}odorous})$ submodels, the number of subcomponents of the $IAQ_{comfort/health}$ model will be more than seven.

(c) Processing of submodel input data to obtain concentration values c_j in $\mu g/m^{-3}$.

(d) Calculation of excess concentration values for each pollutant identified (Table 2) using $\Delta c_j = c_j - c_{ref}$.

(e) Calculation of the sum of excess concentration values for submodels 1–7 via $\sum \Delta c_{1 \ldots 7}$

(f) Determination of the adjusted weights $W_{1, \ldots 7}$ for the equation of the $IAQ_{comfort/health}$ model on the basis of arithmetic means or by adjusting all the values of Δc_j in a given model using Equation (10), only for groups of pollutants with similar concentration values.

(g) Calculation of the values of the ventilating air stream from the total concentration of indoor air pollutants $\sum c_j$ (for instance, Reference [42]), for the environment described by the $\sum IAQ_{index}$ model (Figure 5) in accordance with the requirements of the EN 16798-1:2019 standard for the individual substances emitted and using an alternative method when the concentration in the room has stabilized.

(h) Calculation of the given IAQ input parameters, including concentrations of all pollutants, c_j, assigned to the submodels. The PD values are taken from their sensory equations (Table 3) depending on the percentage of persons dissatisfied, $PD = f(c_j, \ldots)$, or selected from the formulas for determining this function given in Table 3.

(i) Development of $IAQ_{PM2.5}$, IAQ_{PM10}, and $IAQ_{non\text{-}odorous}$ submodels. When it is planned to use the indoor air quality index scale IAQI or a similar scale, it is necessary to convert these to PD^* values in %, in two steps: (1) by reading from the standard curves of the $IAQI = f(c_j)$ all IAQI values for the determined (measured) $VOC_{non\text{-}odorous}$ concentration values and using the converted scale calibration curve, $PD = f(IAQI)$, by reading from the recalibration curve (Equation (16)), the $PD = f(c_j)$ values on the dissatisfaction rating scale from zero to 100 in %, according to Footnote 7 in Table 3.

(j)   Calculation of IAQ values for Pj pollutant submodels from Equation (8) and insertion into the model equation $\sum IAQ_{comfort/health}$ (14) (with weights $W_1, W_2, \ldots, W_7$ or more).

2.8. The IEQ Assessment Equation with $\sum IAQ$ As a Subcomponent

The proposed IAQ model can be a substantial component of the IEQ model; for example, in the case study shown later in the article. The indoor environmental quality index refers to the quality of a building's environment with respect to the occupants' satisfaction in %. The morphology of the IEQ$_{index}$ model used to assess buildings, to determine as an IEQ component the IAQ$_{index}$ and to determine other subcomponents TC$_{index}$—thermal comfort, ACc$_{index}$—acoustic comfort and L$_{index}$—light quality based on measurements of physical properties in each of the submodels—in accordance with the scheme of the Piasecki–Kostyrko model, is presented in Figure 6 [22].

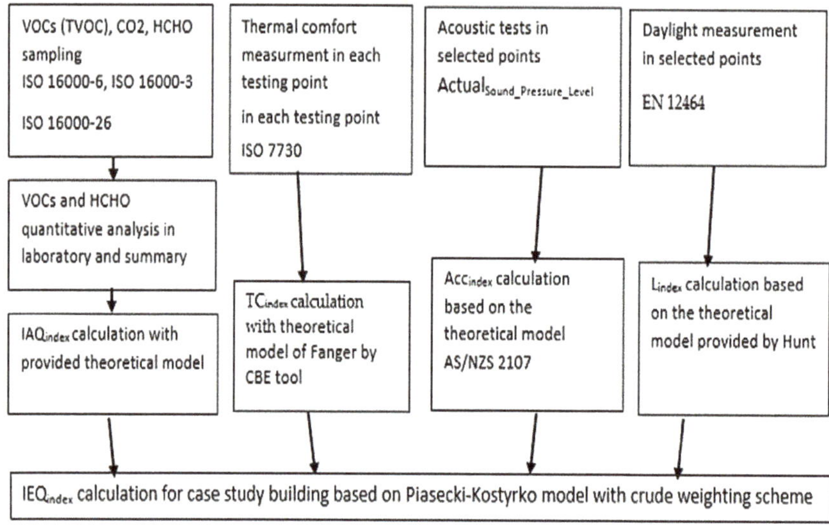

Figure 6. The research steps necessary to determine the IEQ$_{index}$ for buildings, including physical and design parameters of buildings and subcomponent models.

The EN 16798-1:2019 is the reference for IEQ model creation [22,33]. The standard allows complex indoor information to be presented as one overall indicator of indoor environmental quality of the building—IEQ$_{index}$. The model reliability, including the uncertainties of measurements and data for this model, was discussed clearly in Reference [34], where the authors also presented the internal incongruity in the IEQ model structure and the justification for using the crude weights method for each subcomponent. Originally, the IEQ model was expressed as a polynomial equation consisting of four terms by Wong [43]. The IEQ$_{index}$ is composed of the following subcomponents (SI_i): thermal comfort (TC_{index}), indoor air quality (IAQ_{index}), acoustics (ACc_{index}), and lighting quality (L_{index}). Multiplying their weights, W_i, leads to Equation (17).

$$IEQ_{index} = \Sigma \ W_i \cdot SI_i \tag{17}$$

The authors adopted the crude weighting system, where all elements are weighted in the same way (0.25 for W_1–W_4), as shown in Equation (18).

$$IEQ_{index} = 0.25 \cdot TC_{index} + 0.25 \cdot \Sigma IAQ_{index} + 0.25 \cdot ACc_{index} + 0.25 \cdot L_{index} \tag{18}$$

As a consequence of the equation, the subcomponents SI_i (the predicted percentage of those satisfied) can be calculated using Equation (19).

$$SI_i = 100 - PD(SI_i) \tag{19}$$

where PD is the predicted percentage dissatisfied (PPD) and $PD(SI_i)$ is the percentage of persons dissatisfied with the IEQ subcomponent (SI_i) level. The authors' simulations for IEQ_{index} sub-indices and preliminary metrological analysis of the overall IEQ model fitting were performed with Monte Carlo tests.

It is easy to show that the standard deviations of these values are equal:

$$SD(SI_i) = SD(PD(SI_i)) \tag{20}$$

2.9. A Case Study of a Building

The experimental part of this study was performed simultaneously with the BREEAM certification process, including determination of the three primary IAQ pollutants: formaldehyde concentration, CO_2, and VOCs in the indoor air [22]. The building is a high tower, made of a convex concrete–steel structure with a glass facade. The basic information on the assessed building is presented in Table 4. At the time of the test, the building had a standard empty office without furniture (so-called pre-occupancy stage). The walls were plastered and painted, the suspended ceilings were in place, and the floors were finished with synthetic carpets. All building installations were active, including the mechanical ventilation controlled by the Building Management System BMS system with zonal CO_2 concentration sensors. The building was tested a few days after the formal end of finishing works. The tests were made on the 55th and 47th floor.

Table 4. Information on the building in the case study.

Office Building Certificate	Facade View	Indoor View	Life-Stage	Number of Floors	Net Area (m²)	IAQ Assessed	
						Area (m²)	Number of Floors
BREEA Mexcellent			pre-occupant	49	59,000	3000	2

Measurement points in the building were determined based on the analysis of frequencies of designed occupancies of the room and interior finish standards (open spaces). The sampling plan was prepared with the BREEAM assessors conducting the certification process of the facility. The main focus was on the IAQ index of open spaces in which the largest number of people may reside, and these represent the largest occupied usable floor space. According to the detailed design project documents, the building emphasizes the use of materials with known and low emission levels (BREEAM certified).

2.10. The Equipment, Measurements, and Experimental Approach

Standardized CEN and ISO analytical methods were used to determine the VOC concentrations and CO_2 and formaldehyde concentrations in the indoor air of the building. Selection of the sampling points was made with the BREEAM assessor in two representative office zones per tested floor and a minimum of two floors. The building was tested three days after formal final finishing works at the pre-occupancy stage with no users inside. For this office building, the tests were conducted on the 55th and 47th floors. Air samples were collected using an active sampling procedure with an electronic mass flow controller, which controlled the air flow (10 dm³/h for VOC tests and up to 30 dm³/h for formaldehyde tests). Indoor samples were set up in selected representative office locations, approximately 1.5 m above the floor, away from windows, doors, potential emission sources, and direct sunlight. Air samples were tested in accordance with the ISO 16000-6:2011 and ISO 16000-3:2011 standards. The VOCs were assessed using tubes filled with Tenax adsorbent. Then, they were thermally desorbed using a thermal desorption apparatus (TD-20, Shimadzu, Tokyo, Japan). The process of separation and analysis of volatile compounds was achieved using a gas chromatograph equipped with a mass spectrometer (GC/MS) (model: GCMS-QP2010, Shimadzu, Tokyo, Japan). The following GC oven temperature program was applied: initial temperature 40 °C for five min, 10 °C per min to 260 °C, and the final temperature of 260 °C for 1 min. The 1:10 split ratio injection mode was applied. The method used has a limit of quantification of 2 $\mu g/m^3$. The volatile compounds were identified by comparing the retention times of chromatographic peaks with the retention times of reference compounds and by searching the NIST data base (National Institute of Standards and Technology, Gaithersburg, MD, USA) mass spectral database. Identified compounds were quantified using a relative identification factor obtained from standard solution calibration curves. TVOC was calculated by summing identified and unidentified compounds eluting between n-hexane and n-hexadecane. In order to determine volatile aldehydes, air samples were taken via cassettes using a solid absorbent silica gel coated with 2,4-dinitrilophenyl hydrazine (2,4-DNPH), and then subjected to a laboratory test using high-performance liquid chromatography (HPLC) with UV-Vis detection (Dionex 170S, Dionex, Sunnyvale, CA, USA) and an isocratic pump (Dionex P580A, Dionex, Sunnyvale, CA, USA). The described method has a limit of quantification at 2 $\mu g/m^3$.

Other IEQ_{index} components were tested as follows. The acoustic tests confirming the designed values were carried out by the measurement of the equivalent sound levels, LAeq, in the selected locations. The measurements were carried out during the daytime (starting at 11:00). The following equipment was used for the measurements: Brüel&Kjær 4231 acoustic calibrator (Brüel&Kjær, Nærum, Denmark), Nor-121 analyzer (Norsonic, Tranby, Norway), Brüel&Kjær 4165 measuring microphones (Brüel&Kjær, Nærum, Denmark), analyzer with microphone Norsonic-140 (Norsonic, Tranby, Norway). Before the tests were carried out, the calibration of the measuring path was conducted in accordance with the instructions to "check the acoustic measurement channel". The test results were evaluated in relation to the requirements considering permissible sound levels A in rooms intended for human dwellings. Thermal environmental measurements were provided using the microclimate multifunctional instrument HD32.1 and the tests were in accordance with ISO 7726 and ISO 7730. VOCs were tested simultaneously at all points. Visual comfort (Hea 01) was confirmed by using a MAVOLUX 5032C instrument (USB version) with a 3C15683 detector (Gossen, Nürnberg, Germany), in accordance with EN 12464 provisions.

2.11. Additional Explanations

The adaptation of the IAQ model to a practical casestudy was mainly for illustrative purposes in the context of the presented IAQ calculation/aggregation method. We did not focus deeply on discussing the technical or environmental issues of the presented building. Other IEQ subcomponents, such as thermal, acoustic, and visual satisfaction (in %), used to determine the IEQ index, were experimentally determined and partly presented in References [22,33]. Authors do not focus on these results in this article, as they have already been discussed in other papers [22].

3. Results

3.1. Results for the IAQ_{index} and IEQ_{index} Prediction

A previous publication of ours [22] reported on IEQ and IAQ building assessments for a larger number of BREEAM buildings, where IEQ was assessed without calculating the combined $\sum IAQ$ index. The combined model of the $\sum IAQ_{index}$ presented in this paper had not yet been previously developed, and we were limited in determining the IEQ_{index}, thus we only took into account two of the most well-known pollutants (i.e., CO_2 and TVOC) separately. The assessment of the IEQ index was made by adaptation of the measured parameters (complying with the draft EN 16798-1:2019 standard for indoor environments) as the input values for the submodels of the IEQ_{index}. The input values for the case study are presented in Table 5, which provides the input data for determining the IEQ_{index} sub-indices of thermal comfort (TC_{index}), indoor air quality (IAQ_{index}), acoustics (ACc_{index}), and lighting quality (L_{index}) for an office building (47th floor) three days after completion of the finishing work before users were allowed in the building (i.e., pre-occupancy stage).

Table 5. Physical parameters [1] and IEQ_{index} results calculated using Equation (9) separately for an IAQ_{index} with internal air pollution of CO_2 and an IAQ_{index} with internal TVOC air pollution, assuming a realistic uncertainty of parameter measurement for the case study of a building (47th floor; open space) three days after the completion of finishing works.

Sub-Index	Sub-Index $PD(SI_i)$ Models	Input Values	Sub-Index (Satisfied) and ±SD
TC_{index}	PMV (Fanger-CBE-ISO 7730) $PMV = f(t_a, t_r, v_a, p_a, M, I_{cl,})$ $PD_{TC} = f(PMV)$	I_{cl} 0.55 clo t_a 24.0 °C t_r 24.5 °C v_a 0.15 m/s RH 45% M 1.1 met	90% ± 3.2%
IAQ_{index}	$PD_{IAQ(CO_2)} =$ $395 \cdot \exp(-15.15 \cdot C_{CO2}^{-0.25})$ $PD_{IAQ(TVOC)} =$ $405 \cdot \exp(-11.3 \cdot C_{TVOC}^{-0.25})$	450 ppm 787 µg/m³	85.2% ± 0.6% 52.0% ± 18.0%
ACc_{index}	$PD_{ACc} =$ $2 \cdot ($Actual$_{Sound_Pressure_Level}$(dB(A)) $-$ Design$_{Sound_Pressure_Level}$(dB(A))) Actual (background) noise level Design sound level	55 dB(A) 45 dB(A)	80% ± 6.7%
L_{index}	$PD_L = -0.0175 + 1.0361/\{1 + \exp(+4.0835 \cdot (\log_{10}(E_{min}) - 1.8223))\}$	450 lux	98.4% ± 9.0%
$IEQ_{(CO_2)}$ First variant with c_{CO2} as an IAQ_{index} parameter $IEQ_{CO_2} = 92.2\% \pm 5.8\%$ [1]			
IEQ_{TVOC}	Second variant with c_{TVOC} as an IAQ_{index} parameter	$IEQ_{TVOC} = 80.1\% \pm 10.7\%$	

[1] The IEQ and its measurement's uncertainty (with subcomponent standard deviation values) were calculated for IEQ physical parameter values, where t_a is the air temperature (°C), t_r is the mean radiant temperature (°C), v_a is the relative air velocity (m/s), p_a is the water-vapor partial pressure (Pa), M is the metabolic rate (met), and $I_{cl,}$ is the clothing insulation (clo). In addition, c_{CO2} is the concentration in ppm, c_{TVOC} is the highest observed TVOC concentration in µg/m³, actual noise is in dB(A), and E_{min} is the minimum daylight illuminance (lux).

3.2. Results for the ΣIAQ_{index} and IEQ_{index} Assessment Including Identified Pollutants (CO_2, TVOC, and HCHO)

The example of a modified calculation of the collective submodel $\Sigma IAQ_{quality}$ for three basic pollutants, as a component of the IEQ model for determining one project value for this indicator, is provided in two variants. The first variant uses sub-indices of IAQ for two pollutants, CO_2 and TVOC, which are described in Table 5, as well the sub-index of the third pollutant, HCHO (according to Reference [47]), where these differences in the approaches mean one must combine them into one submodel ΣIAQ in order to be used in IEQ calculation. The second variant uses submodels of IAQ for TVOC and HCHO pollutants based on the IAQI system [30] and then converts them into percentages of persons dissatisfied PD* in %. According to the diagram of the model ΣIAQ from Figure 5 and using Equation (12) of the $\Sigma IAQ_{quality}$ model, the submodel weights are calculated as follows.

W_{CO_2} for the submodel IAQ(CO_2) = 0.5 is a component of the polynomial:

$$\sum IAQ_{quality} = 0.5 \cdot IAQ(CO_2) + 0.5 \cdot IAQ(VOC) \tag{21}$$

W_{VOC} for submodel IAQ(VOC) = 0.5 is a weight for combined submodel of the polynomial:

$$IAQ(VOC) = W_{TVOC} \cdot IAQ(TVOC) + W_{HCHO} \cdot IAQ(HCHO) \tag{22}$$

with the terms W_{TVOC} and W_{HCHO} calculated from Equation (14) using the measured values c_j (actual concentration of TVOC and HCHO) and the reference values c_{ref} (Table 6).

Table 6. Calculation of the weights of the W_{TVOC} and W_{HCHO} values for the two sub-indices of the combined IAQ(VOC) model.

Sub-Index	Input Value c_j	Input [1] Value c_{ref}	Excess Concentration Δc_j	W_j
IAQ(TVOC)	787 µg/m^3	300 µg/m^3	487 µg/m^3	0.96
IAQ(HCHO)	18 µg/m^3	0	18 µg/m^3	0.04

[1] EN 16798-1:2019 for a very low-pollution building.

In our case study, the value of the submodel IAQ weight for the model (HCHO) ought to also be calculated from the measured value and the reference value c_{ref}. However, formaldehyde is an unusual pollutant because, although it belongs to VOC$_{odorous}$ compounds, the concentrations found in buildings are many times lower than the HCHO threshold c_{th} = 300 µg/m^3 according to the WHO [7] and lower than the threshold concentrations of HCHO from 60 µg/m^3 to 70 µg/m^3 issued in 2013 by the American Industrial Hygiene Association [44]. The permissible value of c_{ref} = 100 µg/m^3 is also higher than the formaldehyde concentration found in buildings, according to Reference [5] and the standard EN16798-1:2019 [10]. Therefore, the authors propose that in such a case (to avoid a negative value of Δc_j), the value modelling the reference should be taken as zero. Then, the form of the adjusted W_{HCHO} weight in the model described by Equation (11) for air with three pollutants, would be as follows.

$$W_{HCHO} = \frac{\Delta c_j}{\sum\limits_{j=2\ldots3} \Delta c_{2\ldots3}} = \frac{(c_{HCHO} - 0)}{(c_{TVOC} - c_{ELV}) + (c_{HCHO} - 0)} \tag{23}$$

The results of the weights assessment for the two variants of the $\sum IAQ_{quality}$ model are presented in Table 6.

According to the diagram of the model $\sum IAQ$ from Figure 5 and Equation (8), we proposed sensory equations for the percentage of persons dissatisfied %*PD** in two variants.

The submodel $\sum IAQ$'s first variant includes the following:

1. The IAQ submodels used so far in References [22,33] for CO_2 and TVOC pollutants, as shown in Table 5;

2. The IAQ submodel for formaldehyde, using two types of equations depending on the range of HCHO concentrations measured in the building. Formaldehyde concentrations in the air with values above the threshold concentration, c_{th}, for its odor, i.e., above 60 or even 300 µg/m^3, can be used to create IAQ submodels for rooms with volatile and aromatic VOC compounds as well as for the HCHO equation [50].

$$PD_{HCHO} = \frac{\exp(2.14 \cdot OI - 3.81)}{\exp(2.14 \cdot OI - 3.81) + 1} \tag{24}$$

However, in the case study building, the maximum concentration of HCHO was 18 µg/m^3 and, therefore, its concentration in the air was several times lower than the concentration of the odor threshold, c_{th} [44]. The intensity of the formaldehyde odor was undetectable under these conditions, and the sensory equation $PD = f(OI)$, which is appropriate for sensory detection of IEQ, is not applicable for odors below the threshold. Therefore, for small concentrations, we proposed the use of the equation taken from the work of Zhu and Li [47] based on the analysis of "health effects on the human body", derived from "indoor air quality comfort evaluation experiments and the literature".

$$PMV_{HCHO} = 2log\frac{c_{HCHO}}{0.01} \tag{25}$$

This equation links the value of the new unit "the effect of formaldehyde on human comfort", called PMV_{HCHO}, with its c_{HCHO} concentration ($\mu g/m^3$) in the air. It covers the range from 10 $\mu g/m^3$ to 320 $\mu g/m^3$ and, as declared by the authors, this value has the same nature as PMV thermal comfort, which can be converted into a $PD\%$ unit according to the formula in Reference [48], experimentally confirmed for nearly zero energy buildings (NZEBs) by Reference [58].

$$PD_{HCHO} = 100 - 95 \cdot \exp(-0.03353 \cdot PMV^4 - 0.2179 \cdot PMV^2) \tag{26}$$

The submodel $\sum IAQ$'s second variant includes the following.

i.   The IAQ(CO_2) submodel used so far for CO_2 pollution, as shown in Table 5.
ii.   The IAQ submodels for TVOC and HCHO types of pollution used as indoor air quality index ratio values borrowed from the IAQI system [30], which are then converted into percentages of persons dissatisfied (PD* in %) in the following way

(a)   The reference curves of $IAQI = f(c_j)$ [30] for two dependencies of the IAQI index on TVOC and HCHO contamination values must be reconstructed. On the y-axis are the IAQI index values from zero to 200 in the range from "no risk" to "unhealthy" and on the x-axis, the c_j values are presented.
(b)   In accordance with the measured values of c_{TVOC} and c_{HCHO}, the values $IAQI_{TVOC}$ and $IAQI_{HCHO}$ are determined from the functions $IAQI_{TVOC} = f(c)$ and $IAQI_{HCHO} = f(c)$.
(c)   Based on the IAQI system parameters [30] given in Footnote 6 of Table 3, which are presented as data for the functions for indexes, the $IAQI_{TVOC}$ and $IAQI_{HCHO}$ values appropriate for the break pointsi n perceived pollution concentration values are calculated in a range from zero (good) to 200 (unhealthy) using the ordinal scale IAQI = f(c) [27]. This function was converted to $IAQI_{TVOC}$ and $IAQI_{HCHO}$ scales using the concentration function scales $PD*$(TVOC) and $PD*$(HCHO) in the $PD*$ range from 0 to 100% (Figure 8).
(d)   The data used for calculation and conversion of IAQI and $PD*$ scales are presented in Table 7.
(e)   Based on data determined for the new converted scales (Table 7) for the percentage of persons dissatisfied ($PD*$(TVOC) and $PD*$(HCHO) concentration functions), the $PD*(c)$ working graphs were drawn (Figure 7). (To harmonize the concentration scale of pollutants on the y-axis, c_{HCHO} values multiplied by 10 were applied.)
(f)   The interpolation of the percentage of persons dissatisfied ($PD*$(TVOC) and $PD*$(HCHO)) in % for the measured values of pollution concentrations in the air must be made using the graph from Figure 7 or using Equation (16) [31].

Table 7. Recalculation and conversion of IAQI value scales.

*PD** Value	IAQI Value	Evaluation	c_{TVOC}-TVOC (1 h)		c_{HCHO}-HCHO (1 h)	
%	-	-	ppm	µg/m³	ppm	µg/m³
0	0	No risk	0	0	0	0
25	50	Good	0.3 [1]	300	0.01 [2]	12.3
50	100	Moderate	0.9	900	0.04	49.1
75	150	Unhealthy for sensitive	3.0	3000	0.10	122.8
100	200	Unhealthy	4.6	4600	0.75	921.0

[1] Conversion factors TVOC: 0.3 ppm corresponds to 300 µg/m³ [44]. [2] Conversion factors HCHO: 1 ppm corresponds to 1228 µg/m³ [44].

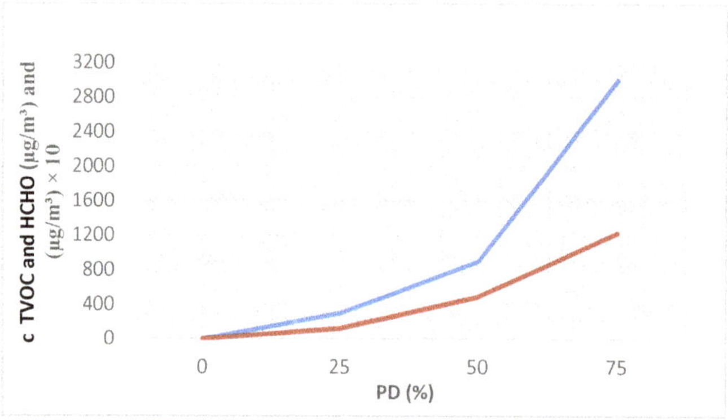

Figure 7. The percentage of persons dissatisfied function of TVOC (blue line) and HCHO (brown line) versus concentration.

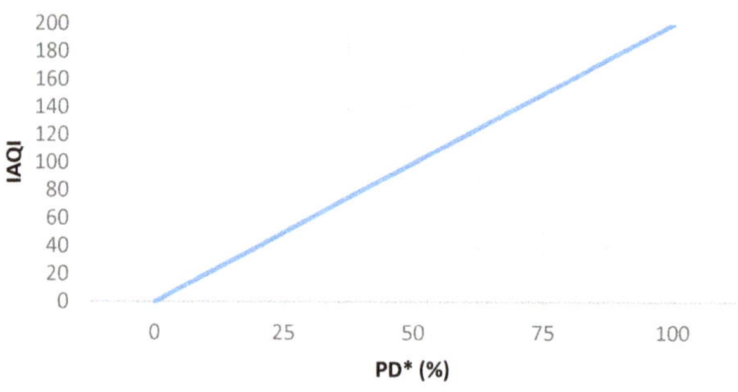

Figure 8. Percentage of persons dissatisfied, %*PD**, in relation to IAQI values.

In the context of the results for the overall IEQ model index when treating both the main pollutants CO_2 and TVOC separately, we present in Table 8a the transformed IEQ calculations with the sub-indices $\sum$IAQ. The results for the individual case study building IAQ subcomponents are taken from Table 5 (for c_{TVOC} = 787 µg/m³ and c_{HCHO} = 18 µg/m³ [22]). The IEQ index with $\sum$IAQ$_{quality}$ values was calculated using two variants—the first conventional and the second borrowed from the IAQI scale [30] (Table 8a).

Standard deviations for each concentration are provided in Table 8b.

Table 8. (a) Physical parameters (Footnote [1] in Table 5) and IEQ results calculated from Equation (9) with $\sum IAQ$; assuming realistic uncertainty of parameter measurements for the case study building (47th floor; open space) a few days after completion of the finishing works. (b) Measured pollutant concentrations c and standard deviations $SD(c)$.

(a)

Sub-Index	Sub-Index $PD(S_I)$ Models	Input Values	Sub-Index (Satisfied) and $\pm$SD
TC_{index}	PMV (Fanger-CBE-ISO 7730) $PMV = f(t_a, t_r, v_a, p_a, M, I_{cl,dyn})$ $PD_{TC} = f(PMV)$	I_{cl} 0.55 clo t_a 24°C t_r 24.5°C v_a 0.15 m/s RH 45% M 1.1 met	90.0% $\pm$ 3.2%
$\sum IAQ_{index}(1)$ Sub-indices First variant	$PD_{IAQ(CO_2)} = 395 \cdot \exp(-15.15 \cdot C_{CO2}^{-0.25})$ $PD_{IAQ(TVOC)} = 405 \cdot \exp(-11.3 \cdot C_{TVOC}^{-0.25})$ $PMV_{HCHO} = 2log \frac{c_{HCHO}}{0.01}$ $PD_{HCHO} = 100 - 95 \cdot \exp(-0.03353 \cdot PMV^4 - 0.2179 \cdot PMV^2)$	$c = 450$ ppm $c = 787$ µg/m^3 $c = 0.018$ mg/m^3	85.2% $\pm$ 0.6% 52.0% $\pm$ 18.0% 65.8% $\pm$ 10.7%%
$\sum IAQ_{index}(1)$	$IAQ_{VOC} = 0.96 \cdot IAQ_{variant1}(TVOC) + 0.04 \cdot IAQ_{variant1}(HCHO)$ $\sum IAQ_{index}(1) = 0.5 \cdot IAQ(CO_2) + 0.5 \cdot IAQ_{variant\ 1}(VOC)$		53.0% $\pm$ 17.3% 69.1% $\pm$ 9.0%
$\sum IAQ_{index}(2)$ Sub-indices [1)] Second Variant	$PD_{IAQ(CO_2)} = 395 \cdot \exp(-15.15 \cdot C_{CO2}^{-0.25})$ PD^*_{TVOC} read from the graph with $PD^*_{TVOC} = f(c_{TVOC})$ or determined from Equation (16) PD^*_{HCHO} read from the graph with $PD^*_{HCHO} = f(c_{HCHO})$ or determined from Equation (16) SD for c_{TVOC} and c_{HCHO} at "break points" $\pm$12%	$c = 450$ ppm $c = 787$µg/m^3$\pm$18% $c = 18$µg/m^3$\pm$12%	85.2% $\pm$ 0.6% 54% $\pm$ 13.8% 71.1%$\pm$ 11.0%
$\sum IAQ_{index}(2)$	$IAQ_{VOC} = 0.96 \cdot IAQ_{variant2}(TVOC)$ $+ 0.04 \cdot IAQ_{variant2}(HCHO)$ $\sum IAQ_{index}(2) = 0.5 \cdot IAQ(CO_2)$ $+ 0.5 \cdot IAQ_{variant\ 2}(VOC)$		54.7% $\pm$ 13% 70.0% $\pm$ 6.5%
ACc_{index}	$PD_{ACc} = 2 \cdot (Actual_{Sound\_Pressure\_Level}(dB(A) -$ $Design_{Sound\_Pressure\_Level}(dB(A))$ Actual (background) noise Design sound level	55 dB(A) 45 dB(A)	80.0% $\pm$ 6.7%
L_{index}	$PD_L = -0.0175 + 1.0361/\{1 + \exp(+4.0835 \cdot (\log_{10}(E_{min}) - 1.8223))\}$	450 lux	98.4% $\pm$ 9.0%

Table 8. *Cont.*

Sub-Index	Sub-Index $PD(S_I)$ Models	Input Values	Sub-Index (Satisfied) and $\pm$SD
$IEQ_{index}(1)$ with $\sum IAQ_{index}(1)$ meas. $IEQ_{index}(1)$ $\pm$SD	$IEQ_{index}\pm SD = W_1 \cdot TC_{index} + W_2 \cdot \sum IAQ_{index}(1) + W_3 \cdot Ac_{cindex} + W_4 \cdot L_{index}$		$84.4\%\pm3.7\%$ $u_{meas} = 2 \cdot 3.7 = \pm7.4\%$
overall $IEQ_{index}(1)\pm u_{overall}$ [2]	$\pm u_{overall}(IEQ) = (\sum (SD_{real}(PD(SI_i))^2 + \sum (SD_{vote}PD(SI_i))^2)^{-2}$ $IEQ_{index}(1)\pm u_{overall}$		$u_{oveall} = \pm16.24\%$ $84.4\%\pm16.24\%$
$IEQ_{index}(2)$ with $\sum IAQ_{index}(2)$ meas$IEQ_{index}(2)\pm$SD	$IEQ_{index}\pm SD =$ $W_1 \cdot TC_{index} + W_2 \cdot \sum IAQ_{index}(2) + W_3 \cdot ACc_{index} + W_4 \cdot L_{index}$		$84.6\%\pm3.3\%$ $u_{meas} = 2 \cdot 3.3 = \pm6.6\%$
overall $IEQ_{index}(2)\pm u_{overall}$ [2]	$\pm u_{overall}(IEQ) = (\sum (SD_{real}(PD(SI_i))^2 + \sum (SD_{vote}PD(SI_i))^2)^{-2}$ $IEQ_{index}(2)\pm u_{overall}$		$u_{overall} = \pm16.15\%$ $84.6\%\pm16.15\%$

(b)

	c_{mes}	$SD(c_{meas})$	c_H	$SD(c_H)$	c_L	$SD(c_L)$	PD^*_H	SD	PD^*_L	SD
TVOC	787	$18\%\Rightarrow141.7$	900	$12\%\Rightarrow108$	300	$12\%\Rightarrow36$	50	12%	25	12%
HCHO	18	$12\%\Rightarrow2.16$	49.1	$12\%\Rightarrow5.89$	12.3	$12\Rightarrow1.48$	50	12%	25	12%

[1] The method of calculation of mean values PD^* and $\pm$ SD(PD^*) for TVOC and HCHO is based on Equation (16) and takes the form (27).

$$PD* = \frac{\left(PD^*_H - PD^*_L\right)}{(c_H - c_L)} \times (c_{meas} - c_L) + PD^*_H \tag{27}$$

Concentrations c and SD(c) are in µg/m^3; PD^*_H, PD^*_L, and SD(PD^*) are in %, and standard deviations of the HCHO concentration of 12% was adopted on the basis of reports from IAQ research conducted as part of BREEAM in 2016 [22]. The assumptions were that c_H and PD^*_H are the coordinates of the upper break point (i.e., high break point) of the converted scale $PD^* = f(c)$, and c_L and PD^*_L are the coordinates of the lower break point (i.e., low break point) of the converted scale $PD^* = f(c)$. Standard deviations were assumed for c_{meas}, c_{TVOC}, and c_{HCHO} as well as for $PD^* = \pm12\%$, as this is half of the transformed segment of the scale, which according to Table 7 covers a range of 25% of PD^*, with one perceived category of air quality, e.g.,"no risk" or "moderate". Therefore, the maximum standard deviation was 12.5% and, according to the literature on AQI and IAQI values, it should be rounded to a total value. [2] SD$_{vote}(PD(SI_i))$ from the $\pm u_{overall}$(IEQ) equation was the standard deviation of a probability distribution of an each. $(SI_i)_{vote}$ and was calculated primary using the $PD(SI_i)$ equation calibration curve [34].

4. Discussion

4.1. Discussion of the $\sum IAQ_{index}$ Theoretical Model

For years, the authors, as accredited laboratory personnel, have conducted IAQ pollution tests in indoor environments for various applications. Based on our experience, it was concluded that the general approach to assessing combined IAQ has not yet been systematized and that there is a global tendency to assess individual IAQ parameters separately or to group them without a justified aggregation method. This is not a good situation from the point of view of building users' needs. This, in our opinion, may lead to incorrect IAQ interpretations in specific building situations. In the context of analyzing the problem in this paper, authors presented a summary of the state-of-the-art methods and also provided a new approach for solving some of these problems. As presented, it is possible to create a $\sum$IAQ index aggregating the results of indoor air analyses, taking into account various representative pollutants. Three levels of comprehensive air quality assessments (with three, five or seven subcomponents), depending on the application of the assessment, were proposed, together with step-by-step procedures. This may be practical, as shown in the evaluation of a case study on a building. We originally selected the main IAQ subcomponent equations and user satisfaction dependences, $\%PD = f(c_j)$, and provided them all in one place (Table 3). We then proposed and justified the weighting schemes for the IAQ total equation. In most of the studies in the literature, the weighting schemes used for IEQ or IAQ assessments are not physically justified or explained. There are known methods of weighting sub-indices, but the problem that was solved in this paper was an effective system for weight adjustments. For the construction of the combined model IAQ_{index}, with a weighting scheme useful for aggregating sub-indices, we proposed the model scheme presented in Figure 5. According to the results, the advantage of the complex model $\sum IAQ_{index}$, in which the input quantities always constitute concentrations of given pollutants, is the ability to use these concentrations to calculate excess pollution concentrations from Equation (10) and generate weighting schemes $W_{1, \ldots n}$ for all three models by adjusting the weights based on the concentration values of excess air pollutants to a value ≤ 1.0 for each IAQ_{index} model. The Δc_j values determine the masses of pollutants that must be removed by ventilation to eliminate the target pollutant effect. They can be determined as differences between the current concentrations of pollutants and the concentration of pollutants at the reference or standard level (e.g., c_{ELV} or c_{LCI}), and in the case of $VOC_{odorous}$, the odor threshold c_{th}. The presented approach may allow planning of air quality for the building.

As discussed, it is important to identify those VOCs with comfort, health, and impacts and focus on the IAQ sub-model choice aspect, briefly defined as the strategy "VOCs—Total vs. Target: Comfort, Irritancy, Odor, and Health Impact". The model from Figure 5 has uniform inputs, i.e., concentration levels c_j and two outputs: (1) weighed (adjusted) and (2) sensory equations, $PD^* = f(c_{TVOC})$, constituting the IAQ submodel equations. These second outputs of submodels (PD^* values) are coefficients of satisfaction from the comfort sensation or lack of "health risk". These are the terms of the equation describing "combined $\sum IAQ$", which meets the requirements of the abovementioned strategy of selecting IAQ sub-models related to IAQ components that have the most impact on the resulting IEQ perception.

Models for subcomponents of IAQ not perceived by humans but influencing health are recommended to be used from the index set in the AQI system [26–28], which was adapted by the authors to assess the quality of indoor air based on, and in accordance with, the concepts of the air quality assessment system used globally by the American EPA. In the context of the subcomponent of the TVOC concentration in Figure 7, the authors provided the relationships of $PD^* = f(c_{TVOC})$ based on Jokl research [49] and resulting from the IAQI scale [30] as converted by the authors. The relationship between PD^* and TVOC concentration in both approaches is strongly correlated, as shown in Figure 9.

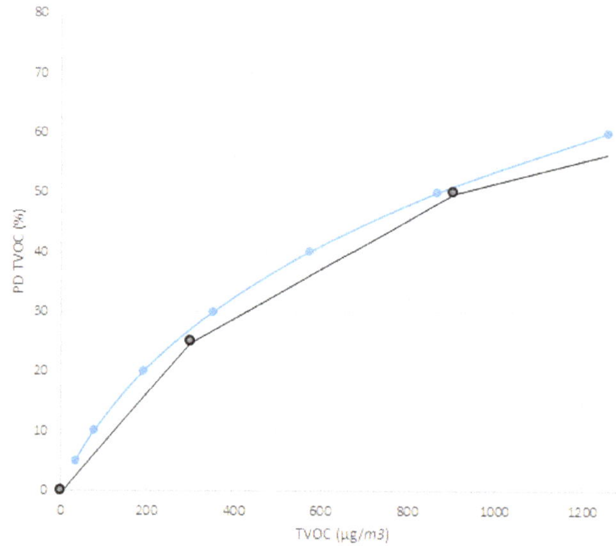

Figure 9. PD_{TVOC} based on conversion from IAQI to the PD* scale and study of Weber-Fechner theory.

The curves obtained from the conversion confirm Jokl's predictions provided in [49] and previously accepted idea. However, the final confirmation of these curves will be done experimentally as panel tests, as planned by the authors for the near future.

4.2. Discussion of Results for the Case Study on a Building

The experimental study was performed in a BREEAM certified building, and included the determination of formaldehyde concentration, CO_2, and VOCs in the indoor air. The example calculation of the combined $\sum$IAQ model for three basic pollutants, as components of the IEQ_{index} model, is presented in two variants but the calculated $PD*_{TVOC}$ values obtained with both calculation methodswere very similar. The first variant of $\sum$IAQ calculation used %PD = $f(c_j)$ curves in % sub-indices of IAQ for three pollutants, and the differences in approach in Tables 5 and 8a meant combining them into one IEQ submodel: the $\sum$IAQ model intended for the IEQ calculation. The second variant used submodels of IAQ for TVOC and HCHO pollutants based on the IAQI system [30] which were then converted into percentages of persons dissatisfied (*PD** in %).

The first conclusion is that CO_2 concentration cannot be used separately for the IAQ_{index} assessment, especially at the pre-occupancy stage (Table 5). The building was polluted with VOC emissions and HCHO from the construction products directly after finishing works were completed.

The authors confirmed that all three pollutions should be a simultaneously integrated part of the IAQ model, because the importance of TVOC is much greater, representing the main source of pollution—the construction and finishing materials.

According to the results, we recognized two variants of the combined $\sum IAQ_{index}$ calculation. For the first variant [22], the combined $\sum$IAQ index of satisfied users was 69.1% and, for the second variant (new approach with converted AQI index), the $\sum$IAQ index was 70.0% satisfied. The results ofthe $IEQ_{index}(1)$ (for Variant 1) were within the interval of combined overall uncertainty,±16.24%, and the results of the $IEQ_{index}(2)$ (for Variant 2) were that the overall uncertainty was ±16.15%. Therefore, the result was convergent, which confirms the credibility of the proposed approach.

The results obtained also showed that, in the period immediately after completion of finishing works in indoor spaces, there may be a temporarily increased concentration of TVOC, which systematically decreases over time, as we have shown in other papers. In the case of a building, the research showed that tests carried out immediately after finishing works gave results that significantly exceeded the BREEAM limits for TVOC at 300 $\mu g/m^3$ (twice as high). It should be expected that an acceptable level should be reached after a minimum of one month from the completion of the work.

For correctness of the obtained calculations, the authors are conducting a model credibility analysis that will be provided in the next article—Indoor Air Quality Model Part II: The Combined Model $\sum IAQ_{index}$ Reliability Analysis. The model uncertainty estimate may be compromised because the model reproduces the discomfort level associated with the dominant component.

Author Contributions: Conceptualization, M.P. and K.B.K.; methodology, M.P. and K.B.K.; formal analysis, M.P. and K.B.K.; investigation, M.P. and K.B.K.; resources, M.P. and K.B.K.; data curation, M.P. and K.B.K.; writing—original draft preparation, M.P. and K.B.K.; writing—review and editing, M.P. and K.B.K.; visualization, M.P. and K.B.K.; supervision, K.B.K.; project administration, M.P.; funding acquisition, M.P.

Funding: This research received no external funding.

Conflicts of Interest: The authors declare no conflict of interest.

References

1. Fanger, P.O. What is IAQ? *Indoor Air* **2006**, *16*, 328–334. [CrossRef] [PubMed]
2. Jones, L.L.; Hashim, A.; McKeever, T.; Cook, D.G.; Britton, J.; Leonardi-Bee, J. Parental and household smoking and the increased risk of bronchitis, bronchiolitis and other lower respiratory infections in infancy: Systematic review and meta-analysis. *Respir. Res.* **2011**, *12*, 5. [CrossRef] [PubMed]
3. ASHRAE. *ASHRAE Guideline No. 10. Interactions Affecting the Achievement of Acceptable Indoor Environments*; ASHRAE: New York, NY, USA, 2011.
4. Kostyrko, K.; Kozicki, M. Trends in odor measurments and detecting MVOC content in the interiors of buildings. In *Conference on Metrology MKM18*; The Scientific Papers of Faculty of Electrical and Control Engineering; Gdańsk University of Technology: Gdansk, Poland, 2018; Volume 59, p. 97.
5. Carrer, P.; de Oliveira Fernandes, E.; Santos, H.; Hänninen, O.; Kephalopoulos, S.; Wargocki, P. On the development of health-based ventilation guidelines: Principles and framework. *Int. J. Environ. Res. Public Health* **2018**, *15*, 1360. [CrossRef] [PubMed]
6. WHO Regional Office for Europe. *Evolution of WHO Air Quality Guidelines: Past, Present and Future*; WHO Regional Office for Europe: Copenhagen, Denmark, 2017.
7. WHO Regional Office for Europe. WHO guidelines for indoor air quality. *Nutr. J.* **2010**, *9*, 454.
8. General Administration of Quality Supervision, Inspection and Quarantine of the People's Republic of China. *GB/T 18883-2002, Indoor Air Quality Standard*; General Administration of Quality Supervision, Inspection and Quarantine of the People's Republic of China: Beijing, China, 2002. (In Chinese)
9. Asikainen, A.; Carrer, P.; Kephalopoulos, S.; Fernandes, E.D.O.; Wargocki, P.; Hänninen, O. Reducing burden of disease from residential indoor air exposures in Europe (HEALTH-VENT project). *Environ. Health A Glob. Access Sci. Source* **2016**, *15* (Suppl. 1), 35.
10. CEN. Part 1: Indoor environmental input parameters for design and assessment of energy performance of buildings addressing indoor air quality, thermal environment, lighting and acoustics. In *EN 16798 Energy Performance of Buildings–Ventilation of Buildings*; CEN: Brussels, Belgium, 2019.
11. Wang, Z.; Bai, Z.; Yu, H.; Zhang, J.; Zhu, T. Regulatory standards related to building energy conservation and indoor-air-quality during rapid urbanization in China. *Energy Build.* **2004**, *36*, 1299–1308. [CrossRef]
12. *China National StandardsGB 3095-2012 Ambient Air Quality*; The Standardization Administration of China (SAC): Beijing, China, 2012.
13. The Government of the Hong Kong Special Administrative Region IAQ Management Group. *A Guide on Indoor Air Quality. Certification Scheme for Offices and Public Places*; The Government of the Hong Kong Special Administrative Region IAQ Management Group: Hong Kong, China, 2019.
14. Piasecki, M.; Kostyrko, K. Indoor Environmental Quality Assessment Model IEQ Developed in ITB. Part 1. Choice of the Indoor Environmental Quality Subcomponent Models. *Ciepłownictwo Ogrzew. Went.* **2018**, *49*, 223–232.
15. Sekhar, S.C.; Tham, K.W.; Cheong, K.W. Indoor air quality and energy performance of air-conditioned office buildings in Singapore. *Indoor Air* **2003**, *13*, 315–331. [CrossRef]

16. Moschandreas, D.J.; Sofuoglu, S.C. The indoor environmental index and its relationship with symptoms of office building occupants. *J. Air Waste Manag. Assoc.* **2004**, *54*, 1440–1451. [CrossRef]

17. Sofuoglu, S.C.; Moschandreas, D.J. The link between symptoms of office building occupants and in-office air pollution: The Indoor Air Pollution Index. *Indoor Air* **2003**, *13*, 332–343. [CrossRef]

18. IEA. *Indoor Air Quality Design and Control in Low-Energy Residential Buildings-Annex 68 Subtask 1: Defining the Metrics in the Search of Indices to Evaluate the Indoor Air Quality of Low-Energy Residential Buildings*; AIVC Contributed Report, 17; Abadie, M.O., Wargocki, P., Eds.; IEA: Paris, France, 2017.

19. Chen, C.; Chiang, C.; Shao, W.; Huang, L. Hazards of VOCs emissions from building material and ventilation effectiveness in Taiwan. In Proceedings of the Indoor Air Conference, Copenhagen, Denmark, 17–22 August 2008; Volume 1.

20. Zhu, C.; Li, N. Study on Grey Clustering Model of Indoor Air Quality Indicators. *Procedia Eng.* **2017**, *205*, 2246–2253. [CrossRef]

21. Salisa, L.C.R.; Abadiea, M.; Wargocki, P.; Rode, C. Towards the definition of indicators for assessment of indoor air quality and energy performance in low-energy residential buildings. *Energy Build.* **2017**, *152*, 492–502. [CrossRef]

22. Piasecki, M.; Kozicki, M.; Firlag, S.; Goljan, A.; Kostyrko, K. The approach of including TVOCs concentration in the indoor environmental quality model (IEQ)- case studies of BREEAM certified office buildings. *Sustainability* **2018**, *10*, 3902. [CrossRef]

23. Toftum, J.; Jørgensen, A.S.; Fanger, P.O. Upper limits of air humidity for preventing warm respiratory discomfort. *Energy Build.* **1998**, *28*, 15–23. [CrossRef]

24. Wang, Z.; Yu, Z. PM2.5 and Ventilation in a Passive Residential Building. *Procedia Eng.* **2017**, *205*, 3646–3653. [CrossRef]

25. Morawska, L.; Ayoko, G.A.; Bae, G.N.; Buonanno, G.; Chao, C.Y.H.; Clifford, S.; Fu, S.C.; Hänninen, O.; He, C.; Isaxon, C.; et al. Airborne particles in indoor environment of homes, schools, offices and aged care facilities: The main routes of exposure. *Environ. Int.* **2017**, *108*, 75–83. [CrossRef]

26. Cornell's Legal Information Institute. Available online: https://www.law.cornell.edu/sfr/text/40/appendix-A_to_part_58 (accessed on 20 May 2019).

27. Saad, S.M.; Shakaff, A.Y.M.; Saad, A.R.M.; Yusof, A.M.; Andrew, A.M.; Zakaria, A.; Adom, A.H. Development of indoor environmental index: Air quality index and thermal comfort index. In *AIP Conference Proceedings*; American Institute of Physics: Melville, NY, USA, 2017; Volume 1808, p. 020043.

28. Mintz, D. *Technical Assistance Document for the Reporting of Daily Air Quality–The Air Quality Index (AQI)*; U.S. EPA: Washington, DC, USA, 2013.

29. United States EPA-Environmental Protection Agency Office of Air Quality Planning and Standards Air Quality Assessment Division Research Triangle Park, NC. Available online: https://airnow.gov/index.cfm?action=airnow.calculator (accessed on 25 June 2019).

30. Wang, H.; Tseng, C.; Hsieh, T. Developing an indoor air quality index system based on the health risk assessment. In Proceedings of the Indoor Air Conference, Copenhagen, Denmark, 17–22 August 2008; Volume 1.

31. United States EPA-Environmental Protection Agency Office of Air Quality Planning and Standards Air Quality Assessment Division Research Triangle Park, NC. *EPA-454/B-18-007- Technical Assistance Document for the Reporting of Daily Air Quality*; United States EPA-Environmental Protection Agency Office of Air Quality Planning and Standards: Research Triangle Park, NC, USA, 2018.

32. Sakellaris, I.A.; Saraga, D.E.; Mandin, C.; Roda, C.; Fossati, S.; de Kluizenaar, Y.; Carrer, P.; Dimitroulopoulou, S.; Mihucz, V.G.; Szigeti, T.; et al. Perceived indoor environment and occupants' comfort in European "modern" office buildings. The OFFICAIR Study. *Int. J. Environ. Res. Public Health* **2016**, *13*, 444. [CrossRef]

33. Piasecki, M.; Kostyrko, K.; Pykacz, S. Indoor environmental quality assessment: Part 1: Choice of the indoor environmental quality subcomponent models. *J. Build. Phys.* **2017**, *41*, 264–289. [CrossRef]

34. Piasecki, M.; Kostyrko, K.B. Indoor environmental quality assessment, part 2: Model reliability analysis. *J. Build. Phys.* **2018**, *42*, 288–315. [CrossRef]

35. Ye, W.; Won, D.; Zhang, X. A Simple VOC Prioritization Method to Determine Ventilation Rate for Indoor Environment Based on Building Material Emissions. *Procedia Eng.* **2015**, *121*, 1697–1704. [CrossRef]

36. Bienfait, D. *ECA Report No. 11 Guidelines for Ventilation Requirements in Buildings*; Commission of the European Communities: Bruxelles, Belgium, 1992.
37. Ye, W.; Won, D.; Zhang, X. A practical method and its applications to prioritize volatile organic compounds emitted from building materials based on ventilation rate requirements and ozone-initiated reactions. *Indoor Built Environ.* **2017**, *92–93*, 324–333. [CrossRef]
38. Persily, A.K. Evaluating building IAQ and ventilation with indoor carbon dioxide. *ASHRAE Trans.* **1997**, *103*, 4072.
39. Gyot, G.; Sherman, M.H.; Walker, I.S.; Clark, J.D. *Residential Smart Ventilation: A Review*; LBNL 2001056; Ernest Orlando Lawrence Berkeley National Laboratory: Berkeley, CA, USA, 2017.
40. Engineering ToolBox. Available online: https://www.engineeringtoolbox.com (accessed on 5 May 2019).
41. Noguchi, M.; Mizukoshi, A.; Yanagisawa, Y.; Yamasaki, A. Measurements of volatile organic compounds in a newly built daycare center. *Int. J. Environ. Res. Public Health* **2016**, *13*, 736. [CrossRef] [PubMed]
42. Haag, W. Rapid continuous TVOC measurements using PPB-level PIDs. In Proceedings of the 10th International Conference on Indoor Air Quality and Climate, Beijing, China, 4–9 September 2005; pp. 938–941.
43. Wong, L.T.; Mui, K.W.; Hui, P.S. A multivariate-logistic model for acceptance of indoor environmental quality (IEQ) in offices. *Build. Environ.* **2008**, *43*, 1–6. [CrossRef]
44. American Industrial Hygiene Association (AIHA). *Odor Thresholds for Chemicals with Established Health Standards*; American Industrial Hygiene Association: Falls Church, VA, USA, 2013.
45. Jokl, M.V. Evaluation of indoor air quality using the decibel concept based on carbon dioxide and TVOC. *Build. Environ.* **2000**, *35*, 677–697. [CrossRef]
46. Tahsildoost, M.; Zomorodian, Z.S. Indoor environment quality assessment in classrooms: An integrated approach. *J. Build. Phys.* **2018**, *42*, 336–362. [CrossRef]
47. Zhu, C.; Li, N. Study on indoor air quality evaluation index based on comfort evaluation experiment. *Procedia Eng.* **2017**, *205*, 2246–2253. [CrossRef]
48. European Committe for Standardization. *CEN/CR 1752:1998 Ventilation for Buildings. Design Criteria for the Indoor Environment*; International Organization for Standardization: Geneva, Switzerland, 1998.
49. Jokl, M.V. Indoor Air Quality Assessment Based on Human Physiology–Part 1. New Criteria Proposal. *Acta Polytech.* **2003**, *43*, 31–37.
50. Wargocki, P.; Knudsen, H.N.; Krzyżanowska, J. Some methodological aspects of sensory testing of indoor air quality. In Proceedings of the CLIMA 2010, Antalya, Turkey, 9–12 May 2010.
51. ISO. *ISO 16000-28 Determination of Odor Emissions from Building Products Using Test Chambers*; ISO: Geneva, Switzerland, 2013.
52. Kim, K.H.; Kim, Y.H. Composition of key offensive odorants released from fresh food materials. *Atmos. Environ.* **2014**, *205*, 2246–2253. [CrossRef]
53. Wu, C.; Liu, J.; Zhao, P.; Piringer, M.; Schauberger, G. Conversion of the chemical concentration of odorous mixtures into odor concentration and odor intensity: A comparison of methods. *Atmos. Environ.* **2016**, *127*, 283–292. [CrossRef]
54. Yeganeh, B.; Haghighat, F.; Gunnarsen, L.; Afshari, A.; Knudsen, H. Evaluation of building materials individually and in combination using odor threshold. *Indoor Built Environ.* **2006**, *15*, 583–593. [CrossRef]
55. Gunnarsen, L.; Fanger, P.O. Adaptation to indoor air pollution. *J. Environ. Int.* **1992**, *18*, 43–54. [CrossRef]
56. Fang, L.; Clausen, G.; Fanger, P.O. Impact of temperature and humidity on the perception of indoor air quality. *Indoor Air* **1998**, *8*, 80–90. [CrossRef]
57. Polednik, B. *Zanieczyszczenia a Jakość Powietrza Wewnętrznego w Wybranych Pomieszczeniach*; Biblioteka Cyfrowa Politechniki Lubelskiej: Lublin, Poland, 2013. (In Polish)
58. Piasecki, M.; Fedorczak-Cisak, M.; Furtak, M.; Biskupski, J. Experimental Confirmation of the Reliability of Fanger's Thermal Comfort Model—Case Study of a Near-Zero Energy Building (NZEB) Office Building. *Sustainability* **2019**, *11*, 2461. [CrossRef]

Article

An Accident Model with Considering Physical Processes for Indoor Environment Safety

Zhengguo Yang *, Yuto Lim and Yasuo Tan

School of Information Science, Japan Advanced Institute of Science and Technology 1-1 Asahidai, Nomi, Ishikawa 923-1292, Japan; ylim@jaist.ac.jp (Y.L.); ytan@jaist.ac.jp (Y.T.)
* Correspondence: yangzg@jaist.ac.jp

Received: 15 September 2019; Accepted: 1 November 2019; Published: 6 November 2019

Abstract: Accident models provide a conceptual representation of accident causation. They have been applied to environments that have been exposed to poisonous or dangerous substances that are hazardous in nature. The home environment refers to the indoor space with respect to the physical processes the of indoor climate, e.g., temperature change, which are not hazardous in general. However, it can be hazardous when the physical process is in some states, e.g., a state of temperature that can cause heat stroke. If directly applying accident models in such a case, the physical processes are missing. To overcome this problem, this paper proposes an accident model by extending the state-of-the-art accident model, i.e., Systems-Theoretic Accident Model and Process (STAMP) with considering physical processes. Then, to identify causes of abnormal system behaviors that result in physical process anomalies, a hazard analysis technique called System-Theoretic Process Analysis (STPA) is tailored and applied to a smart home system for indoor temperature adjustment. The analytical results are documented by a proposed landscape genealogical layout documentation. A comparison with results by applying the original STPA was made, which demonstrates the effectiveness of the tailored STPA to apply in identifying causes in our case.

Keywords: STAMP; STPA; physical process; indoor environment safety; smart home systems

---

1. Introduction

Accident models provide a conceptual representation of accident causation [1]. Their state-of-the-art development is on the phase of systemic models [1–3], i.e., accident models based on system theory rather than reliability. They were specifically applied to understand accidents in industrial areas, e.g., deepwater well control [4], railway [5], and aviation [6]. Some others relate to places that have been exposed to poisonous or dangerous substances, e.g., oil transportation [7] and nuclear power plants [8]. These poisonous or dangerous substances are hazards in nature, which can directly cause harm when leaked or released in workplaces. The workplace is a strictly managed environment for work. Safety-critical systems are taken as preventative measures for leakage and release.

The home environment refers to the indoor space with respect to physical processes of indoor climate, e.g., temperature change, which is different from safety-critical environments in workplaces. Hereafter, we use physical process to represent the physical process of indoor climate. The home environment is not a hazardous place in general. However, it can become hazardous when the physical process transfers from a normal state (e.g., temperature for thermal comfort), then through some intermediate states (e.g., temperature for thermal discomfort), finally reaching a hazardous state (e.g., temperature for heat stroke). The home environment is a place for everyday living, and it is not as strictly managed as that in workplaces. Smart home systems are developed to maintain the home environment in desired states, not only as preventative measures.

In systemic accident models [1–3], accidents are the result of the violation of a set of constraints on the behaviors of the system components, i.e., management, humans, and technology. If directly applying a systemic accident model to the smart home system, the information of physical processes is missing. For example, a smart home system violated its constraint on adjusting the indoor temperature for thermal comfort, and resulted in high temperature for heat stroke. The physical process of temperature change from one of thermal comfort, to some intermediate states for discomfort, then to a high-temperature state that can cause, e.g., heat stroke, is missing. This process is important. First, it assists in understanding how system behaviors could result in accident through intermediate state(s). Second, we need the anomalies' information of the physical process and their causes to deploy reactions and precautionary measures. We can take advantage of hazard analysis techniques to identify the causes in the system to the anomalies. When an intermediate abnormal state or hazardous state of the physical process is detected, with considering the corresponding causes in the system, an effective precautionary or reaction measures can be selected. Generally, if something undesired is not considered in the very beginning of risk analysis, the causes in the following analysis cannot be identified [2,9]. Therefore, it is necessary to extend the systemic accident model by including the physical process.

A newly developing systemic accident model, i.e., Systems-Theoretic Accident Model and Process (STAMP) [2], is considered in this paper, as its underlying rationale has been widely acknowledged by comparing with other systemic accident models [10]. It is based on general system theory for understanding accident causality of sociotechnical systems. A brief introduction of it is presented in Section 3.1. We extend it by considering the following facets. First, the home environment is not inherently hazardous. It can ensure a comfortable life in some physical process states and cause harms in others. Therefore, we take the physical process into account in understanding accident formation. Second, the indoor environment is greatly affected by the behaviors of smart home systems. This is because physical processes are the result of smart home systems and the outdoor climate. However, in a limited period, e.g., days, the outdoor climate can be considered with no big changes. Third, the role of people in the home environment. The characteristics of workers in workplaces and occupants in the home environment are different.

In this paper, we extend the STAMP model with considering physical processes (hereafter denoted by STAMP-PP) to understand accident formation. Smart home systems interact with the home environment through its behaviors, e.g., warm up and cool down. Thus, the STAMP-PP connects the physical world through the behaviors of the systems. Under this consideration, accidents are the result of the violation of a set of constraints on the behaviors of the systems to cause abnormal changes in physical processes, and finally result in personal harm. The extended STAMP-PP model demonstrates accident formation with respect to system behaviors and physical processes. The system behaviors can be controlled either by the smart home system directly or by occupants indirectly.

The information related to physical processes is important, and the abnormal behaviors of systems under specific operation scenarios must be known to select the appropriate reactions and precautionary measures. To this end, hazard analysis techniques [9,11] that can assist in analyzing potential causes of accidents are required. We adopted a hazard analysis technique to identify causes of abnormal system behaviors under related operation scenarios, which can result in abnormal changes in physical processes. A new approach to hazard analysis, called System-Theoretic Process Analysis (STPA) [2,9], is based on the STAMP model, and it is tailored and applied to the smart home system [12] for adjusting the indoor temperature, to demonstrate how to identify causes to abnormal system behaviors that result in physical process anomalies. The STPA can be used to identify unsafe control actions of a controller and are also the reasons why unsafe control actions can happen under specific scenarios. As abnormal behaviors of smart home systems that result in intermediate states of physical processes are also considered, the STPA is then tailored also for identifying causes to these abnormal behaviors. Landscape Genealogical Layout Documentation (LGLD) is proposed for documenting the analytical results, and the relations among the results are clearly and straightforwardly represented by comparing with conventional ways of documentation, i.e., tables and lists. We compared the results with that of applying the original STPA, which demonstrate the effectiveness of the tailored STPA in identifying causes of abnormal system behaviors and the LGLD documentation in representing the relations among the results.

The contributions of this paper are as follows.

- We discussed the characteristics of the smart home in the viewpoint of occupants and the safety of the home environment.
- The concept of the Performers System, which emphasizes the behaviors performed by various home appliances, is proposed.
- We propose the STAMP-PP model for understanding accident formation, i.e., abnormal system behaviors that result in abnormal changes in physical processes and cause hazards.
- We tailored the STPA and applied it to a smart home system to identify inappropriate and unsafe control actions that cause abnormal system behaviors, which result in abnormal changes in physical processes, and hazards.
- An LGLD approach is proposed for documenting the STPA analytical results.

This paper extends a conference paper [13] that introduces the STAMP-PP model while considering physical processes. The application of the tailored STPA and the new way of documenting the results in this paper are novel.

The rest of this paper is organized as follows. Section 2 discusses some knowledge about smart home and home environment safety. Section 3 introduces the proposed accident model STAMP-PP. Then, the hazard analysis technique STPA is tailored and applied to a smart home system for indoor temperature adjustment in Section 4. In Section 5, a discussion is given. Section 6 introduces the related work. Finally, Section 7 concludes this paper and points out the future work.

2. Preliminaries

In this section, we discuss the characteristics of the smart home and home environment safety before introducing the STAMP-PP model and the application of the STPA.

2.1. Smart Home

A home is a place for people like individual or family members, etc. to live. It is the sum of the place where people live permanently and the social unit—family. Since the 20th century, with the introduction of electricity, informtion, and communication technologies, great changes have taken place in the home [14]. One representation of this change is the development of the concept of the smart home since the 1990s [15]. The primary objective of the smart home is to increase occupants' comfort and make daily life easier. An example of possible techniques that make a home smart is machine learning [16]. The smart home has certain characteristics [15,17], e.g., adaptability, connectivity, controllability, and computability. These are discussed from the viewpoint of technology. This section discusses the characteristics of the smart home from the viewpoint of occupants.

- Partial Automation: Although home life has been automated a lot more than ever before due to the development of electricity, electronics, and network technologies, there are still elements of our lifestyles that have been left unchanged in practice. For example, pots and pans are used for cooking with gas in everyday meals. Thus, contemporary homes are only partially automated in practice.
- Application Area: The home is a place where people live. Various off-the-shelf products are used to improve the quality of life. These are manufactured by different manufacturers for different purposes. Occupants are not professional in understanding their rationale, particularly that of high-tech products. Occupants only learn to use them through product instructions or other occupants.
- Complexity: The complexity of a home is owing to three aspects, i.e., variety of appliances and products; variety of occupants in terms of age, health condition, knowledge, gender, etc.;

and outdoor environment. Off-the-shelf appliances and products indoors have various purposes and are produced by different manufacturers. In smart homes, they are connected together by the smart home network to enable a variety of services [15,17]. Different from workers in workplaces which rely on skills, rules, and knowledge for a specific job [18], occupants are usually reliant on their own life experiences to lead their lives with respect to various indoor items. Outdoor climate, air quality, etc. can affect the indoor environment, which in turn impacts the working of indoor appliances or devices. All these add to the complexity of the smart home.

2.2. Home Environment Safety

If we refer to a dictionary, the definition of safety could be the condition of being protected from or unlikely to cause harm, injury or loss. In system safety [2], safety is taken as an emergent property of systems. It is defined as freedom from conditions that can cause death, injury, occupational illness, damage to or loss of equipment or property, or damage to the environment [19]. However, it cannot be freedom from the conditions in practice. Safety is, thus, the condition of risk that has been reduced to an acceptable level, e.g., as low as reasonably practicable [20]. Home environment safety could be understood as that the risk of indoor climate has been reduced to a level of no harm to the health of occupants.

As is known to all, indoor climate is affected by outdoor climate. Bad weather such as heatwaves has been occurring frequently in recent years due to global warming and weather anomalies around the world. For example, many places globally experienced intense heat in the summer of 2018. It was observed that the highest temperature record in Japan was broken and reached a new level of 41.1 °C (Japan Meteorological Agency [visited 2018.10.03] http://www.jma.go.jp/jma/index.html). Indoor climate can thus be endangered by outdoor climate anomalies.

The indoor climate is adjusted by dedicated home appliances, e.g., use of air-conditioner to adjust indoor temperature. Different home appliances are integrated via home networks that yield value-added integrated services [21,22]. In order to ensure thermal comfort, an indoor temperature adjustment service is an example of the integrated service, which can potentially adopt a window, curtain, and air-conditioning unit. Each involved home appliance may have safety instructions. However, due to the complexity of the smart home, they still could be used in scenarios that cause safety problems. For example, the heating mode of an air-conditioner was used when it should not be. Since the smart home is an application area, appliances inside it may be replaced from time to time. Once an appliance is introduced into the smart home, it also brings about risk. For example, the predefined integrated service may not be aware of the new item and cause safety problems when using it. Therefore, home environment safety depends on the proper use of the indoor climate adjustment service with respect to related home appliances if they were properly designed and manufactured.

To understand home environment safety, we also need to discuss how people relate to the smart home. This is mostly because home appliances are produced and operated by people. One group of people are professionals. These relate to the activities of design, manufacture, transport, installation, and disposal of appliances, home networks, and so on. One distinguishing characteristic is that they have expertise in a certain field. The other group is occupants who are non-professionals. They are the customers who use the various home appliances. Both groups of people can affect home environment safety in different ways. Professionals are responsible for designing, manufacturing, etc. safe systems and home appliances. Occupants care more about operational safety, since they are more error-prone in operations. We talk about occupants in this paper.

3. Accident Model

As the smart home is an application area, the STAMP-PP model is discussed with respect to system operations rather than system development. It aims to understand how accident formation relates to system behaviors in adjusting the home environment, i.e., how abnormal system behaviors cause indoor climate anomalies. Concrete information about indoor climate anomalies can be used for indoor climate anomaly detection [23], and abnormal system behaviors under operation scenarios can further be used in selecting reactions and precautionary measures when the corresponding indoor climate anomaly is detected.

The STAMP-PP model starts with system behaviors to describe accident formation. The behaviors can result in abnormal changes in physical processes, which can cause discomfort or harm. The connection between smart home systems and the home environment is the system behaviors. In this section, we first give a brief introduction of the STAMP model, then discuss in detail the proposed STAMP-PP model.

3.1. STAMP

In systems theory, systems are viewed as interrelated components kept in a state of dynamic equilibrium through feedback control loops. Figure 1 presents a standard contrl loop [2]. The STAMP model is based on system theory rather than the reliability that traditional accident models are grounded on. Safety, in STAMP, is an emergent property of systems. Accidents are the result of the lack of or inappropriate constraints imposed on the system design and operations. The STAMP model consists of three building blocks, i.e., safety constraints, a hierarchical safety control structure, and process models.

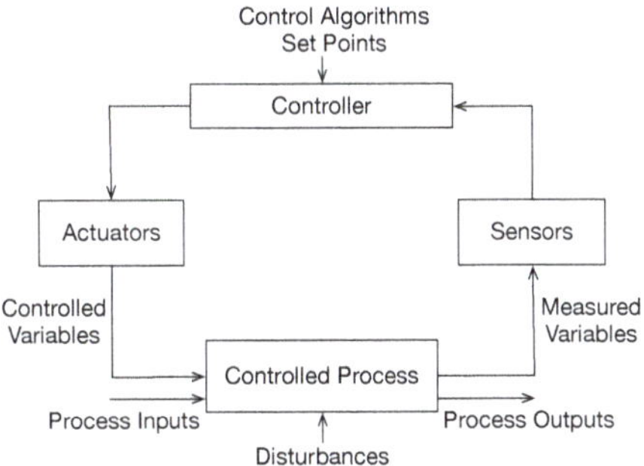

Figure 1. A standard control loop.

Safety constraints are a basic concept in the STAMP model. Losses occur only because safety constraints were not successfully enforced. Systems, in system theory, are viewed as hierarchical structures, where each level imposes constraints on the activity of the levels beneath it. Constraints are enforced by control actions of a higher-level system component (controller) to the lower-level one (controlled process).

The hierarchical safety control structure presents all stakeholders with their internal structures within the system under analysis, and the control actions and feedbacks that link the independent stakeholders and their internal components [6]. Control processes operate between levels of a system to control the processes at lower levels. The feedbacks provide information about how effectively the control actions ensure the constraints are enforced. The higher level uses the feedbacks to adapt future controls to more readily achieve its goals. An accident occurs when control processes provide inadequate control that violates safety constraints. Inadequate control comes from missing constraints, inadequate safety control commands, commands that were not executed correctly at a lower level, and inadequately communicated or processed feedback about constraint enforcement.

Process models are used by the controller to determine appropriate control actions. It is up to the type of the controller. For an automated controller, the process model is embedded in the control logic. For a human controller, the process model is the mental model. In both situations, it contains information of the required relationship among the system variables, the current system state, and the ways the process can change state. There are four conditions required to control a process, i.e., goal, action condition, observability condition, and model condition. Accidents related to component interaction can usually be explained in terms of an incorrect process model. The process model used by the controller does not match the controlled process that results in interaction accidents.

In the STAMP model, safety is achieved when the behaviors of components of a system appropriately ensured safety constraints. Accidents are the results of flawed processes involving interactions among people, societal and organizational structures, engineering activities, and physical system components that lead to violating the system safety constraints. The process leading up to an accident is described in terms of an adaptive feedback function that fails to maintain safety as system performance changes over time to meet a complex set of goals and values.

3.2. STAMP-PP

Since the STAMP-PP model focuses on the behaviors of systems to affect physical processes, we first define the concept of systems that emphasize behaviors, i.e., Performers System. The reason for choosing the word "performer" is to highlight that the behaviors are performed by the systems. Then, based on the Performers System, we can describe accident formation considering physical processes.

3.2.1. Performers System

The indoor environment is adjusted by indoor environment adjustment services with respect to various home appliances. To differentiate the home appliances with other indoor items, e.g., router or furniture, we define the concept of Performer.

Definition 1 (Performer). *A performer is a network-enabled home appliance that can adjust the indoor environment independently.*

There are two points to explain this definition. First, a Performer has networking capability so that it can be used by indoor environment adjustment services. Second, it has functions of adjusting the indoor environment, e.g., adjusting indoor temperature, or dehumidification.

There are two types of Performers based on the way it adjusts the indoor environment. One is direct adjustment, e.g., an air-conditioner which heats or cools indoor air directly; another is indirect adjustment, e.g., an electric window which adjusts, e.g., indoor temperature by introducing an air flow or solar radiation of the outdoor environment. In the latter case, the outdoor climate is passively used to adjust the indoor environment. Then, concept of Performer is used to define the Performers System.

Definition 2 (Performers System). *It is a system of all installed Performers in a house that are connected to the same home network.*

By connecting to the same home network, we can ensure that the Performers System can be used by the same indoor environment adjustment service. The Performers System has a goal prescribed by the indoor environment adjustment service. The goal is achieved by taking advantage of the functions of the Performers of the Performers System. Each Performer is taken as a subsystem of the Performers System. However, there is no need to have all related Performers working at the same time. Figure 2 shows an example of the Performers System. It consists of an air-conditioner, an electric window, and an electric curtain, which are taken as Performers. They connect to the same network and have the ability to adjust the indoor temperature. When adjusting the indoor temperature for thermal comfort (the goal), the indoor temperature adjustment service could use any combination of the Performers, but not necessarily all of them. The indoor temperature adjustment service is executed in the smart home system core. The Performers can of course be operated by occupants who are also the beneficiaries of the adjustment.

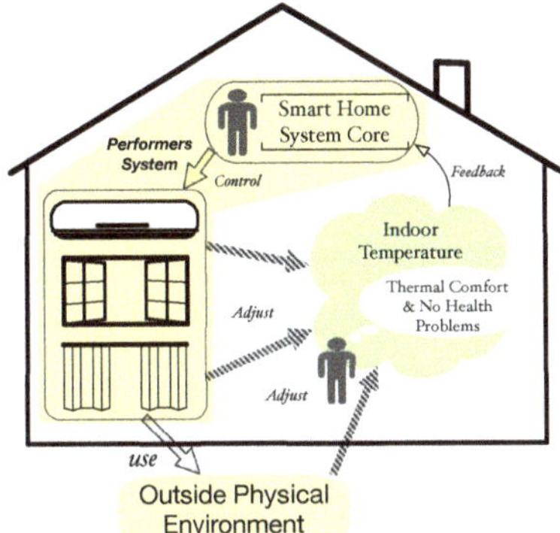

Figure 2. An example of the Performers System.

The Performers System can adjust physical processes with respect to various physical properties, e.g., temperature and humidity, by utilizing the functions the Performers provide. These physical processes may have different forms, for example, increasing or decreasing indoor temperature to the prescribed temperature level for thermal comfort.

3.2.2. Service

In this section, we discuss the behaviors of the Performers System, and when the behaviors can be taken as Services.

Definition 3 (Behavior of the Performers System). *The way the* Performers System *behaves to adjust the indoor environment.*

The Behaviors of the Performers System are the representation of the functions of related Performers. For example, the Performers System in Figure 2 has the ability to cool the temperature down (behavior), which could be achieved by setting a lower temperature level under the cool mode of the air-conditioner (function). The representation of the Behaviors is the physical process, e.g., temperature change.

Definition 4 (Service). *The* Behavior of the Performers System *exhibited in order to fulfill occupants' comfort requirement.*

One example of the Service can be the Performers System in Figure 2, which increases the indoor temperature to 22 °C for thermal comfort. Comfort means psychological and physical satisfaction with the state of the indoor environment, e.g., thermal comfort. It is the way to evaluate the Behaviors of the Performers System, and thus implicitly constrain the Behaviors. The comfort has different contents for different goals of the Performers System, e.g., thermal comfort; comfort in terms of humidity levels.

There are two ways to evaluate comfort. Let us take thermal comfort as an example. The first is based on the perception of occupants. If occupants feel uncomfortable, one can manually set a desired temperature level to the Performers System. It is easy and accurate but limited in some scenarios. For example, babies and elderly people may not be sensitive to temperature change due to their nervous system not being well developed or being degenerated. The second way is for comfort to be automatically evaluated by the smart home system core. This depends on various indices [24,25] for evaluating the physical environment for thermal comfort. For example, the PMV–PPD (Predicted Mean Vote and Predicted Percentage of Dissatisfied) index [25] is used to evaluate thermal comfort with respect to environmental factors and personal factors. PMV and PPD are short for predicted mean vote and predicted percentage of dissatisfaction, respectively, and are calculated based on Fomulas (1) and (2), where TS denotes the thermal sensation transfer coefficient and is determined by metabolic rate; MV means internal heat production in the human body and is determined by the metabolic rate and external work; and $HL1$, $HL2$, $HL3$, $HL4$, $HL5$, and $HL6$ represent heat losses through skin, sweating, latent respiration, dry respiration, radition and convection, respectively. Then combine with the ASHRAE (American Society of Heating, Regrigerating and Air-Conditioning Engineers) thermal sensation scale as shown in Table 1 to determine the comfortableness. The details are referred to in [25].

$$PMV = TS \times (WM - HL1 - HL2 - HL3 - HL4 - HL5 - HL6) \tag{1}$$

$$PPD = 100 - 95 \times e^{-0.03353 \times PMV^4 - 0.2179 \times PMV^2} \tag{2}$$

Table 1. ASHRAE thermal sensation scale.

Hot	Warm	Slightly Warm	Neutral	Slightly Cool	Cool	Cold
+3	+2	+1	0	−1	−2	−3

Environmental factors like humidity can be acquired through humidity sensors. Personal factors like metabolic rate can be roughly evaluated based on the occupant's activities, e.g., sedentary. Another way is to take advantage of wearable devices to measure personal information and send these data to the smart home system core to evaluate comfortableness. However, as far as the author is aware, cloth insulation cannot be evaluated through wearable devices for now. It can only be roughly evaluated through scenarios like in hot summer, where the value of cloth insulation is small, and an average value is assigned for the summer season.

Definition 5 (Service Manner). *The* Service *under a specific condition is called a Service Manner.*

The condition can be personal or environmental. For a Service, e.g., adjust the indoor temperature for thermal comfort by an air-conditioner, changes in conditions result in different Service content. For example, compare a sweating man in summer and a sedentary man in winter. The conditions in the former are high metabolic rate, high outdoor temperature, etc., and he needs for the temperature to be cooled down. For the latter, the conditions may be low metabolic rate, low outdoor temperature, etc., and he needs for the temperature to go up.

Definition 6 (Critical Service Manner). *The* Service Manner *in demand is called Critical Service Manner.*

This definition is given from the viewpoint of people who need the Service. In the example from Definition 5, the adjustment of indoor temperature to achieve a cool environment for thermal comfort is the Critical Service Manner for the sweating man; the adjustment of indoor temperature to achieve a warm environment for thermal comfort is the Critical Service Manner for the sedentary man.

3.2.3. Accident Formation

Accidents can be understood as the resilience [26] of the Performers System, i.e., adjusting indoor environment anomalies to maintain a normal performance has failed and resulted in undesired consequences. A Service may fail and result in uncomfortableness, then further evolve into hazards and cause harm to occupants. Accident formation is to describe how Services may fail and further evolve into a hazard to cause accidents. The causes on the system part, i.e., the Performers System, can be understood by the STAMP model. The relation between the Behaviors of the Performers System and the physical processes in accident formation is discussed in this section.

There are some considerations about physical processes from the viewpoint of engineering. Before a hazard is detected, physical process anomalies should be detected as early as possible so as to trigger precautionary measures. The information of hazards is also important for triggering reaction measures. All in all, accident formation, which is shown in Figure 3 focuses on the physical process anomalies resulting from abnormal Behaviors.

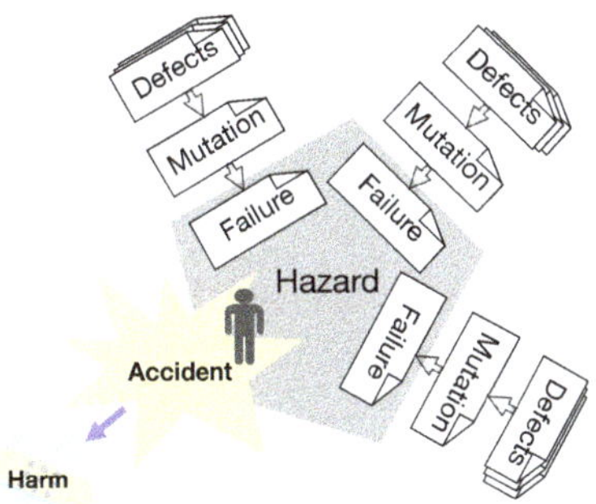

Figure 3. The accident causality model.

Next, let us discuss the terms and related rationales of the accident causality model in Figure 3.

Definition 7 (Defect). Defects *are the direct causes of a* Mutation.

A Mutation (Definition 8) relates to abnormal Behaviors of the Performers System. Abnormal Behaviors are due to unexpected control actions of the Performers System. Thus, Defects are inappropriate control actions (ICAs) and unsafe control actions (UCAs) of the Performers System to the abnormal Behaviors in adjusting the home environment. ICAs occur when Mutations result in Service Failures (Definition 9), and UCAs occur when Service Failures evolve into Hazards (Definition 10). For example, an ICA can be wrongly setting the heating mode of a Performer in hot summer, which will cause thermal discomfort. To differentiate from causes introduced in the STAMP model, Defects refer to more superficial reasons. As the home environment is an application area, the deeper causes for system development defects are not considered here. When physical process anomalies are detected, the home environment is expected to react to them immediately with respect to Defects. Therefore, the Defects are the direct causes and should be controllable, e.g., through reconfiguration.

As discussed at the beginning of Section 3, the STAMP-PP model relates to system operations. The occurrence of Defects is under the scenarios of system operations. The operations can be categorized into three types. One is operations by occupants. Another is by some controllers, e.g., Performers. The other is a mixture of by occupants and the controller. To react to a physical process anomaly efficiently, it is necessary to know the Defect with its corresponding operation scenario. To identify Defects with respect to the scenarios, we applied the hazard analysis technique STPA [2,9] in Section 4.

Definition 8 (Mutation). *A Mutation is the violation of constraint on the Behavior of the* Performers System.

Abnormal changes in physical processes are the representation of the abnormal Behavior of the Performers System. Therefore, the constraint on the Behavior of the Performers System is the prescription of changes in physical processes that satisfy the comfort purpose. Thus, the Mutation is the change of physical processes under adjustment unacceptably deviating from the prescribed curve(s), which results in uncomfortableness. For example, the amplitude of temperature fluctuation should not exceed a threshold value for thermal comfort [23]. The Mutation in this case is the amplitude of temperature fluctuation exceeding the threshold value.

The Mutation is a state of physical processes between the state that brings about comfort and the state that results in a hazard. This concept is important not only for the understanding of accident formation, but also because the information can trigger precautionary measures. First, a Mutation indicates the current indoor environment adjustment is inefficient. Second, by combining the *Mutation* with Defects under certain operation scenarios, one can select appropriate precautionary measures to restore the state of physical processes that bring about comfort.

Definition 9 (Service Failure). *The* Behavior of the Performers System *exhibits failed to fulfill occupants' comfort requirement.*

The concept of Mutation is related to physical processes, while Service Failure refers to both physical processes and the perception of occupants. The occurrence of Service Failure is when occupants perceive the uncomfortableness brought about by the Mutation.

There are two ways to determine whether a Service has failed. One is directly determined by the perception of occupants. Another one resorts to various indices [24,25], by which the smart home system core can conclude whether the home environment satisfies the comfort requirement with occupants on the scene.

Definition 10 (Hazard). *It is an indoor environment state that will cause harm to occupants.*

A Hazard will harm the health condition of occupants who are on the scene. One or more Service Failures will form or further evolve into a hazardous situation. For example, a Service Failure for thermal discomfort evolves into a hazard if the indoor temperature reaches a level that can cause heat stroke if occupants are present.

Taking thermal related hazards as an example, the evaluation of a hazard depends on heat stress indices [27] and cold stress indices [28]. These indices are sophisticated techniques to represent thermal sensations to hot and cold conditions. Heat stress [27] is defined as the net heat load to which an occupant is exposed from the combined contributions of metabolic heat, environmental factors, and clothing that results in an increase in heat storage in the body. Cold stress [28] is the climatic condition under which the body heat exchange is just equal to or too large for heat balance at the expense of significant heat and sometimes heat debt. Similarly to the PMV–PPD index, they also have a complex relations with environmental and personal factors, which can be measured in practice.

Definition 11 (Accident). *It is an unintentional event where a* Hazard *results in harm of occupants.*

It involves both the home environment and occupants. The Hazard has harmed occupants. The Accident can be detected by evaluating the Hazard and the health condition of the occupants. The latter can be measured by taking advantage of wearable devices.

Definition 12 (Harm). *Death, physical injury or damage to the health of occupants.*

It is the consequence of the Accident. It varies with respect to Hazards and the health conditions of occupants. For example, it may cause heat illnesses or even death to elderly people due to heat exposure [29]; it may also affect sleep and the circadian rhythm that can cause cardiac autonomic response during sleep due to cold exposure [30].

4. Application of STPA

To identify Defects under operation scenarios, the hazard analysis technique STPA is adopted. In this section, we first introduce the STPA steps, then discuss the way to tailor it and a new way of documenting the analytical results, and finally illustrate the application results and compare them with the application of the original STPA.

4.1. STPA

The Systems-Theoretic Process Analysis (STPA) [2,9] is a new hazard analysis technique based on the STAMP model. The goal is to identify causes that lead to hazards and result in losses so they can be eliminated or controlled. The STPA has three steps, the latter two of which are taken as the main steps.

The first step is to establish the system engineering foundation. Three things should be done in this step. The first is to define the interested accident and its related system hazards. They should be specific and concise. Then, the system level safety requirements and design constraints need to be specified to prevent system hazards from occurring. The last is to define the safety control structure that takes the system level safety requirements and design constraints as inputs.

The second step is to identify potentially unsafe control actions (UCAs). Every controller of systems usually has one or more control actions. System hazards thus result from inadequate control or enforcement of the safety constraints. Four taxonomies are provided to look for the UCAs:

1. A control action required for safety is not provided or not followed;
2. An unsafe control action is provide;
3. A potentially safe control action is provided too early or too late (at the wrong time or in the wrong sequence);
4. A control action required for safety is stopped too soon or applied too long.

Then, we translate the identified UCAs into safety requirement and designed constraints on system component behaviors.

The third step determines how each UCA identified in step two could occur. For each UCA, examine the parts of the control loop as shown in Figure 4 to see if they can cause the UCA under some scenarios. Then, design controls and mitigation measures if they do not already exist or evaluate existing measures.

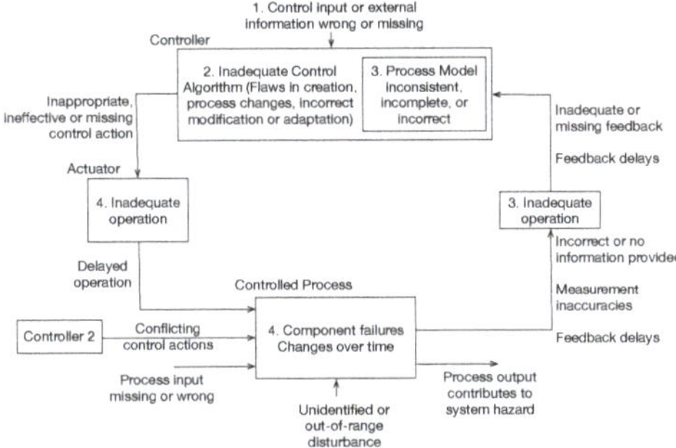

Figure 4. A classification of controls flaws leading to system hazards.

4.2. Tailor

The understanding of accident formation within the indoor environment has extended the STAMP model, i.e., the STAMP-PP model, by also considering physical processes. As disscussed in Section 3.2.3, abnormal Behaviors result in Service Failures, and abnormal Behaviors are due to inappropriate control actions of the Performers System. Thus, it is important to know what these inappropriate control actions are. Therefore, the STPA need to be tailored for this main purpose.

In the first step of STPA, in addition to the definition of accident, system hazard, and safety requirement, information about the Service Failure and requirements of no failure, i.e., the reliability of the Performers System to deliver Services, should also be provided. Design constraints are not required as this work is not for implementing a safe system.

When a Service Failure occurs, which means the corresponding control action is inappropriate, then precautionary measure(s) should be provided. Thus, in the second step of STPA, ICAs should be identified with regard to the taxonomies provided in step two of STPA. Namely, given a state of physical processes, under a specific operation scenario, we need to consider whether a control action with respect to the taxonomies will cause a Service Failure. Precautionary measures are determined by considering the context information which consists of the ICAs, operation scenarios, and the Service Failure. The process of the determination heavily depends on the expertise of concerned areas, which is directed by the reliability requirements. The precautionary measures are also control actions. Ineffective precautionary measures will result in the occurrence of a Hazard (then, reaction measures are required). For each precautionary measure, UCAs are identified by considering the taxonomies provided in step two of STPA under the conditions of operation scenarios and of a state of physical processes. Safety requirements to the UCAs also need to be identified to guide the selection of reaction measures.

The third step of STPA is not necessary. This is because the analysis in our case is to identify Defects under operation scenarios, which can be utilized in selecting appropriate precautionary and reaction measures, but not in designing and manufacturing a system.

STPA adopted tables and lists for documenting analytical results [2,9]. After applying the tailored STPA, the results, i.e., control actions, ICAs, and their related reliability requirements and operation scenarios; and precautionary measures, UCAs, and their related safety requirements and operation scenarios, could be documented by that used by the STPA. However, the relations among them are not clearly represented. In this paper, we propose a Landscape Genealogical Layout Documentation (denoted as LGLD) for documenting the results, which is illustrated in Figure 5. The ancestor is a control action. The first generation illustrates ICAs and their related reliability requirements and operation scenarios. The second generation represents precautionary measures. The third generation represents UCAs and their related safety requirements and operation scenarios. These results can be numbered for better reference, e.g., ICA-m for an ICA, which means the inappropriate control action m. This way of documentation also implies the analysis direction, i.e., from control actions to ICAs, then to precautionary measures, and finally to UCAs.

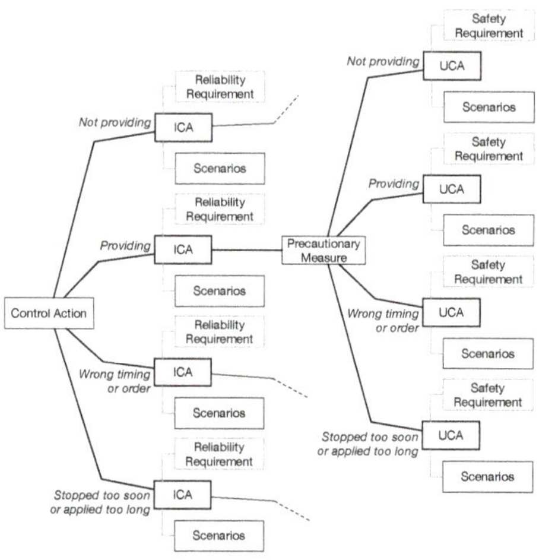

Figure 5. Documenting the analytical results by the Landscape Genealogical Layout Documentation (LGLD) approach.

For each control action, the taxomomies provided in step two of STPA to list the ICAs must be considered. Every ICA is attached with the reliability requirements and operation scenarios. Then, each ICA is connected with a precautionary measure. The precautionary measure is to prevent its connected ICA from changing the home environment from a Service Failure to a Hazard. For each precautionary measure, it is important to list the UCAs attached with safety requirements and operation scenarios.

4.3. Results

As discussed in Section 2.2, weather anomalies affect indoor environment, e.g., the heatwave of summer 2018. Thus, we consider the example of the high-temperature results in heat stroke [31,32]. Heat stroke is clinically diagnosed as a severe elevation in body temperature (a core body temperature of 40°C or higher) that occurs in the presence of central nervous system dysfunction and a history of environmental heat exposure or vigorous physical exertion. It can be classified into nonexertional (classic) heat stroke and exertional heat stroke. The former occurs in very young or older people, or those with chronic illness when the environmental temperature is high. The latter happens to young fit people and involves prolonged excessive activities like sports. This paper focuses on the former case.

Heat stroke can be assessed by heat stress indices, among which the most widely accepted and used one is the wet bulb globe temperature (WBGT) [33]. The Ministry of the Environment of Japan (Ministry of the Environment, Japan: http://www.wbgt.env.go.jp/en/) has recommended a criterion for thermal conditions based on the WBGT as shown in Table 2.

Table 2. Recommended criteria for thermal conditions.

WBGT (°C)	Threat Level
~21	Almost Safe
21~25	Caution
25~28	Warning
28~31	Severe Warning
31~	Danger

We applied the tailored STPA to a Performers System that is for indoor temperature adjustment for thermal comfort. The structure of the system consists of four parts, i.e., home, home gateway, service intermediary, and service provider [12], as shown in Figure 6. The home is the place that occupants live in, which is equipped with different kinds of items, e.g., home appliances, to meet everyday living requirements. The home gateway is the gateway of the networked Performers to the outside world of the home. The indoor temperature adjustment service is executed here to issue commands for the control of Performers. The service intermediary aggregates various services from different service providers and maintains them locally. It can also respond to service subscriptions from home gateways. The service provider designs, publishes, and updates concrete services to the service intermediary for future use. We applied STPA to investigate the Behaviors of the Performers System to elicit possible Defects under related operation scenarios. Thus, we focus on executing the indoor temperature adjustment service in the home gateway to adjust the indoor temperature. The home gateway, service intermediary, and service provider can be taken as the smart home system core.

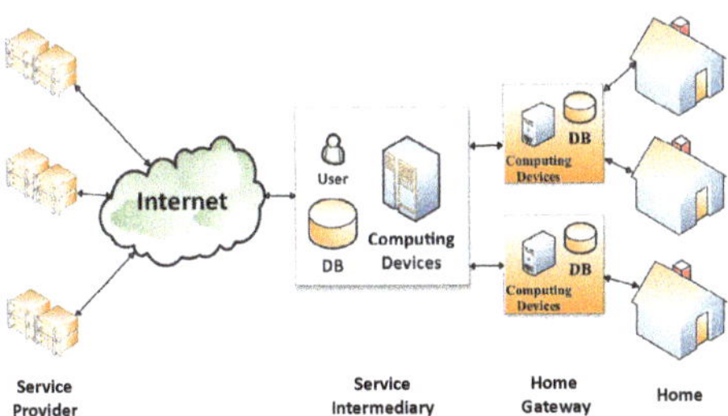

Figure 6. The structure of the Performers System.

In order to apply STPA, an assisting tool, i.e., STAMP Workbench (the STAMP Workbench is an open-source, free, easy-to-use tool for people who are interested in system safety analysis by using the STAMP/STPA. It was developed by the IT Knowledge Center of Information-Technology Promotion Agency, Japan. https://www.ipa.go.jp/english/sec/complex_systems/stamp.html) is adopted for the analysis. The STAMP Workbench is claimed to include features such as concentration on thinking and help analysis and is not just an editing tool, or guide analysis procedure, but an unlimited and intuitive operation. Analytical results can be exported into Excel files and images.

In the first step of STPA, we prepared some concepts for further analysis as shown in Table 3. For demonstration, Service Failure is defined as when the indoor WBGT temperature is adjusted within (25,28] °C, and the Hazard is when the indoor WBGT temperature is adjusted over 28 °C. The reliability requirement corresponds to the Service Failure, which represents the requirement to ensure a Service will not fail.

Table 3. Preparation for the tailored System-Theoretic Process Analysis (STPA).

Accident	Physical harm of occupants due to heat stroke
Service Failure	Indoor WBGT temperature is within (25,28] °C
Reliability Requirement	Indoor WBGT temperature should be adjusted bellow 25 °C
Hazard	Indoor WBGT temperature is over 28 °C
Safety Requirement	Indoor WBGT temperature should be adjusted bellow 28 °C

Figure 7 illustrates the safety control structure for indoor temperature adjustment. The Home Gateway is the controller, which is responsible for executing the indoor temperature adjustment service. The controlled process is the Home Environment. Performers are taken as actuators. We consider an air-conditioner and a window here as Performers. Empirically, for energy saving, these two cannot work at the same time. The control actions are listed on the arrow from the Home Gateway to the Performers. The control action "set to X °C" means to set Performers to adjust the indoor temperature to X °C. The feedback to the Home Gateway is the indoor temperature. Indoor temperature can be adjusted by the indoor temperature adjustment service that is executed in the Home Gateway, or by occupants to issue commands, i.e., control inputs to the Home Gateway.

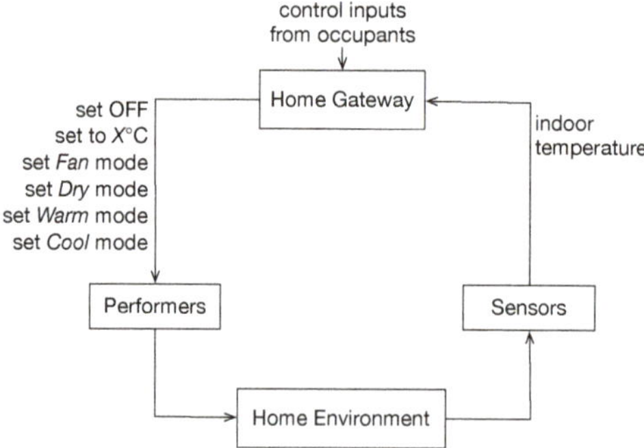

Figure 7. The safety control structure for indoor temperature adjustment.

Next, based on the results of the first step, we identifies ICAs and UCAs and elicited their related requirements and operation scenarios. Part of the results are shown in Figures 8 and 9, which adopted the LGLD approach introduced in Section 4.2 for the documentation. They illustrate the results of analyzing the control actions "set OFF" and "set to X °C". "N/A" denotes the taxonomy is not applicable to the corresponding control action. In Figure 8, the second "set OFF" can be considered as a reconfiguration compared with the first one. The precautionary measure "set Cool mode" can be deployed for the two ICAs that relate to the "set OFF". One reason could be that a different configuration has a higher possibility to restore the physical process to a comfortable state, as the "set

OFF" has caused the ICA. For the second operation scenario, not-providing ICA, it is because people are not sensitive to temperature change and thus did not issue the "set OFF" command manually.

For the results as shown in Figure 9, X satisfies $X < 25$ °C. There are two reasons for the not-providing ICA and UCA to say that the Performers System is working in the Fan mode. The first is if it works in the Warm mode, which in this case is inappropriate or hazardous, and we cannot make a solemn vow to conclude that not providing "set to X °C" is inappropriate or hazardous. Second, the other working modes, i.e., Dry and Cool, have a cooling effect based on our experience, which may not be an ICA or UCA even when "set to X °C" is not provided. One more thing that needs to be explained is that providing "set to X °C" at the "wrong time" is neither inappropriate nor hazardous. If it is not provided in time, the home environment would experience Service Failure or Hazard for some time. However, providing "set to X °C" at a different time should not be inappropriate or hazardous. Conversely, since "set to X °C" is provided, the home environment could be restored to a safe level, even though later than expected. In this case, providing "set to X °C" can be thought of as a precautionary or reaction measure, rather than an inappropriate or hazardous control action.

Figure 8. The analysis results for the control action "set OFF".

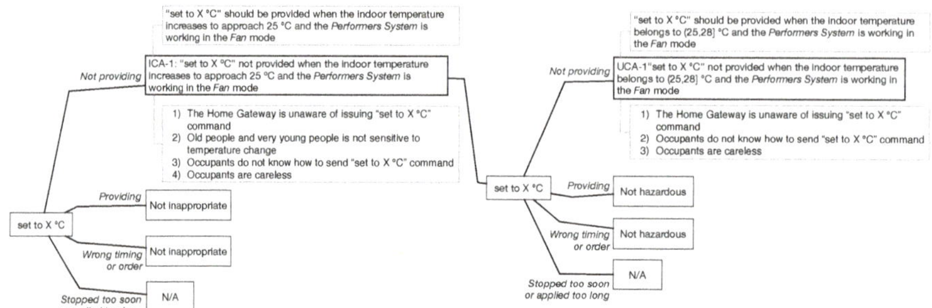

Figure 9. The analysis results for the control action "set to X °C".

4.4. Comparsion of Results

Originally, the STPA is the only hazard analysis technique was based on the STAMP model [2]. Thus, in this section, let us compare the results presented in Section 4.3 with those by adopting the original STPA. Since the third step of STPA was not taken into account, the comparison only considers the results that derived from the first two steps of STPA. For comparability, the temperature issue discussed in Section 4.3 is still part of our focus when adopting the original STPA.

As discussed, the goal of STPA is to identify causes that lead to hazards and result in losses, so they can be eliminated or controlled. The causes to be identified are UCAs, and flaws in the control loop (as shown in Figure 4) under some scenarios which are different from our case. The elimination or control usually resorts to designing and implementing a safe system, while in our case, the aim is to select appropriate precautionary and reaction measures which can restore a safe Service delivery.

The system engineering foundation is given first. The prepared definitions are illustrated in Table 4. The design constraint is system level constraint and is expected to further decompose into constraints that can be assigned to system components as the analysis evolves. Compared with what is shown in Table 3, Service Failures and reliability requirements to the system are gone, which indicates ICAs will not be identified afterward. This is because ICAs are supposed to result in Service Failure, and reliability requirements of ICAs can be taken as the decomposition of the system level reliability requirement. The safety control structure as shown in Figure 7 can also be used here.

Table 4. Preparation for the STPA analysis.

Accident	Physical harm of occupants due to heat stroke
Hazard	Indoor WBGT temperature is over 28 °C
Safety Requirement	Indoor WBGT temperature should be adjusted bellow 28 °C
Design constraint	The Performers System is capable of adjusting the indoor WBGT temperature bellow 28 °C

Next, the UCAs identified in step two of STPA are shown in Tables 5 and 6 for control actions "set OFF" and "set to X °C", respectively. Then, for each UCA, safety requirements and design constraints can be derived. For example, the safety requirement for UCA-1 in Figure 5 could be:

- "set OFF" should be provided when the indoor temperature belongs to (25,28] °C and the Performers System is working in the Warm mode.

The design constraints for UCA-1 could be:

- The Performers System should be accurately aware of the indoor temperature change;
- "set OFF" should be provided when needed.

Table 5. UCAs for the control action "set OFF".

	Hazard: Indoor WBGT temperature is over 28 °C			
Control Action	Not Providing	Providing	Wrong Timing or Order	Stopped Too Soon or Applied Too Long
set OFF	UCA-1: "set OFF" not provided when the indoor temperature belongs to (25,28] °C and the Performers System is working in the Warm mode	UCA-2: "set OFF" provided when the Performers System is working in the Cool mode and the indoor temperature belongs to (25,28] °C	UCA-3: "set OFF" provided after the Performers System has been working in the Warm mode for some time when the indoor temperature belongs to (25,28] °C	N/A

Table 6. UCAs for the control action "set to X°C".

	Hazard: Indoor WBGT Temperature Is over 28 °C			
Control Action	Not Providing	Providing	Wrong Timing or Order	Stopped Too Soon or Applied Too Long
set to X °C	UCA-4: "set to X °C" not provided when the indoor temperature belongs to (25,28] °C and the Performers System is working in the Fan mode	Not hazardous	Not hazardous	N/A

There are some differences in the comparison with the results derived in step two of STPA. The ICAs, operation scenarios for ICAs, reliability requirements, precautionary measures, and operation scenarios for UCAs cannot be obtained by adopting the original STPA. However, the design constraints can be derived. The UCAs identified by adopting the original and tailored STPA are equivalent. The results documented by tables and lists are separated. It is a trivial problem when checking the relations between the results that were documented by the conventional approach. The LGLD approach integrated the results and overcame this problem. We discuss the advantages of the tailored STPA at the end of Section 5.

5. Discussion

As one of our everyday living experiences, physical processes could be in hazardous states that cause harm to occupants [29,30]. The occurrence of hazardous states can be transferred from a normal state of a physical process that brings about comfortableness, through some intermediate states for uncomfortableness. The normal state is maintained by the smart home system, which takes advantage of home appliances or the outdoor climate. When the behaviors of the smart home system deviate from expectations, the above transformation will occur. Thus, this living experience validates the proposed STAMP-PP model.

To validate the terms defined in the STAMP-PP model, let us consider the eight abnormal indoor temperature situations that were discussed in our previous work [23]. Mutations can be undesired fluctuation, constantly cooler/warmer than expectation, undesired duration of the temperature that results in discomfort, and Service Failures when combining the Mutations with the feeling of occupants on the scene. Hazards can be unbearable hot/cold and undesired duration in hot/cold situations.

Mutations and Hazards provide detectable evidence of physical process anomalies. The information of Service Failure combines with ICAs under related operation scenarios—e.g., as depicted in Figures 8 and 9, one can predict whether Hazards would happen before providing precautionary measures, because Service Failures and the ICAs under related operation scenarios can be the context, in which a Hazard could occur if time goes on. This context information can be acquired from various sensors (e.g., temperature sensor), system states, etc. For example, in Figure 9, we can know whether the temperature is increasing to approach 25 °C through temperature sensors. Then, precautionary measures can be selected under the direction of reliability requirements. If a Hazard is unfortunately detected, reaction measures have to be selected based on the UCAs under operation scenarios and the Hazard, which are guided by safety requirements.

The purpose of precautionary measures is to restore Service delivery. This is achieved by reconfiguration of the Performers System. The reconfiguration has two forms. One is to reset the current working Performer(s), e.g., the precautionary measure "set OFF" (the second one) as shown in Figure 8, which is achieved by resetting the OFF command to the same Performer. Another is to reconfigure ready-to-use Performers to achieve the goal, e.g., the "set Cool mode" in Figure 8, which could be reconfigured to a standby Performer to the Cool mode. The purpose of reactions is to restore a safe home environment state, which could be a Service Failure (that needs further interference) or the process of normal Service delivery. It is also achieved by reconfiguring the Performers System, which has the same contents as those introduced for precautionary measures. The reactions should have another content that precautionary measures do not have, i.e., a warning mechanism. The warning mechanism is triggered when a Hazard is detected that implies precautionary measures have failed. In the very beginning, the warning signals will be sent to occupants who could leave the scene or do something else to ensure safety. If the reconfiguration in the reaction stage cannot restore the system to a safe situation, the warning mechanism will inform an emergency department, e.g., a hospital, through networks.

Accident models can be roughly classified into three categories [1,2], i.e., sequential models, epidemiological models, and systemic models. Sequential models describe accidents as the result of time-ordered sequences of discrete events. Epidemiological models view accidents as a combination of latent and active failures within a system, which is analogous to the spreading of a disease. Latent conditions, e.g., management practices or organizational culture, can lie dormant within a system for a long time, which can finally create conditions at a local level to result in active failures. The STAMP model is based on system theory and so is the STAMP-PP model. It describes physical process anomalies that cause uncomfortableness and health problems as a result of abnormal Behaviors of the Performers System. Furthermore, to better understand the causal relation between the Performers System and physical processes and better represent the time order of physical process evolvement, the terms in the STAMP-PP model can be taken as events.

Hazard analysis techniques are devised based on accident models. The STPA was invented based on the STAMP model to identify causes that conventional approaches overlooked [2,9]. We found it is an efficient tool for assisting in the identification of Defects, i.e., ICAs and UCAs (e.g., the results shown in Figures 8 and 9). However, hazard analysis heavily depends on expertise in a specific area. Accident models and hazard analysis techniques are to assist in understanding accidents and guiding for causal analysis. The STAMP-PP model offers a way to understand physical process anomalies with respect to the behaviors of smart home systems. To apply the tailored STPA to that in our case (some results of which are shown in Figures 8 and 9), one may need to possess knowledge in at least the fields of computer science, software engineering, computer networks, and even physics. This makes the authors feel that safety research in different areas is like different research fields.

In the first step of STPA, the accidents and system hazards provided mostly depend on the interest of an organization or the government [2]. For example, in this paper, Accident is defined as physical harm of occupants due to heat stroke, as illustrated in Table 3. It can also, however, encompass other symptoms aside from heat stroke, e.g., severe cold. Concrete accidents are different due to the variety of environments, e.g., workplaces and the home environment. It is also determined by budget, severity of occurrence, and frequency of occurrence [2].

In step two of STPA, originally, the taxonamies were provided to identify UCAs with respect to control actions. Physical process anomalies (not Hazards) are the result of abnormal Behaviors of the Performers System. The behaviors are the representations of the functions which are achieved by the control actions issued by the Performers Sytem. For example, stopping the Performers System from working is achieved by the control action "set OFF" as shown in Figure 8. Thus, the taxonomies are applicable for identifying ICAs. This ensures that all possible ICAs can be effectively identified.

The states of physical process change from a state of comfort to one of discomfort, then to one of hazard. When a state of discomfort is detected, precautionary measures are adopted to prevent the physical process from transferring into a state of hazard. Precautionary measures refer to control actions achieved by reconfigurations. Thus, the taxonomies are also applicable to the precautionary measures. For example, the control action "set OFF" in Figure 8 can both be the original control action and the precautionary measure. The same applies to the control action "set to X °C" as shown in Figure 9.

The purpose of adopting the hazard analysis technique is to identify Defects, so as to select appropriate precautionary and reaction measures for adjusting home environment anomalies to maintain a normal performance. By checking the comparison presented in Section 4.4, we found the tailored STPA can satisfy this purpose, while the original STPA cannot. First, ICAs under operation scenarios and the reliability requirements of the ICAs can be identified by the tailored STPA. This is due to Service Failure, and the system level reliability requirement is provided in the first step of STPA. Naturally, precautionary measures are not necessarily identified in the original STPA, because precautionary measures are selected for ICAs. Second, even though UCAs can be identified by both the original and tailored STPA, the operation scenarios have different contents. For the tailored STPA, the operation scenarios refer to occupants or controllers or both as discussed when introducing the concept of Defect. For the original STPA, the operation scenarios refer to the control flaws as shown in Figure 4, which are identified in step three of STPA.

As discussed in Section 4.2, the STPA takes advantage of tables and lists to document the analytical results (see the results shown in Tables 5 and 6). We proposed the LGLD approach to document the analytical results. The advantage of comparing tables and lists is that it can clearly represent the relations among the results in a straightforward way. For example, in Figures 8 and 9, the relations among control actions, ICAs, precautionary measures, and UCAs are clear. Further, it is clear to see the (reliability or safety) requirements attached to each ICA and UCA, and the related operation scenarios under which the ICA and UCA can occur. This kind of relations is not explicitly represented using tables and lists such as the ones presented in Section 4.4. To build such documentation, one can build along the way of analysis, because the analysis starts from control actions to identify ICAs under related operation scenarios, and reliability requirements, then to determine precautionary measures, and finally to identify UCAs under related operation scenarios, and safety requirements.

When applying the STAMP-PP model to understand accident formation, the Performers System has to fully control the home environment, or its behaviors may not be the "only" reason to cause Service Failure, then Hazard. This can be the limitation of the proposed STAMP-PP model.

6. Related Work

This section includes two parts. The first is about accident models based on system theory. The second discusses safety-related research in the smart home environment.

6.1. Accident Models

The systems approach is considered as the dominant paradigm in safety research [34]. It views accidents as the unexpected interactions among system components, i.e., technical, social, and human elements. Among various systemic models, there are three most cited models [10], that is, STAMP [2], Functional Resonance Analysis Method (FRAM) [3], and Accimap [35,36].

The concept of STAMP has been briefly introduced in Section 3.1. According to [10], over half of the reviewed papers were STAMP-related, which indicates a pervasive acknowledgement of its underlying rationale. Its application has attracted researchers from a broad field. The authors of [6] adopted it to investigate aircraft rapid decompression events. The authors of [4] applied it in the analysis of deepwater well control safety. In the field of railway, the authors of [5] investigated railway accidents and accident spreading by taking the China Jiaoji railway accident as the example. It has also been applied to the field related to poisonous or dangerous substances. The authors of [7] adopted it in analyzing the China Donghuang oil transportation pipeline leakage and explosion accident. The authors of [8] applied it to the Fukushima Daiichi nuclear disaster and to promote safety of nuclear power plants. Smart home systems are generally not considered as safety-critical systems. However, as weather anomalies, e.g., heatwaves, occur regularly due to global warming, smart home systems for indoor environment adjustment, in this context, can be taken as safety-critical. Thus, the STAMP model can be adopted for understanding accident formation in the home environment.

FRAM was developed to act as both an accident analysis and risk assessment tool [10]. The FRAM model graphically describes systems as interrelated subsystems and functions that will exhibit varying degrees of performance varation. Accidents result from the fact that the emergent variation produced from the performance variability of any system component to "resonate" with that of the rest of the elements is too high to control. It has been discussed that the FRAM and STAMP approaches focus better on qualitative modeling and description of systemic behavior and accidents [37]. FRAM also has applications in different fields, e.g., the authors of [38] applied it to railway traffic supervision to investigate interdisciplinary safety analysis of complex sociotechnological systems. The authors of [39] extended FRAM by including a framework with steps to support hazard analysis. Some efforts have been tried to quantify it, e.g., the authors of [40] developed a semiquantitative FRAM based on a Monte Carlo simulation.

The Accimap method is a graphical representation of a particular accident scenario that relates to systemwide failures, decisions, and actions [3,36]. Accimap is a generic approach and does not use taxonomies (that is different than that of the STPA [2,9]) of failures across the different levels of considered [41]. The Accimap produces less reliable accident analysis results compared to STAMP [42].

The selection of accident analysis techniques depends on the system characteristics, i.e., manageability and coupling [1]. The systemic approaches are usually adopted by systems with low manageability and tight coupling. Systemic approaches related to complex sociotechnical systems have their own strengths. For one such system, it is better to adopt multiple approaches which supplement each other, even though STAMP is considered much more effective and reliable in understanding accidents and hazard analysis [10,41,42].

6.2. Smart Home Safety

In the past, safety research inside the home environment used to be based on events or chains-of-events. With the emergence of the so-called smart homes, safety research inside the home environment also has new forms. Some of it refers to monitoring the home environment. With the purpose of detecting safety problems of indoor climate abnormal variations, the authors of [12] proposed a CPS (cyberphysical system) home safety architecture to support an event-based detection. The authors of [43] presented a method that maps the real home connection to a virtual home environment, together with related policies to ensure remote monitoring, to ensure home safety. Elderly safety in the smart home environment was achieved by analyzing and inferring locations, time slots, and periods of stay of elderly people [44]. Robot techniques were also employed, e.g., the authors of [45] developed a robot which can, for example, sense gas leakage and shut off the gas valve. Others focus on a specific part of the home. The authors of [46] proposed risk analysis and assessment when cooking to prevent potential risks. This is because the kitchen is also prone to safety problems like gas leakage and fire accidents. Electricity is also an important risk factor. The authors of [47] adopted an alert circuit with a voltage level indicator to prevent the smart solar home system from being overloaded and damaged. With cloud computing techniques becoming pervasive in implementing smart home systems, risks like cloud service unavailability have also been introduced. To overcome this, the authors of [48] discussed home resilience in the presence of possible unavailability and proposed RES-Hub, i.e., a standalone hub to ensure the continuity of required functionalities.

Most of the studies like those discussed above focus on implementing systems to deal with home safety problems. If not properly designed and implemented, the system itself can be a risk factor. Thus, requirement elicitation becomes critical. Conventional safety-related techniques are applied to safety-critical areas, e.g., aviation [6]. Our work employed these techniques to the smart home systems.

7. Conclusions and Future Work

We extended the accident model STAMP by considering physical processes in the home environment. As the home environment is adjusted by behaviors of the smart home system, we first proposed the concept of Performers System that emphasizes the behaviors performed by various home appliances. Then, based on the Behavior of the Performers System, we proposed accident formation with respect to physical processes going from normal state to some intermediate states that result in uncomfortableness, and finally to states that cause harm. In order to identify the Defects, i.e., ICAs and UCAs that result in abnormal system behaviors, the hazard analysis technique STPA was tailored and applied to the smart home system for indoor temperature adjustment. After comparison with the results derived by adopting the original STPA, we found that the tailored STPA is an efficient tool to assist in identifying the Defects. The analytical results of applying STPA were used to adopt tables and lists for documentation, but the relations among the results were not straight-forward and unclear. We then proposed the LGLD approach, whose advantages are demonstrated by the comparison of results.

For future work, we aim to map the context information in which physical process anomalies occurred in the cyber world, such that precautionary measures and reactions can be effectively selected in real time. This brings about the problem of how to parameterize the STPA analytical results, i.e., ICAs, operation scenarios, and reliability requirements; and UCAs, operation scenarios, and safety requirements.

Author Contributions: Conceptualization, Z.Y., Y.L., and Y.T.; methodology, Z.Y. and Y.T.; validation, Z.Y. and Y.T.; writing—original draft preparation, Z.Y.; writing—review and editing, Z.Y. and Y.T.; supervision, Y.T.

Funding: This research received no external funding.

Acknowledgments: The authors would like to thank Toshiaki Aoki who gave constructive comments.

Conflicts of Interest: The authors declare no conflict of interest.

References

1. Underwood, P.; Waterson, P. *Accident Analysis Models and Methods: Guidance for Safety Professionals*; Loughborough University: Loughborough, UK, 2013; p. 28.
2. Leveson, N.G. *Engineering a Safer World: Systems Thinking Applied to Safety*; The MIT Press: Cambridge, MA, USA; London, UK, 2011.
3. Hollnagel, E. *FRAM: The Functional Resonance Analysis Method: Modelling Complex Socio-technical Systems*, 1st ed.; CRC Press: Boca Raton, FL, USA; London, UK; New York, NY, USA, 2012.
4. Meng, X.; Chen, G.; Shi, J.; Zhu, G.; Zhu, Y. STAMP-based analysis of deepwater well control safety. *J. Loss Prev. Process Ind.* **2018**, *55*, 41–52. [CrossRef]
5. Ouyang, M.; Hong, L.; Yu, M.H.; Fei, Q. STAMP-based analysis on the railway accident and accident spreading: Taking the China—Jiaoji railway accident for example. *Saf. Sci.* **2010**, *48*, 544–555. [CrossRef]
6. Allison, C.K.; Revell, K.M.; Sears, R.; Stanton, N.A. Systems Theoretic Accident Model and Process (STAMP) safety modelling applied to an aircraft rapid decompression event. *Saf. Sci.* **2017**, *98*, 159–166. [CrossRef]
7. Gong, Y.; Li, Y. STAMP-based causal analysis of China-Donghuang oil transportation pipeline leakage and explosion accident. *J. Loss Prev. Process Ind.* **2018**, *56*, 402–413. [CrossRef]
8. Daisuke, U. STAMP Applied to Fukushima Daiichi Nuclear Disaster and the Safety of Nuclear Power Plants in Japan. Master's Thesis, School of Engineering, Massachusetts Institute of Technology, Cambridge, MA, USA, 2016.
9. Leveson, N.G.; Thomas, J.P. *STPA Handbook*; Massachusetts Institute of Technology: Cambridge, MA, USA, 2018.
10. Underwood, P.; Waterson, P. A critical review of the STAMP, FRAM and Accimap systemic accident analysis models. In *Advances in Human Aspects of Road and Rail Transportation*; CRC Press: Boca Raton, FL, USA, 2012; pp. 385–394.

11. Ericson, C.A., II. *Hazard Analysis Techniques for System Safety*; John Wiley and Sons, Inc.: Hoboken, NJ, USA, 2005; Chapter 3, pp. 31–54.

12. Yang, Z.; Lim, A.O.; Tan, Y. Event-based home safety problem detection under the CPS home safety architecture. In Proceedings of the 2013 IEEE 2nd Global Conference on Consumer Electronics (GCCE), Tokyo, Japan, 1–4 October 2013; pp. 491–495. [CrossRef]

13. Yang, Z.; Lim, Y.; Tan, Y. A risk model for indoor environment safety. In Proceedings of the 2017 IEEE 6th Global Conference on Consumer Electronics (GCCE), Nagoya, Japan, 24–27 October 2017; pp. 1–5. [CrossRef]

14. Harper, R. *Inside the Smart House*; Springer: Berlin/Heidelberg, Germany, 2003.

15. Lobaccaro, G.; Carlucci, S.; Löfström, E. A Review of Systems and Technologies for Smart Homes and Smart Grids. *Energies* **2016**, *9*, 1–33. [CrossRef]

16. Djenouri, D.; Laidi, R.; Djenouri, Y.; Balasingham, I. Machine Learning for Smart Building Applications: Review and Taxonomy. *ACM Comput. Surv.* **2019**, *52*, 24:1–24:36. [CrossRef]

17. Toschi, G.M.; Campos, L.B.; Cugnasca, C.E. Home automation networks: A survey. *Comput. Stand. Interfaces* **2017**, *50*, 42–54. [CrossRef]

18. Rasmussen, J. Skills, rules, and knowledge; signals, signs, and symbols, and other distinctions in human performance models. *IEEE Trans. Syst. Man Cybern.* **1983**, *SMC-13*, 257–266. [CrossRef]

19. MIL-STD-882E. *Department of Defense Standard Practice: System Safety*; Standard, Department of Defense: Washington, DC, USA, 2012.

20. Defence Standard 00-56. *Safety Management Requirements for Defence Systems, Part 1, Requirements*; Standard, Ministry of Defence: London, UK, 2007.

21. Chemishkian, S. Building smart services for smart home. In Proceedings of the 2002 IEEE 4th International Workshop on Networked Appliances (Cat. No.02EX525), Gaithersburg, MD, USA, 15–16 January 2002; pp. 215–224. [CrossRef]

22. Nakamura, M.; Tanaka, A.; Igaki, H.; Tamada, H.; Matsumoto, K. Constructing Home Network Systems and Integrated Services Using Legacy Home Appliances and Web Services. *Int. J. Web Serv. Res. (IJWSR)* **2008**, *5*, 82–98. [CrossRef]

23. Yang, Z.; Aoki, T.; Tan, Y. Multiple Conformance to Hybrid Automata for Checking Smart House Temperature Change. In Proceedings of the 2018 IEEE/ACM 22nd International Symposium on Distributed Simulation and Real Time Applications (DS-RT), Madrid, Spain, 15–17 October 2018; pp. 1–10. [CrossRef]

24. Anderson, G.B.; Bell, M.L.; Peng, R.D. Methods to Calculate the Heat Index as an Exposure Metric in Environmental Health Research. *Environ. Health Perspect.* **2013**, *121*, 1111–1119. [CrossRef]

25. ANSI/ASHRAE Standard 55-2010. *Thermal Environmental Conditions for Human Occupancy*; Standard, American Society of Heating, Refrigerating and Air-Conditioning Engineers, Inc.: Atlanta, GA, USA, 2010.

26. Woods, D.D.; Hollnagel, E.; Leveson, N. *Resilience Engineering: Concepts and Precepts*, 1st ed.; CRC Press: Boca Raton, FL, USA, 2006.

27. Jacklitsch, B.; Williams, W.J.; Musolin, K.; Coca, A.; Kim, J.H.; Turner, N. *NIOSH Criteria for a Recommended Standard: Occupational Exposure to Heat and Hot Environments*; Recommended Standard 2016-106; U.S. Department of Health and Human Services, Centers for Disease Control and Prevention, National Institute for Occupational Safety and Health: Cincinnati, OH, USA, 2016.

28. ISO 11079. *Ergonomics of the Thermal Environment—Determination and Interpretation of Cold Stress When Using Required Clothing Insulation (IREQ) and Local Cooling Effects*; Standard, International Organization for Standardization: Geneva, Switzerland, 2007.

29. van Loenhout, J.; le Grand, A.; Duijm, F.; Greven, F.; Vink, N.; Hoek, G.; Zuurbier, M. The effect of high indoor temperatures on self-perceived health of elderly persons. *Environ. Res.* **2016**, *146*, 27–34. [CrossRef] [PubMed]

30. Okamoto-Mizuno, K.; Mizuno, K. Effects of thermal environment on sleep and circadian rhythm. *J. Physiol. Anthropol.* **2012**, *31*, 14. [CrossRef] [PubMed]

31. Leon, L.R.; Bouchama, A. Heat Stroke. In *Comprehensive Physiology*; American Cancer Society: Atlanta, GA, USA, 2015; pp. 611–647, doi:10.1002/cphy.c140017. [CrossRef]

32. Gaudio, F.G.; Grissom, C.K. Cooling Methods in Heat Stroke. *J. Emerg. Med.* **2016**, *50*, 607–616. [CrossRef] [PubMed]

33. Budd, G.M. Wet-bulb globe temperature (WBGT)—Its history and its limitations. *J. Sci. Med. Sport* **2008**, *11*, 20–32. [CrossRef] [PubMed]
34. Salmon, P.; Williamson, A.; Lenné, M.; Mitsopoulos-Rubens, E.; Rudin-Brown, C. Systems-Based Accident Analysis in the Led Outdoor Activity Domain: Application and Evaluation of a Risk Management Framework. *Ergonomics* **2010**, *53*, 927–39. [CrossRef] [PubMed]
35. Svedung, I.; Rasmussen, J. Graphic representation of accident scenarios: Mapping system structure and the causation of accidents. *Saf. Sci.* **2002**, *40*, 397–417. [CrossRef]
36. Rasmussen, J. Risk management in a dynamic society: A modelling problem. *Saf. Sci.* **1997**, *27*, 183–213. [CrossRef]
37. Bjerga, T.; Aven, T.; Zio, E. Uncertainty treatment in risk analysis of complex systems: The cases of STAMP and FRAM. *Reliab. Eng. Syst. Saf.* **2016**, *156*, 203–209. [CrossRef]
38. Belmonte, F.; Schön, W.; Heurley, L.; Capel, R. Interdisciplinary safety analysis of complex socio-technological systems based on the functional resonance accident model: An application to railway traffic supervision. *Reliab. Eng. Syst. Saf.* **2011**, *96*, 237–249. [CrossRef]
39. Tian, J.; Wu, J.; Yang, Q.; Zhao, T. FRAMA: A safety assessment approach based on Functional Resonance Analysis Method. *Saf. Sci.* **2016**, *85*, 41–52. [CrossRef]
40. Patriarca, R.; Gravio, G.D.; Costantino, F. A Monte Carlo evolution of the Functional Resonance Analysis Method (FRAM) to assess performance variability in complex systems. *Saf. Sci.* **2017**, *91*, 49–60. [CrossRef]
41. Salmon, P.M.; Cornelissen, M.; Trotter, M.J. Systems-based accident analysis methods: A comparison of Accimap, HFACS, and STAMP. *Saf. Sci.* **2012**, *50*, 1158–1170. [CrossRef]
42. Filho, A.P.G.; Jun, G.T.; Waterson, P. Four studies, two methods, one accident—An examination of the reliability and validity of Accimap and STAMP for accident analysis. *Saf. Sci.* **2019**, *113*, 310–317. [CrossRef]
43. Yang, L.; Yang, S.H.; Yao, F. Safety and Security of Remote Monitoring and Control of intelligent Home Environments. In Proceedings of the 2006 IEEE International Conference on Systems, Man and Cybernetics, Taipei, Taiwan, 8–11 October 2006; Volume 2, pp. 1149–1153. [CrossRef]
44. Kim, S.C.; Jeong, Y.S.; Park, S.O. RFID-based Indoor Location Tracking to Ensure the Safety of the Elderly in Smart Home Environments. *Pers. Ubiquitous Comput.* **2013**, *17*, 1699–1707. [CrossRef]
45. Lee, K.H.; Seo, C.J. Development of user-friendly intelligent home robot focused on safety and security. In Proceedings of the ICCAS 2010, Gyeonggi-do, Korea, 27–30 October 2010; pp. 389–392. [CrossRef]
46. Yared, R.; Abdulrazak, B.; Tessier, T.; Mabilleau, P. Cooking Risk Analysis to Enhance Safety of Elderly People in Smart Kitchen. In *Proceedings of the 8th ACM International Conference on PErvasive Technologies Related to Assistive Environments*; ACM: New York, NY, USA, 2015; pp. 12:1–12:4. [CrossRef]
47. Hasan, T.; Nayan, M.F.; Iqbal, M.A.; Islam, M. Smart Solar Home System with Safety Device Low Voltage Alert. In Proceedings of the 2012 UKSim 14th International Conference on Computer Modelling and Simulation, Cambridge, UK, 28–30 March 2012; pp. 201–204. [CrossRef]
48. Doan, T.T.; Safavi-Naini, R.; Li, S.; Avizheh, S.; K., M.V.; Fong, P.W.L. Towards a Resilient Smart Home. In Proceedings of the 2018 Workshop on IoT Security and Privacy, Budapest, Hungary, 20–20 August 2018; ACM: New York, NY, USA, 2018; pp. 15–21. [CrossRef]

applied sciences

Article

A Contactless Measuring Method of Skin Temperature based on the Skin Sensitivity Index and Deep Learning

Xiaogang Cheng [1,2], Bin Yang [3,4,*], Kaige Tan [5], Erik Isaksson [5], Liren Li [6], Anders Hedman [5], Thomas Olofsson [4] and Haibo Li [1,5]

1   College of Telecommunications and Information Engineering, Nanjing University of Posts and Telecommunications, Nanjing 210003, China; chengxg@njupt.edu.cn (X.C.); haiboli@kth.se (H.L.)
2   Computer Vision Laboratory (CVL), ETH Zürich, 8092 Zürich, Switzerland
3   School of Building Services Science and Engineering, Xi'an University of Architecture and Technology, Xi'an 710055, China
4   Department of Applied Physics and Electronics, Umeå University, 90187 Umeå, Sweden; thomas.olofsson@umu.se
5   KTH Royal Institute of Technology, 10044 Stockholm, Sweden; kaiget@kth.se (K.T.); erikis@kth.se (E.I.); ahedman@kth.se (A.H.)
6   School of computer science and technology, Nanjing Tech University, Nanjing 211816, China; dtspsxzhj@njtech.edu.cn
*   Correspondence: yangbin@xauat.edu.cn

Received: 16 December 2018; Accepted: 15 March 2019; Published: 1 April 2019

Featured Application: The NISDL method proposed in this paper can be used for real time contactless measuring of human skin temperature, which reflects human body thermal comfort status and can be used for control HVAC devices.

Abstract: In human-centered intelligent building, real-time measurements of human thermal comfort play critical roles and supply feedback control signals for building heating, ventilation, and air conditioning (HVAC) systems. Due to the challenges of intra- and inter-individual differences and skin subtleness variations, there has not been any satisfactory solution for thermal comfort measurements until now. In this paper, a contactless measuring method based on a skin sensitivity index and deep learning (NISDL) was proposed to measure real-time skin temperature. A new evaluating index, named the skin sensitivity index (SSI), was defined to overcome individual differences and skin subtleness variations. To illustrate the effectiveness of SSI proposed, a two multi-layers deep learning framework (NISDL method I and II) was designed and the DenseNet201 was used for extracting features from skin images. The partly personal saturation temperature (NIPST) algorithm was use for algorithm comparisons. Another deep learning algorithm without SSI (DL) was also generated for algorithm comparisons. Finally, a total of 1.44 million image data was used for algorithm validation. The results show that 55.62% and 52.25% error values (NISDL method I, II) are scattered at (0 °C, 0.25 °C), and the same error intervals distribution of NIPST is 35.39%.

Keywords: contactless measurements; skin sensitivity index; thermal comfort; subtleness magnification; deep learning; piecewise stationary time series

---

1. Introduction

Higher economic growth drives increasing energy consumption, and 50% of housing consumption is generated by heating, ventilation and air conditioning (HVAC) systems [1,2]. Furthermore, one of the most important reasons for energy waste is that the actual thermal requirements of indoor occupants are ignored, with the result that overheating and overcooling occur often. Fortunately, real-time thermal comfort perception can provide useful signals to HVAC systems for achieving energy saving and human-centered intelligent control. Therefore, many researchers have been studying thermal comfort measurements for indoor environments in recent decades. Many methods were generated, including the questionnaire survey method [3–5], environmental measurement method [6,7] and contact measuring method of human body physiological parameters [8–18]. In recent years, the semi-contact measuring method [19,20] and contactless measuring method [21–23] for human body physiological parameters were also generated. For example, in references [19,20], an infrared sensor was fixed on the frame of eyeglasses in order to measure skin temperature. In reference [21], a normal vision sensor was also used for measuring skin temperature and two non-linear models were trained. In references [22,23], Kinect was used for recognizing human poses or indoor locations, and then human thermal comfort and dynamic metabolism were estimated, respectively. All these methods are meaningful attempts. However, due to the challenges of measuring thermal comfort which are (1) skin subtleness variation [24], (2) inter-individual differences [14,25] and (3) temporal intra-individual differences [14,26], there is still no satisfactory method for perceiving human thermal comfort.

To overcome the aforementioned challenges, the skin sensitivity index (SSI) was defined in this paper. The SSI is strongly related to skin temperature. A contactless measuring method of skin temperature based on SSI and Deep Learning was proposed, hereinafter referred to as NISDL. Two different deep learning methods of NISDL have been designed and trained, respectively, which are NISDL methods I and II. The main difference between them is that the location of SSI participation in the neural network training is not the same. A total of 1.44 million images were collected for 16 Asian female subjects, and this 'big data' was used for algorithm validation.

The main contributions of this paper are:

(1)   The skin sensitivity index (SSI) was proposed for describing individual sensitivity of thermal comfort, and the index was combined with skin images for deep learning network training.
(2)   A novel contactless measuring algorithm (NISDL) based on SSI was proposed, with two different frameworks of NISDL having been designed for real-time thermal comfort measurement.
(3)   A deep learning algorithm without SSI was also generated and trained. Two comparisons were made: (1) comparison between data-driven methods (deep learning) and model-driven methods (linear models); (2) comparison of measuring effects in the case of SSI participation in training and non-participation in training.

The rest of this paper is organized as follows. Section 2 introduces the related work about thermal comfort. In Section 3, the research methods, including SSI computation, subjective experiments and NISDL methods, are introduced. The results and discussion are shown in Sections 4 and 5. Finally, Section 6 gives the conclusion.

2. Related Work

Since the 1970s, Fanger has explored human thermal comfort and conducted many kinds of subjective experiments. Based on this, he eventually established what is known as Fanger's theory [27]. From then on, many studies about thermal comfort were carried out.

Questionnaire surveys are good as a method to understand the inner feeling of an occupant. With the development of the internet, online questionnaire surveys can also be generated [3,4]. However, it is inconvenient and also difficult to guarantee that occupants will continue to give feedback based on their personal thermal feelings [5]. Therefore, the environment measurement method was also adopted in the building industry [6]. With this kind of method, some objective parameters, such as indoor temperature, airflow and humidity, are often measured. Unfortunately, the goal of the environment measurement method is to meet the thermal comfort needs of a majority of indoor occupants. Therefore, the thermal feelings of a minority were ignored. To overcome this drawback, a kind of nonlinear autoregressive network, still belonging to the environment measurement method, was generated to predict indoor temperature [7]. In fact, human thermal comfort is complicated, and with constant indoor parameters it is difficult to meet each individual's requirements for thermal comfort. As such, some researchers study physiological measurement methods, including the contact measuring method, semi-contact measuring method and contactless measuring method.

For the contact measuring method, skin temperature and heart rate are usually the measured parameters. Wang and Nakayama [8,9] made early attempts at measuring the skin temperature around the human body. Liu [10] conducted subjective experiments and a total of 22 subjects were invited. The data, being local skin temperatures and electrocardiograms, were collected. Based on these collected data, 26 measuring methods of mean skin temperature were assessed. Takada [11] presented a multiple regression equation to predict skin temperature in non-steady state. The multiple regression equation was considered as a function of mean skin temperature. Wrist skin temperatures and upper extremity skin temperatures were also adopted to estimate human thermal sensation, respectively [12,13]. Chaudhuri [14] presented a predicted thermal state (PTS) model, and the capture of peripheral skin temperature. Furthermore, body surface area and clothing insulation were used for analyzing inter- and intra-individual differences. As for the thermal comfort study using heart rate, based on physiological experimentation, Yao [15] investigated the relationship between heart rate variation (HRV) and electroencephalograph (EGG). The results show that HRV and EEG are useful for thermal comfort studies, but further data validation is needed. Moreover, Dai, Chaudhuri and Kim [16–18] combined machine learning with contact measurement, and they are all meaningful attempts. However, as the linear kernel of SVM was used for predicting human thermal sensation in different experiments, sometimes overfitting can happen.

For the semi-contact measuring method, Ghahramani [19] used an infrared sensor to estimate skin temperature of different face points. The infrared sensor was mounted on the frame of eyeglasses. Based on this, a hidden Markov model was constructed to capture personal thermal sensation [20].

In practical application, contact and semi-contact measurement are both difficult to apply widely. The reason is that an occupant needs to wear a sensor, which is uncomfortable and is also not in line with the goal of human-oriented intelligent buildings. For this reason, a kind of contactless measuring method was studied in reference [21]. Based on vision sensors, Cheng [21] extracted the saturation (S) channel from skin images and constructed two saturation-temperature models to estimate skin temperature. The two models are the contactless measuring method of thermal comfort based on saturation-temperature (NIST) and the contactless measuring method of thermal comfort based on partly saturation-temperature (NIPST). Alan [22] proposed a contactless measuring method based on human poses. A total of 12 poses of thermal comfort were defined and Kinect was adopted to estimate human skeleton and poses. Further, Dziedzic [23] also used Kinect to predict human thermal sensation and dynamic metabolic rates.

With the development of machine learning (ML) and computer vision (CV), some thermal comfort perception methods based on ML and CV were proposed. Support Vector Machine (SVM) are often used for analyzing existing databases (RP-884) and captured environmental parameters [24,28,29]. Further, Peng [26] use unsupervised and supervised learning to predict occupants' behavior, applied to three types of offices which are single person offices, multi-person offices, and meeting rooms. The results show that the average energy savings for the entire space is 21% in experimental condition. Li [30] proposed a fuzzy model to predict thermal sensation, skin temperature and heart rate considered as objective parameters. For avoiding overheating, Cosma [31] extracted data from multiple local body parts and analyzed them with four kinds of machine learning algorithms, including SVM, Gaussian process classifier (GPC), k-neighbors classifier (KNC) and random forest classifier (RFC).

The kinds of machine learning adopted in references [24,26,28–31] are traditional algorithms. In recent years, the use of deep neural networks is on the rise [32,33]. In addition, a kind of subtleness magnification technology was presented [34,35]. These provide new directions and opportunities for the measurement of human thermal comfort. Based on this technology, a novel contactless measuring method was generated which will be introduced as follows.

3. Research Methods

3.1. Subjective Physiological Experiments

3.1.1. Subjects Data and Chamber Environments

16 human subjects were invited for experiments and the resulting data volume is 1.44 million images. The experiments were conducted in a chamber with controllable indoor air temperature and relative humidity. The corresponding dry-bulb air temperature is $22.2 \pm 0.2\,^{\circ}$C and the relative humidity is $36.9 \pm 2.5\%$. The resolution of vision sensor used for capturing video is 1280×720. The iButton, model DS192H with uncertainty $\pm 0.125\,^{\circ}$C, was used for measuring skin temperature from the back of subject's hand. All the subjects are Asian females with an average age of 23.9 ± 3.9 years, average weight of 52.2 ± 6.5 kg, and body mass index (BMI) 19.9 ± 2.2 kg/m^2.

3.1.2. Experimental Procedures

The experiment was conducted during winter in Sweden. There are generally three steps in subjective physiological experiments. (1) Preparation stage: The indoor environment parameters were measured and controlled to a suitable level. When the subjects came into the chamber, they should rest for 10 min for adaptation. At the same time, warm water with constant temperature ($45\,^{\circ}$C) was prepared. (2) Thermal stimulus: After 10 min adaptation, subjects were asked to immerse hands into the water with $45\,^{\circ}$C. The whole thermal stimulus process lasted for 10 min. (3) Big data collection: After 10 min of stimulus, subjects were asked to sit next to the data collection desk and put her pairs of hands under the vision sensor. The back of the hand is faced up and the data is collected for 50 min. At the same time, skin temperature sensor (iButton) was attached to the back of one hand. The corresponding sampling interval is 1 min. It should be noted that, based on piecewise stationary time series analysis [36], linear interpolation was adopted in this paper, and 11 points were interpolated into 1 min for real skin temperature captured by iButton.

3.2. Skin Sensitivity Index

3.2.1. SSI Definition

When human body encounters thermal stimuli, blood circulation will change which will also be reflected in skin's color and texture. In reference [21], based on the HSV (hue, saturation, value) color space, the S channel was extracted and a linear ST (saturation-temperature) model was established.

$$T = k_i \times S + b_i \tag{1}$$

where i denotes subject number, the k reacts to the change rate of skin temperature, and b denotes the intercept. S and T are skin saturation and temperature respectively. In this paper, k is defined as the skin sensitivity index (SSI). SSI is a high weight coefficient in skin temperature changes and SSI reflects the skin sensitivity level to external thermal stimuli.

3.2.2. SSI Computing

Based on subjective physiological experiments, real skin temperature can be obtained by iButton. The images were also collected from subjects' hands. Therefore, the SSI can be calculated. The steps are as follows: (1) Extracting each frame from captured video; (2) Segment region of interest (ROI); (3) Extracting S channel from ROI images and computing mean values of S for each ROI image; (4) Search SSI value based on real skin temperature and S for each subject.

3.3. NISDL Algorithm

In this paper, considering that SSI is a high weight coefficient for contactless thermal comfort measurement, it will improve the prediction accuracy of skin temperature. Based on SSI, the NISDL algorithm was introduced in this paper. Furthermore, to validate the effectiveness of SSI, two kinds of deep learning frameworks (NISDL method I, II) have been constructed. The main difference between NISDL methods I and II is that the location where the network invokes SSI to participate in the model training is different. The NISDL algorithm constructed is introduced as follows.

3.3.1. Video Pre-Processing

In fact, skin texture variation is subtle and is difficult to perceive. In this paper, for magnifying this kind of subtleness variation, an image subtleness magnification technology known as Euler Video Magnification (EVM) is adopted [34,35]. Based on EVM, let c (x, t) denote skin images which are subtly varied with time t. Suppose that the variation function is formula (2) [34,35].

$$C\,(x, t) = F\,(x + h\,(t)) \tag{2}$$

where, $h\,(t)$ is variation degree, F is a function which constructs the relationship between C (x, t) and h (t). If the skin image C (x, t) is magnified, and the first-order Taylor series expansion can be handed to $F\,(x + h\,(t))$.

$$C\,(x, t) = F(x + (1 + \zeta) * h(t)) \tag{3}$$

where ζ is the magnification coefficient which can be set based on practical application. According to Formula (3), only the variation part was magnified to a magnitude of $1 + \zeta$, while the other part of skin texture is not magnified. Therefore, the invisible texture variation is made to be visible.

It should be noted that de-noise processing should be handled before EMV processing. Further, after the video is magnified, ROI is selected and cropped from each frame. The ROI images are imported into NISDL method I and II for model training.

3.3.2. NISDL Method I

As shown in Figure 1, SSI values are used as input data and imported into the deep learning network at the very beginning. The ROI images were also combined with SSI values in the first step. According to the size of ROI images, the SSI value of each ROI image was expanded into a matrix. The matrix is considered as a channel and combined with the 3 channels of ROI images.

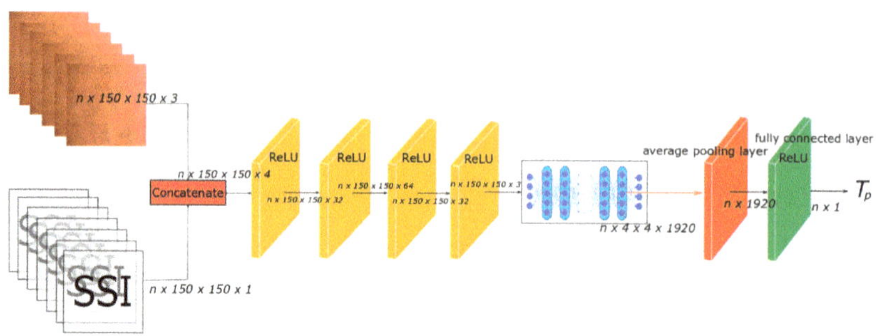

Figure 1. NISDL method I. (SSI values were combined with skin images firstly.).

The merged data between ROI images and SSI values above is inputted into four convolution layers, which are used for dimensionality reduction. In NISDL method I, the DenseNet201 [33] is adopted for features extraction. The last two layers of DenseNet201 are not suitable for skin temperature measurement, hence the two layers are removed. The reason for this is that the activation function of the last layers is softmax. Instead of these two layers, an average pooling layer and a fully connected layer are added behind denseNet201. Based on the deep learning networks designed above, n ROI images and n SSI values were inputted into NISDL method I. Therefore, n skin temperatures can be obtained.

3.3.3. NISDL Method II

Figure 2 is the deep learning framework of NISDL method II. In this method, the ROI images and SSI values were processed for features extraction. Subsequently, the two kinds of features were combined in the second half of the whole framework. An average pooling layer and the DenseNet201 (excluding the last two layers) were also used for features extraction of ROI images. For SSI values, a convolution layer and an average pooling layer were adopted for feature extraction and dimensionality reduction. After features combination, three fully connected layers were constructed in NISDL method II. Therefore, the skin temperature can be obtained. The algorithm in detail, including NISDL method I and II, is shown in Table 1.

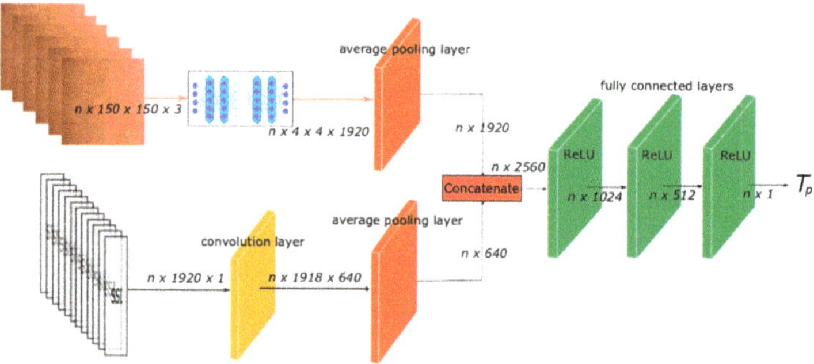

Figure 2. NISDL method II. (SSI values and skin images were processed by deep learning networks. Two kinds of features were extracted and the features were combined.).

Table 1. Contactless measuring method of skin temperature based on skin sensitivity index.

Algorithm: the NISDL algorithm

Output: NISDL model (*.h5*), skin temperature (°C)
Step:

1. Video pre-processing

    (1) De-noise and handle subtleness magnification for captured video.
    (2) Magnification coefficient ξ is 10 (formula (3)).
    (3) Extracting region of interest (ROI) from each frame of video, the size is 150×150.

2. Making label

    (1) Making numerical interpolation for skin temperatures captured by iButton.
    (2) Uniform interpolation is adopted, plus 11 points/min.
    (3) Establish a correspondence table between ROI images and skin temperatures (after interpolation).

3. Algorithm training

    (1) Commonality between NISDL method I and II

         1) Training set and test set ratio: 12: 4.
         2) Validation set: 500 images.
         3) During network training, 32 images/batch, epoch is 8.
         4) Training ~30000 images, validate once.
         5) Activation function: ReLU

    (2) NISDL method I

         1) SSI values were concatenated with ROI images in the first steps.
         2) Convolutional kernel: 1×1.

    (3) NISDL method II

         1) Features are extracted from SSI values and ROI images, respectively.
         2) Concatenating the two kind of features in the second half of network.
         3) Convolutional kernel: 3×1.

4. Optimizing model parameters

3.3.4. Evaluation Metric

For assessing NISDL algorithm constructed in this paper, the absolute error is adopted.

$$Error = |\,T_p(i) - T_r(i)\,|\ i = 1, 2, 3, \ldots \tag{4}$$

where $T_p(i)$ is the prediction values of skin temperature and obtained from the proposed NISDL algorithm. $T_r(i)$ is the real value of skin temperature and captured by iButton. The parameter i denotes the particular ROI image.

3.3.5. Algorithms for Comparison

Two algorithms are used for comparison in this paper: (1) DL algorithm. The commonality between NISDL method I and II is that they all use SSI for model training. For validating the effectiveness of NISDL algorithm (with SSI), we remove the SSI and corresponding hidden layers for SSI features extraction from NISDL method II, so that it will be another deep learning network (without SSI) and is named the DL algorithm hereinafter. (2) NIPST algorithm. DL algorithm, NISDL method I and II are all nonlinear methods and data driven methods with deep learning networks. For further validating NISDL method I and II, the NIPST algorithm is also used for algorithm comparison, which is a linear and model driven method.

4. Results

16 subjects were invited for subjective physiological experiments and a total of 1.44 million images were captured. Based on this, the NISDL algorithm was validated and compared with the NIPST algorithm and the DL algorithm.

4.1. Hardware Parameters

For this paper, a computer with a GPU was used for images processing and algorithm validation. The GPU is GeForce GTX TITAN X, the CPU is Intel core i5-4460 CPU@3.2Ghz X 4, the RAM is 16G and the word size is 64bit.

4.2. Training of NISDL method I

The size of ROI images are $n \times 150 \times 150 \times 3$, and the size of expanded SSI vlues are $n \times 150 \times 150 \times 1$. The SSI matrix was considered as a channel and concatenated with ROI images, so that the result is $n \times 150 \times 150 \times 4$. The activation function of four convolution layers, shown in Figure 1, are Rectified Linear Units (ReLU) and the size of convolution kernel is 1×1. DenseNet201 was used for feature extraction and its output is a matrix with size of $n \times 4 \times 4 \times 1920$. Based on this, two hidden layers are constructed and the size of the last layer, being a fully connected layer, is $1920 \times n$.

4.3. Training of NISDL Method II

The size of the expanded SSI matrix is $n \times 1920 \times 1$, which differs from that of NISDL method I. The corresponding convolution kernel is 3×1. The features of SSI were extracted by two hidden layers, and the features of ROI images were extracted by DenseNet201. As shown in Figure 2, in the second half of framework, the two kinds of features are concatenated. In order to ensure that the SSI features have a suitable influence on network training (moderate, not too big or too small), the size of ROI image features is set as $n \times 1920$, and size of SSI features is set as $n \times 640$ (triple relationship). Finally, the size of last three hidden layers is 2560×1024, 1024×512 and 512×1, respectively.

4.4. Commonality between NISDL Method I and II

During network training, the same parameters of NISDL method I and II are shown as follows. Based on data of 1.44 million images, the ratio of training set and test set is 12:4, the number of validation set is 500. The epoch is 8, which means that the training set was trained 8 times. The input data batch is 32. When the error of validation set is less than 0.46 °C, the corresponding model (*.h5) will be saved. Further, when 30,0000 images of training set were trained, the corresponding model (*.h5) will also be saved. After the generation of the model, the test set images were inputted into generated model, so that the prediction values of skin temperatures could be obtained.

4.5. Quantitative Comparison

The prediction values of skin temperature are shown in Figure 3. The set of values obtained from iButton is ground truth. The corresponding error statistics, including mean, median, are shown in Figure 4, which is a box-whisker plot. The mean values of NISPT, DL, NISDL method I and II in °C are 0.579, 0.359, 0.335 and 0.265, respectively. In addition, the median values of them in °C are 0.343, 0.309, 0.238 and 0.228, respectively. It was shown that deep learning methods (DL, NISDL method I and II) are all better than the nonlinear model (NIPST) and further that the method with SSI (NISDL method I and II) is better than the method without SSI (DL).

In this paper, the error distributions are given in Figure 5 and Table 2. The errors of DL, NISDL method I and II are mainly concentrated in the range of 0 °C and 0.75 °C. NISDL is better than DL, because two error percentages of NISDL corresponding to [0, 0.25) are 52.25% and 55.62%. In addition, the error percentages of NISDL corresponding to [0.25, 0.5) and [0.5, 0.75) are less than that of DL.

The error percentage of NIPST is increased from the interval of [0.75, 1), meaning that the performance of NIPST is worse than DL and NISDL methods I and II.

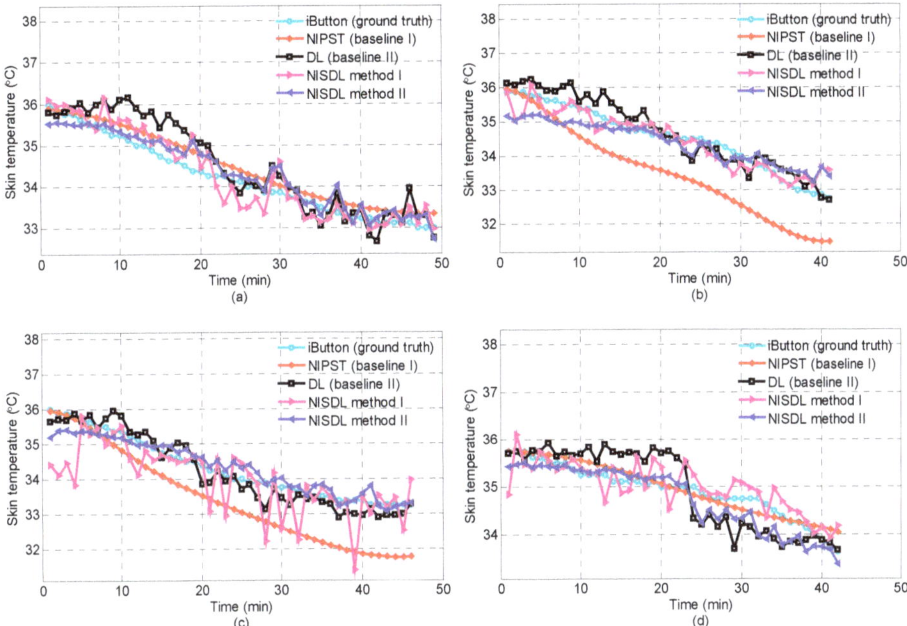

Figure 3. Skin temperatures comparison between ground truth, baseline and NISDL (Images of 12 subjects (0.96 million images) were training set; images of 4 subjects (0.32 million images) were test set. Sub-figures (**a**), (**b**), (**c**) and (**d**) are the comparison results of 4 test set (data of 4 subjects)).

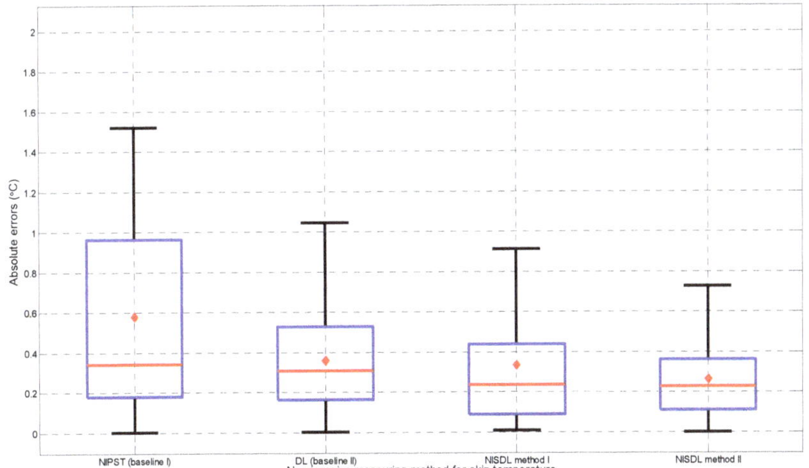

Figure 4. Error statistics (box-whisker plot) comparison between baseline and NISDL (**a**). NIPST was published in references [21]. (**b**). Skin images were trained directly by DenseNet201 to obtain a model and predict skin temperature, and SSI was not involved. (**c**). NISDL method I and NISDL method II all belong to NISDL. The main difference is that SSI values and skin images are combined at different times.

Table 2. Absolute error distribution.

Absolute Error (°C)	NIPST (Baseline, %)	DL (Baseline, %)	NISDL Method I (Figure 1, %)	NISDL Method II (Figure 2, %)
[0, 0.25)	35.39	37.64	52.25	55.62
[0.25, 0.5)	24.16	35.96	28.09	30.34
[0.5, 0.75)	5.06	19.10	10.11	11.80
[0.75, 1.0)	11.24	5.62	3.37	2.250
[>1.0)	24.16	1.69	6.18	0

5. Discussion

5.1. Situation of Overcoming Challenges

NISDL has overcome the three challenges mentioned in Section 1 to some extent. A kind of subtleness magnification technology, which is Euler Video Magnification (EVM), was used for magnifying the skin texture variation, so that the challenges '(1)' given in Section 1 can be overcome. For overcoming challenges '(2)' which is inter-individual difference, the skin sensitivity index (SSI) was proposed and SSI is related with skin saturation. Figures 4 and 5 shows that the performance of NISDL algorithm with SSI is better than that of DL without SSI. In practical application, skin images will be captured in real-time (30 frames/s), the skin temperature variation can always be obtained. Therefore, challenges '(3)' proposed in Section 1 can be overcome. Furthermore, piecewise stationary time series analysis was adopted in this paper for overcoming challenges '(3)'. Considering operability, the breakpoint interval of piecewise stationary signal is set to 5 s, supposing, e.g., that the skin temperature has a constant value during 5 s.

5.2. The Deep Learning Framework

In this paper, NISDL method I and II all belong to the deep learning method. In addition, DL generated for algorithm comparison is also a deep learning method. The error distribution comparison is shown in Figures 4 and 5 and the performance of NISDL II is encouraged. From the perspective of deep learning, the main reason for this is that big data is adopted.

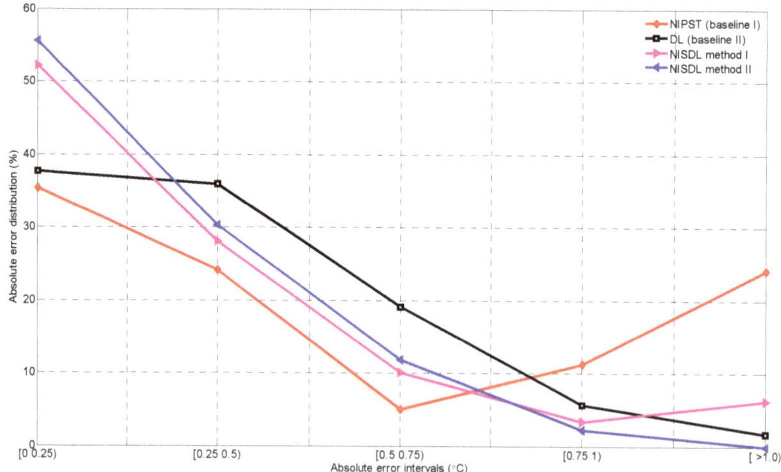

Figure 5. Error distribution comparison between baseline and NISDL (**a**). NIPST was published in [21]. (**b**). Skin images were used in training directly by DenseNet201 to obtain a model and predict skin temperature, and SSI was not involved. (**c**). NISDL method I and NISDL method II all belong to NISDL. The main difference is that SSI values and skin images are combined at different times.).

5.3. The Proposed SSI

Although NISDL and DL are all better than NIPST, there are still a big gap between NISDL and DL. Figures 4 and 5 and Table 2 also show that NISDL is better than DL. The main reason is that SSI is used in NISDL. The NISDL method II is also shown to be better than NISDL method I. Further, when SSI features are extracted and concatenated with ROI images features in the second half of network, the performance will be better.

5.4. Reasons of Designing Two Frameworks for NISDL

Some researchers may ask, why do we design NISDL methods I and II together? The main reason is that we want to extensively confirm the effectiveness of SSI. In this paper, SSI participates in network training from different locations, and the results are all good. When the SSI values are removed from network (DL), the corresponding performance decreased significantly. Based on this, we can know that SSI is helpful for predicting skin temperature through deep learning networks.

5.5. Practical Application

Some researchers may argue that the method proposed in this paper still cannot be applied in practice right now. In fact, a method is always being gradually improved. For example, the NISDL proposed in this paper is better than NIPST which was proposed in 2017. In addition, when more diverse data is captured and used for model training, the performance of NISDL will be better.

Some researchers may argue that the infrared sensor can also be used for measuring skin temperature, so why do we use a vision-based method? In fact, the study [19,20] focused on thermal comfort measurement with an infrared sensor. However, the measurement accuracy is limited. Beyond this, there are other drawbacks in the infrared based measuring method: (1) Distance. The infrared sensor should be placed close to occupant. (2) Cost. The infrared sensor with high accuracy is expensive and the accuracy of an infrared sensor with low cost is also low. (3) Information is limited. From the perspective of the human sensory system, the infrared sensor is 'touch' and vision-based method is 'sight'. The data captured by vision-based methods is much more than that collected by infrared sensor. e.g., human poses can be captured by vision sensor rather than infrared sensor for analyzing human thermal comfort. Based on these three drawbacks, the infrared-based method is difficult to widely apply in practice.

Furthermore, some other researchers may say that vison-based contactless measuring method may have concerns related to personal privacy. In fact, there are at least two options to protect personal privacy issues: (1) Switch button. Based on this switch button, any customer can choose to accept or reject the implementation of real-time personal service of thermal comfort. 92) Information selection. In future practical applications, only the information about human thermal comfort be processed and saved, while other information will be discarded in real-time. (3) Data protection. In order to avoid data protection issues, processed data related to thermal comfort can be directly transferred to the HVAC system instead of being saved. Therefore, from the perspective of human-centeredness, the NISDL algorithm proposed in this paper is helpful.

In future practical applications, the processing capacity of HVAC system should be considered. The HVAC system is usually equipped with computer server. The framework proposed in this paper can be embedded on the computer server directly and GPU is required in computer server. Based on occupant number in the building, it is necessary to prepare one or more GPUs. Further, the best ratio between GPU number and occupant number should be validated and tuned in practical application and will not be considered in this paper.

5.6. Exceptions

While human physiology and human thermal comfort remains a complex issue, the possibility to measure temperature distributions on the body's surface can provide valuable indicators. Therefore, in this paper, we just focus on the prediction of skin temperature. However, some exceptions still should be mentioned. (1) There are some exceptions to the close relationship between skin temperature and thermal comfort. e.g., while sweating occurs, as the reason of sweat evaporates and heat absorption, the skin surface temperature will drop. However, the human perception could be hot. (2) There are some special cases between thermal sensation and thermal comfort. e.g., Occupant sometimes has a warm sensation, but he or she is very comfortable.

5.7. Others

Some potential limitations should be noted. (1) In this paper, all subjects are Asian females. Therefore, maybe we only can say that the NISDL is applicable to Asian women at the present time. Further, more data validation is required. (2) The subject acclimatization time is 10 min in this paper. The result of a longer time, e.g., more than 30 min, will be better than that of 10 min. When the acclimatization time is 30 min, the subjects are more likely to reach a stable starting state. (3) Relative constant parameters were set in an experiment chamber. This means that only one thermal condition was tested. If we handle the physiological experiment in different indoor parameters (e.g., indoor temperature), more valuable data and conclusions can be obtained.

6. Conclusions

In this paper, a kind of contactless measuring method based on skin sensitivity index for thermal comfort (NISDL) is proposed. For validating the effectiveness of SSI, two different deep learning frameworks with SSI were designed. A total of 1.44 million images were used for algorithm validation. The conclusions can be summarized as follows.

(1) SSI is a good and high weight parameter in contactless measurement of skin temperature based on a deep learning network.
(2) The location of SSI participation in NISDL network training has little impact on measuring the performance of skin temperature. Of course, if the SSI features are extracted firstly, and then merged with the features of ROI images, the corresponding effect is slightly better.
(3) The NISDL method proposed in this paper can be used for measuring thermal comfort and more diverse data can help it to improve the measuring accuracy.

In practical application, the inter-difference is very large. How to define and calculate suitable SSI will affect the measuring results. Further, more diverse data comparison is required to improve the algorithm robustness. These areas will be our research directions in the near future.

Author Contributions: Conceptualization, X.C. and B.Y.; Formal analysis, X.C. and B.Y.; Methodology, X.C., B.Y., T.O. and H.L.; Software, X.C. and L.L.; Validation, B.Y., K.T. and E.I.; Writing—original draft, X.C.; Writing—review & editing, B.Y. and A.H.

Funding: This study was supported by the International Postdoctoral Fellowship Program from China Postdoctoral Council (No. 20160022), the National Natural Science Foundation of China (No. 61401236), the Jiangsu Postdoctoral Science Foundation (No. 1601039B), the Key Research Project of Jiangsu Science and Technology Department (No. BE2016001-3).

Acknowledgments: The authors thank William T. Freeman (MIT) for providing his MATLAB code of Euler Video Magnification (EVM).

Conflicts of Interest: The authors declare no conflict of interest.

References

1. U.S. Energy Information Administration. *International Energy Outlook 2017*; IEO2017 Report; U.S. Energy Information Administration (EIA): Washington, DC, USA, 2017.
2. U.S. Energy Information Administration. *Energy Implications of China's Transition toward Consumption-Led Growth*; IEO2018 Report; U.S. Energy Information Administration (EIA): Washington, DC, USA, 2018.

3.   Zagreus, L.; Huizenga, C.; Arens, E.; Lehrer, D. Listening to the occupants: A web-based indoor environmental quality survey. *Indoor Air* **2004**, *14*, 65–74. [CrossRef]

4.   Zhao, Q.; Zhao, Y.; Wang, F.; Wang, J.; Jiang, Y.; Zhang, F. A data-driven method to describe the personalized dynamic thermal comfort in ordinary office environment: From model to application. *Build. Environ.* **2014**, *72*, 309–318. [CrossRef]

5.   Ghahramani, A.; Tang, C.; Becerik-Gerber, B. An online learning approach for quantifying personalized thermal comfort via adaptive stochastic modeling. *Build. Environ.* **2015**, *92*, 86–96. [CrossRef]

6.   Liu, W.; Lian, Z.; Zhao, B. A neural network evaluation model for individual thermal comfort. *Energ. Build.* **2007**, *39*, 1115–1122. [CrossRef]

7.   Afroz, Z.; Urmee, T.; Shafiullah, G.M.; Higgins, G. Real-time prediction model for indoor temperature in a commercial building. *Appl. Energ.* **2018**, *231*, 29–53. [CrossRef]

8.   Wang, D.; Zhang, H.; Arens, E.; Huizenga, C. Observations of upper-extremity skin temperature and corresponding overall-body thermal sensations and comfort. *Build. Environ.* **2007**, *42*, 3933–3943. [CrossRef]

9.   Nakayama, K.; Suzuki, T.; Kameyama, K. Estimation of thermal sensation using human peripheral skin temperature. In Proceedings of the IEEE International Conference on Systems, Man and Cybernetics (SMC2009), San Antonio, TX, USA, 11–14 October 2009; pp. 2872–2877.

10.   Liu, W.; Lian, Z.; Deng, Q.; Liu, Y. Evaluation of calculation methods of mean skin temperature for use in thermal comfort study. *Build. Environ.* **2011**, *46*, 478–488. [CrossRef]

11.   Takada, S.; Matsumoto, S.; Matsushita, T. Prediction of whole-body thermal sensation in the non-steady state based on skin temperature. *Build. Environ.* **2013**, *68*, 123–133. [CrossRef]

12.   Sim, S.Y.; Koh, M.J.; Joo, K.M.; Noh, S.; Park, S.; Kim, Y.H.; Park, K.S. Estimation of thermal sensation based on wrist skin temperatures. *Sensors* **2016**, *16*, 420. [CrossRef] [PubMed]

13.   Wu, Z.; Li, N.; Cui, H.; Peng, J.; Chen, H.; Liu, P. Using upper extremity skin temperatures to assess thermal comfort in office buildings in Changsha, China. *Int. J. Environ. Res. Pu* **2017**, *14*, 1092–1109.

14.   Chaudhuri, T.; Zhai, D.; Soh, Y.C.; Li, H.; Xie, L. Thermal comfort prediction using normalized skin temperature in a uniform built environment. *Energ. Build.* **2018**, *159*, 426–440. [CrossRef]

15.   Yao, Y.; Lian, Z.; Liu, W.; Jiang, C.; Liu, Y.; Lu, H. Heart rate variation and electroencephalograph - the potential physiological factors for thermal comfort study. *Indoor Air* **2009**, *19*, 93–101. [CrossRef] [PubMed]

16.   Dai, C.; Zhang, H.; Arens, E.; Lian, Z. Machine learning approaches to predict thermal demands using skin temperatures: Steady-state conditions. *Build. Environ.* **2017**, *114*, 1–10. [CrossRef]

17.   Chaudhuri, T.; Soh, Y.C.; Li, H.; Xie, L. Machine learning based prediction of thermal comfort in buildings of equatorial Singapore. In Proceedings of the IEEE International Conference on Smart Grid and Smart Cities, Singapore, 23–26 July 2017; pp. 72–77.

18.   Kim, J.; Zhou, Y.; Schiavon, S.; Raftery, P.; Brager, G. Personal comfort models: Predicting individuals' thermal preference using occupant heating and cooling behavior and machine learning. *Build. Environ.* **2018**, *129*, 96–106. [CrossRef]

19.   Ghahramani, A.; Castro, G.; Becerik-Gerber, B.; Yu, X. Infrared thermography of human face for monitoring thermoregulation performance and estimating personal thermal comfort. *Build. Environ.* **2016**, *109*, 1–11. [CrossRef]

20.   Ghahramani, A.; Castro, G.; Karvigh, S.A.; Becerik-Gerber, B. Towards unsupervised learning of thermal comfort using infrared thermography. *Appl. Energ.* **2018**, *211*, 41–49. [CrossRef]

21.   Cheng, X.; Yang, B.; Olofsson, T.; Li, H.; Liu, G. A pilot study of online non-invasive measuring technology based on video magnification to determine skin temperature. *Build. Environ.* **2017**, *121*, 1–10. [CrossRef]

22.   Meier, A.; Dyer, W.; Graham, C. Using human gestures to control a building's heating and cooling System. In Proceedings of the International Conference on Energy-Efficient Domestic Appliances and Lighting (EEDAL 2017), Irvine, CA, USA, 13–15 September 2017.

23.   Dziedzic, J.W.; Yan, D.; Novakovic, V. Real time measurement of dynamic metabolic factor (D-MET). In Proceedings of the 9th International Cold Climate Conference Sustainable new and renovated buildings in cold climates (Cold Climate HVAC 2018), Kiruna, Sweden, 12–15 March 2018.

24.   Peng, B.; Hsieh, S. Data-driven thermal comfort prediction with support vector machine. In Proceedings of the ASME 2017 12th International Manufacturing Science and Engineering Conference (MSEC), Los Angeles, CA, USA, 4–8 June 2017; pp. 1–8.

25. Wang, Z.; de Dear, R.; Luo, M.; Lin, B.; He, Y.; Ghahramani, A.; Zhu, Y. Individual difference in thermal comfort: A literature review. *Build. Environ.* **2018**, *138*, 181–193. [CrossRef]
26. Peng, Y.; Rysanek, A.; Nagy, Z.; Schlüter, A. Using machine learning techniques for occupancy-prediction-based cooling control in office buildings. *Appl. Energ.* **2018**, *211*, 1343–1358. [CrossRef]
27. Fanger, P.O. *Thermal Comfort: Analysis and Application in Environmental Engineering*; Danish Technical Press: Copenhagen, Denmark, 1970.
28. Farhan, A.A.; Pattipati, K.; Wang, B.; Luh, P. Predicting individual thermal comfort using machine learning algorithms. In Proceedings of the IEEE International Conference on Automation Science and Engineering (CASE), Gothenburg, Sweden, 24–28 August 2015; pp. 708–713.
29. Megri, A.; Naqa, I. Prediction of the thermal comfort indices using improved support vector machine classifiers and nonlinear kernel functions. *Indoor Built Environ.* **2016**, *25*, 6–16. [CrossRef]
30. Li, W.; Zhang, J.; Zhao, T.; Wang, J.; Liang, R. Experimental study of human thermal sensation estimation model in built environment based on the Takagi-Sugeno fuzzy model. *Build. Simul.* **2018**. [CrossRef]
31. Cosma, A.C.; Simha, R. Machine learning method for real-time non-invasive prediction of individual thermal preference in transient conditions. *Build. Environ.* **2019**, *148*, 372–383. [CrossRef]
32. LeCun, Y.; Bengio, Y.; Hinton, G. Deep learning. *Nature* **2015**, *521*, 436–444. [CrossRef] [PubMed]
33. Huang, G.; Liu, Z.; Maaten, L.V.D. Densely connected convolutional networks. In Proceedings of the IEEE Conference on Computer Vision and Pattern Recognition (CVPR), Honolulu, HI, USA, 21–26 July 2017.
34. Wu, H.; Rubinstein, M.; Shih, E.; Guttag, J.; Durand, F.; Freeman, W. Eulerian video magnification for revealing subtle changes in the world. *Acm T Graph.* **2012**, *31*, 1–8. [CrossRef]
35. Wadhwa, N.; Wu, H.; Davis, A.; Rubinstein, M.; Shih, E.; Mysore, G.; Chen, J.; Buyukozturk, O.; Guttag, J.; Freeman, W.; et al. Eulerian video magnification and analysis. *Commun. Acm* **2017**, *60*, 87–95. [CrossRef]
36. Cheng, X.; Li, B.; Chen, Q. Online structural breaks estimation for non-stationary time series models. *China Commun.* **2011**, *7*, 95–104.

Article

Keeping Doors Closed as One Reason for Fatigue in Teenagers—A Case Study

Anna Mainka * and Elwira Zajusz-Zubek

Department of Air Protection, Silesian University of Technology, 22B Konarskiego St., 44-100 Gliwice, Poland
* Correspondence: Anna.Mainka@polsl.pl; Tel.: +48-322-371-060

Received: 9 July 2019; Accepted: 22 August 2019; Published: 28 August 2019

Featured Application: Presented results are only a case study and the authors are not able to relate them to the population of children and teenagers in Poland; however, it is worth underlining the positive effects of opening doors during sleep, since the CO_2 concentration decreases 55–64% without a reduction in thermal comfort. This simple action applied by parents can decrease the contribution of low indoor environment quality (IEQ_{IV}) in children's and teenager's bedrooms to approximately 1% during the night.

Abstract: (1) Background: Healthy teenagers are often sleepy. This can be explained by their physiology and behavioral changes; however, the influence of CO_2 concentration above 1000 ppm should not be neglected with respect to sleep dissatisfaction. (2) Methods: CO_2 concentrations were measured in two similar bedrooms occupied by girls aged 9 and 13 years old. The scheme of measurements included random opening and closing of the bedroom doors for the night. Additionally, the girls evaluated their sleep satisfaction in a post-sleep questionnaire. (3) Results: During the night, the CO_2 concentration varied from 402 to 3320 ppm in the teenager's bedroom and from 458 to 2176 ppm in the child's bedroom. When the bedroom doors were open, inadequate indoor air quality (IEQ_{III} and IEQ_{IV} categories) was observed in both the teenager's and child's bedroom during 11% and 25% of the night, respectively; however, closing the doors increased the contribution of moderate (IEQ_{III}) and low (IEQ_{IV}) categories of air to 79% and 86%, respectively. The girls were dissatisfied only when the bedroom door was closed. The satisfied category of sleep was selected only by the younger girl. (4) Conclusions: Opening the bedroom door during the night can decrease the CO_2 concentration 55–64% without reducing thermal comfort.

Keywords: teenagers; children; bedroom; IEQ; CO_2

1. Introduction

There is a generally accepted opinion in society that teenagers are sleepy. This notion indicates that sleepiness is a negative issue; however, this particular group of young people has several reasons for being tired. From a physiological perspective, teenagers undergo the processes of puberty and rapid increases in height [1], and, from a psychological aspect, they have problems with low self-esteem, seeking their role in society, and getting easily involved in personal conflicts at school or within the family [1,2]. The duration of sleep varies among individuals, with an average teenager attending school requiring at least 8 h of sleep per night; however, the maturational changes combined with a cell phone or computer use within the hour prior to trying to fall asleep lead to shorter sleeping hours or sleep deprivation [3]. Sleep issues have drawn worldwide attention in recent years. Sleep problems threaten health and quality of life for up to 45% of the world's population, and 35% of individuals feel that they do not get enough sleep, which has negative effects on both physical and mental health [4]. Rafihi-Ferreira et al. [5] referenced several studies, in which the association between sleep quality and behavioral problems was investigated, demonstrated a relationship between externalizing and internalizing problems and sleep problems in children. This association becomes even more serious when considering that childhood sleep problems may persist and that they constitute a greater risk for other behavioral problems. In comparison with psychological state, body condition, and circadian rhythm, indoor environment quality (IEQ) including thermal conditions and indoor air quality (IAQ) can improve sleep through environmental control. Matricciani et al. [6] in the meta-review underlined that sleep duration is associated with adiposity in children, while sleep quality, timing and variability appear important for children's health, but further research is needed. Children and teenagers are more adversely affected by indoor air pollution than adults since they breathe a greater volume of air relative to their body weight, which may lead to a greater burden of pollutants on their bodies [7,8]. The quality of the indoor environment not only affects health and comfort, but it may also impair learning ability. Poor IEQ, as a result of inadequate thermal conditions, and IAQ can cause symptoms such as being too hot or too cold as well as feeling restless or sleepy [7].

A powerful remedial to improve IAQ is to increase the outdoor air supply rate; however, in recent years, effort has been focussed on decreasing energy consumption. Within houses, the building airtightness is maximized and the ventilation rates are minimized, leading to a general deterioration in indoor air quality. Consequently, by focussing on optimal indoor thermal comfort, air infiltration and dilution of indoor air pollutants are lowered.

Generally, in passive stack ventilation systems, to create optimal air quality conditions for the maintenance of health and comfort of the occupants, windows should be open slightly. In moderate climate zones such as Poland, this is the best solution from spring to fall seasons; however, during winter, unsealed windows decrease thermal comfort. The Upper Silesia region, in comparison with other Polish regions and European Union countries, is characterized by relatively high particulate matter concentrations. Although the last three decades of economic changes have forced the greatest drop in Polish industrial air pollution by closure of the old steelworks, cookeries, coal mines, and coking plants, emissions from small-scale combustion utilities, such as domestic boilers, together with re-suspension processes from urban surfaces and road traffic, have become particularly dangerous. Investigations performed during the winter season confirm that the hazard of domestic sources originates from the low quality of fuels (coal, biomass, culm, or even refuse) used for heating [9,10]. According to the dominant role of coal combustion in the region, residents are even advised not to open windows during winter, particularly at low wind speed, since it decreases IAQ.

IAQ is a multi-disciplinary phenomenon and is determined by many pathways in which chemical, biological, and physical contaminants eventually become a portion of the total indoor environmental composition. There is spatial and temporal heterogeneity of these contaminants, and the determination of exposure is difficult due to the diversity of time that occupants spend within the space [11,12]. Among indoor air pollutants, carbon dioxide (CO_2) is considered a useful and easily measurable indicator of the ventilation and air quality in indoor environments. Although this pollutant itself does not cause serious health issues at lower concentrations and, for short durations, higher concentrations can indicate a lower ventilation level and possible air contamination with other pollutants [13].

In residential non-smoking areas, with the exception of kitchens, occupants are a major source of CO_2 through exhalation. Mean concentrations of CO_2 during cooking times are significantly higher ($p < 0.01$) than those during non-cooking times, with no significant differences among fuel types [14]. While the CO_2 generation rate per person varies as a function of age, activity, and diet, on average, children aged 6–11 years old produce 448 mg/min/p and adults generate 763 mg/min/p [15]. Persily and de Jonge in [16] presented CO_2 generation rates according to age and gender. During sleep, the level of physical activity (MET) is 1.0; thus, the calculated metabolic rate in children aged 6–11 years old is 295 mg/min/p and 271 mg/min/p for boys and girls, respectively, while, for 11–16-year-old males and females, it is 401 mg/min/p and 342 mg/min/p, respectively. Teenagers generate more CO_2 according to their higher body weight, and they demand more privacy, therefore keeping their door closed all day and night. The metabolic activity itself influences air quality in bedrooms by reducing the concentration of oxygen and increasing the level of CO_2.

For CO_2, the exposure limits have been derived exclusively on the basis of health considerations. The Exposure Guidelines for Residential Indoor Air Quality [17] suggest that CO_2 concentrations above 1800 mg/m^3 (1000 ppm) are indicative of an inadequate supply of fresh air, although complaints have been documented at concentrations as low as 1100 mg/m^3 (600 ppm) [17]. Assumptions regarding these levels are in accordance with ASHRAE standards, which previously (62-1989) considered 1000 ppm as the highest acceptable concentration for a minimum sanitary requirement and 8 l/p/s as the minimum ventilation rate. More recent standards (ASHRAE 62-1999, 62-2001, and 62-2004) recommend that the indoor-outdoor differential concentration should not exceed 700 ppm [15]. Generally, a sufficient margin to protect against undesirable changes in the acid-base balance and subsequent adaptive changes such as the release of calcium from bones is a level of 6300 mg/m^3 (3500 ppm) [17]. However, a higher maximum exposure concentration is recommended for direct physiological effects of exposure to CO_2 as opposed to subjective symptoms. Subjective symptoms such as fatigue, headaches, and an increased perception of warmth and unpleasant odors have been associated with CO_2 levels of 900–5800 mg/m^3 (500–3200 ppm) [17].

Polish legal acts do not specify permissible concentrations of CO_2 in ambient air or in rooms intended for the permanent residence of individuals, i.e., apartments and houses. The regulations of the Ministry of Family, Work, and Social Policy define the highest permissible concentrations and intensities of agents harmful to health in the work environment [18]. The highest permissible CO_2 concentration is 9000 mg/m^3. Thus far, the Polish Committee for Standardisation followed general guidelines concerning the quality of air inside non-residential buildings (PN-EN 13779 standard [19]), but, in May 2019, the Polish Committee for Standardisation accepted EN-16798-1 (in English) [20] as a national standard and in CEN/TR 16798-2 Technical Report [21] including recommended criteria for the CO_2 calculation for demand-controlled ventilation in occupied living rooms and bedrooms. Design ΔCO_2 concentration for bedrooms (ppm above outdoors) are within the limits of $\leq$380 ppm (IEQ$_I$), 380–550 ppm (IEQ$_{II}$), 550–950 ppm (IEQ$_{III}$), and $\geq$950 ppm (IEQ$_{IV}$) corresponding to high (I), medium (II), moderate (III), and low (IV) indoor environment quality, respectively.

IEQ evaluation based on CO_2 concentration and possible effects on occupants comfort has been examined [11,22–27]. Polish research on CO_2 levels and possible occupant fatigue includes a few publications [9,28–31]; however, no research regarding IEQ in the bedrooms of Polish children and teenagers have been performed. Such studies can help to better define how behavioral patterns influence the possible exposure levels of air pollutants and occupant fatigue. The present paper presents the results of the measurement of CO_2 levels and the corresponding IAQ linked to the possible tiredness and lethargy of a teenager as compared with a child.

2. Materials and Methods

The concentration of CO_2 was continuously measured in two bedrooms: one occupied by a teenage girl aged 13 and a second by a younger girl aged 9. The choice of girls instead of boys was intentional since, among children and teenagers, more males than females exhibit excessive nightly use of computer games and consequent sleep disturbances [1]. Moreover, Karjalainen [32] suggested that female occupants should primarily be used as subjects when examining indoor thermal comfort requirements, since, if women are satisfied, then it is highly probable that men are also satisfied. Participation in the present study was voluntary. Prior to starting the measurements, the participants' parents gave informed consent.

The home of the girls was located in the suburbs of the western part of the Upper Silesia region. The distance from the urban area was approximately 10 km (Figure 1). The house was built in 2008 and was located in the third row of buildings, 100 m from the local street. It was a detached house (Figure 1) with an attic and no basement. The house area was 98 m^2 with a glass terrace of 12 m^2. The measurements were performed in the South and North bedrooms, which each had an area of approx.

12 m^2 and were occupied by the older and younger girl, respectively; detailed information has been added in Supplementary Materials Table S1.

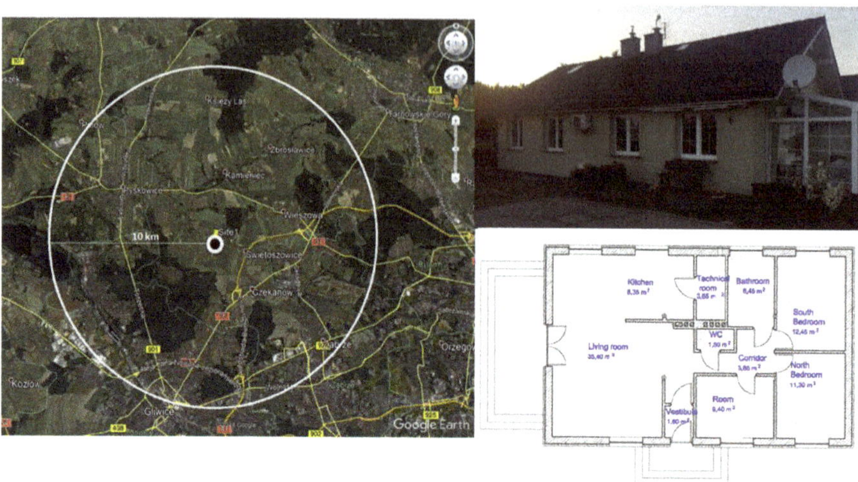

Figure 1. Location, south view and ground floor plan of the house.

The sampling strategy for CO_2 included indoor CO_2 concentration, measurement, and questionnaires. Precautions necessary to avoid measurements in air directly exhaled by building occupants were performed as described in [33]. One automatic portable monitor (model 77535, Az Instruments International Ltd., Hong Kong, China) connected to a PC with RS232 software installed was used in the present study. According to Mahyuddin et al. [34], one CO_2 sensor in a room with <100 m^2 floor area had significant *p*-value relationships, and the breathing zone within the occupied space was considered to be between 1.0 and 1.2 m in the middle of a zone as a representative location. The monitor was equipped with a non-dispersive infrared sensor. The precision of measurements ranged between +0 and +10,000 ppm CO_2: ±100 ppm CO_2 or ±3% at a concentration below 100 ppm. The monitor also displayed and recorded in real-time the measurements of air temperature and relative humidity in the bedroom, allowing for logged data to be downloaded for analysis. The selected sampling interval was 60 s.

To estimate the parameters influencing CO_2 concentrations in the bedrooms during the measurements, parents were asked to note whether each daughter left the room for a longer period during the night. They provided a diary in which it was specified whether the windows and doors were closed, ajar, or fully open during the monitored days and nights. Additionally, the girls were asked to evaluate their sleep satisfaction as categorized into five levels [4]: very dissatisfied (1), dissatisfied (2), moderate (3), satisfied (4), and very satisfied (5). However, for the analysis, three levels were used: very dissatisfied and dissatisfied, defined as dissatisfied; moderately satisfied, defined as moderate; and satisfied and very satisfied, defined as satisfied. The evaluations of sleep quality were included in the post-sleep questionnaire. In total, 137 questionnaires were acquired, with 121 being valid.

During the study, the family was asked to maintain their regular routine regarding the opening of doors and windows. It should be underlined that, during daylight hours, the bedroom windows were fully open only twice for cleaning and ajar for a maximum of 1 h less than 10 times. The measurements were performed between 16 September 2018 and 2 March 2019. For the final analysis, 102 nights from 9:00 p.m. to 7:00 a.m. (51 nights per room) were included.

3. Results

As mentioned above, proper ventilation controls IAQ. Among the techniques adopted to evaluate ventilation rates, the one based on the measurement and analysis of the indoor CO_2 concentration and trends is the most common approach; however, it could be improved by the integration of an electronic nose for odor detection [35]. Acceptable ventilation conditions can be easily achieved in mechanically ventilated rooms, but it is not equally simple to maintain CO_2 and odor levels under control in indoor environments that are naturally ventilated [35]. Natural ventilation is typical for detached buildings in Poland, where the ventilation and the IAQ are controlled only by means of air infiltration through cracks and openings. The IAQ gets worse during the winter season when the desire for thermal comfort and acceptable IAQ are in conflict. For the characterization of IAQ by measuring CO_2 concentration, several indicators and criteria can be used [25]. We selected the average CO_2 concentration and a time fraction over a limit of ΔCO_2 values spent during the night according to the PN-EN 16798 standards [20,21].

Figure 2 shows the field measurements in the teenager's and child's bedrooms, including the CO_2 concentration range, when the doors were open and closed.

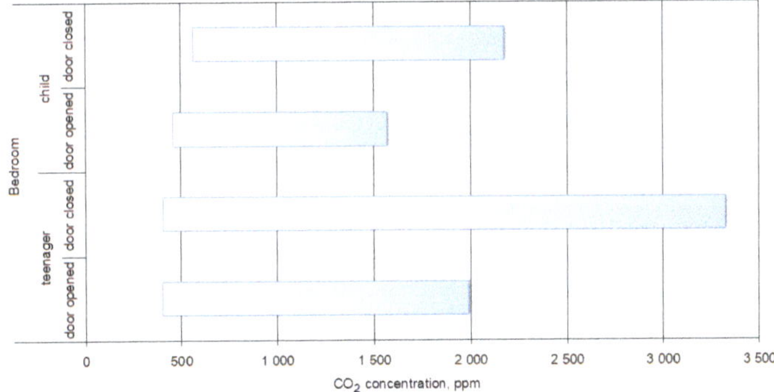

Figure 2. CO_2 concentration inside the teenager's and child's bedrooms during the night.

During the night, the short-term (60 s average) concentrations varied from 402 to 3320 ppm in the teenager's bedroom and from 458 to 2176 ppm in the child's bedroom. Thus, the highest concentrations exceeded the limits characteristic of subjective symptoms such as fatigue, headaches, and an increased perception of warmth and unpleasant odors [14]. If we compare the data separately by day of the week (Table 1), it is clear that, despite similar average concentrations, the maximum concentrations were significantly higher in the teenager's bedroom.

Table 1. Average, minimum, and maximum CO_2 concentrations according to the day of the week.

Parameters	CO₂ Concentrations, ppm					
	Teenager's Bedroom			Child's Bedroom		
	Average	Minimum	Maximum	Average	Minimum	Maximum
Monday	1127	402	3102	1828	1828	1828
Tuesday	998	597	2336	1114	554	1855
Wednesday	1088	452	2897	1144	458	1855
Thursday	1199	411	2695	1118	688	1778
Friday	1195	479	3320	1076	563	1757
Saturday	3061	3061	3061	1262	599	2176
Sunday	1191	405	3188	1993	1993	1993

Distinguishing between open and closed doors in the teenager's and child's bedrooms, the average nightly concentrations were not significantly different (p = 0.76; 1077 and 1103 ppm, respectively); however, there was a statistically significant difference in CO_2 concentration between bedrooms with open and closed doors (Figure 3). In the teenager's bedroom, opening the doors during the night decreased the average CO_2 concentration by 55%, and in the child's bedroom, the observed decrease was 64%.

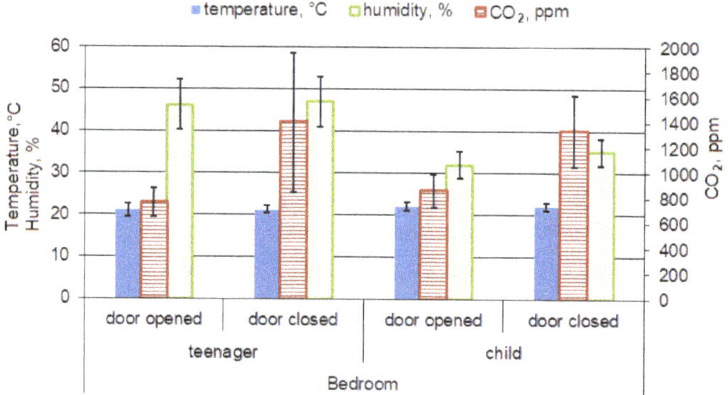

Figure 3. Mean temperature, relative air humidity, and CO_2 concentration inside the teenager's and child's bedrooms during the night.

Moreover, room temperature and humidity are important factors that may influence perceived environment quality. They both affect the thermal balance of the human body through effects on the skin and respiratory organs. Changes in air temperature trigger a sympathetic reflex via the skin that strengthens with lower air temperature. High temperatures and humidity require the human body to respond by increasing heat loss through the skin surface via blood circulation. The relation between indoor temperature and humidity depends on the season and indoor conditions [36]. In our study in the younger girl bedroom, the temperature and humidity correlation coefficient were −0.10 and −0.08 (p < 0.05) with opened or closed doors, respectively. In the older girl bedroom, the temperature and humidity were weakly positively correlated in the case of opened doors 0.26 and negatively correlated −0.33 in the case of closed doors. The mean room temperature differed between the teenager's and child's bedrooms significantly (p < 0.01). The mean temperature in the teenager's bedroom (21.1 ± 1.11 °C) was lower than that in the child's bedroom (22.05 ± 0.59 °C); however, the influence of opening and closing the doors to each bedroom during the night was not statistically significant (p > 0.05). The mean nightly temperature in the teenager's bedroom was 21.07 ± 1.39 °C and 21.01 ± 0.76 °C with open and closed doors, respectively, while, in the child's bedroom, the mean nightly temperature was 22.06 ± 0.62 °C and 22.03 ± 0.57 °C, respectively. In the case of humidity, a significant difference (p < 0.01) between the teenager's and child's bedrooms was also observed (46.46 ± 5.79% and 33.39 ± 3.41%, respectively). In the teenager's bedroom, the humidity did not differ between nights with respect to open and closed doors (46.04 ± 5.97% and 46.89 ± 5.68%, respectively), while, in child's bedroom, this difference was statistically significant (p < 0.01; 31.83 ± 3.20% and 34.88 ± 2.95%, respectively).

Based on WHO regulations [37] and the PN-EN 16798-1 and CEN/TR 16798-2 standards [20,21], the increase in CO_2 concentration in relation to that of outdoor air (ΔCO_2) was measured in both bedrooms. During the night, the general IEQ in the bedrooms included 32% in the high category (I), 28% in the medium category (II), 19% in the moderate category (III), and 21% in the low category (IV) in the older girl's bedroom and 14%, 29%, 32%, and 24%, respectively, in the younger girl's bedroom. This indicates that, for 10 h of rest at night, the teenager spent approximately 3 h in proper air quality, while the child spent only approximately 1.5 h. The indoor concentrations of CO_2 showed a higher contribution of inadequate air quality in the III and IV categories, corresponding to moderate and low IEQ, in the younger girl's bedroom. When the bedroom doors were open, inadequate IEQ was observed in the teenager's and child's bedrooms for 11% and 25% of the night, respectively (Figure 4); however, closing the doors increased the contribution of the moderate (III) and low (IV) categories to 79% and 86%, respectively.

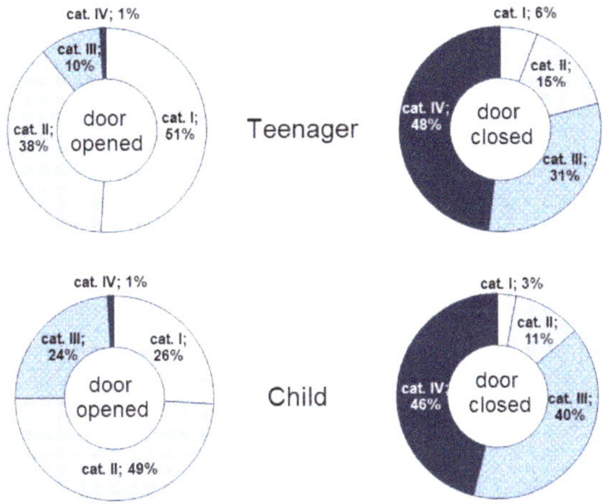

Figure 4. Categories of the indoor environment inside the teenager's and child's bedrooms during the night.

Despite the fact that the teenage girl spent an extra 1.5 h in the high and medium quality environment (IEQ_I and IEQ_{II}), this did not correspond to sleep satisfaction. In the older girl's opinion, her sleep during study was never satisfying. During 51 nights of measurement, the teenage girl evaluated her sleep as moderate (3) for 36 nights, 13 nights as dissatisfied (2), and only two nights as very dissatisfied (1). The younger girl made a more positive assessment, evaluating seven nights as satisfied (4) and one night as very satisfied (5), eight nights categorized as dissatisfied (2), while one night as very dissatisfied (1), and 34 nights as moderate (3). Following the small contribution of very dissatisfied (1) and very satisfied (5) categories, we gathered categories (1) and (2) into one group defined as low sleep satisfaction, in addition to categories (4) and (5) in one group defined as high sleep satisfaction, as presented in Figure 5.

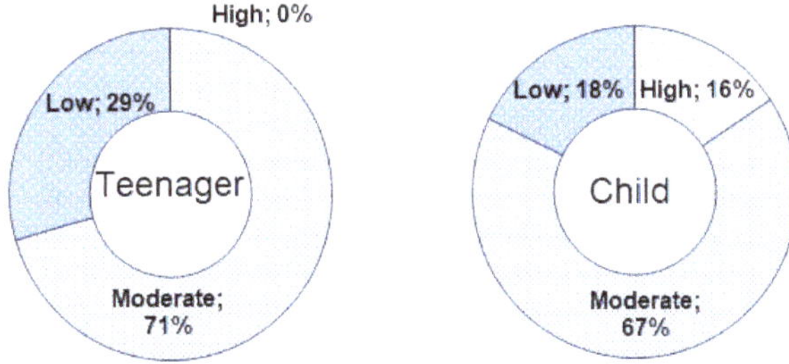

Figure 5. Low, moderate, and high sleep satisfaction of the teenager and child corresponding to categories: Very dissatisfied and dissatisfied (1) + (2); moderate (3), as well as satisfied and very satisfied (4) + (5).

Deeper analysis reveals that the girls were dissatisfied only when the bedroom door was closed. It is interesting that the average CO_2 concentration during the nights evaluated as dissatisfied had a similar average value of 1691 ppm and 1623 ppm in the older and younger girl's bedroom, respectively. Moreover, the average concentration ranges were not statistically different (1070 ppm and 1079 ppm to 2308 ppm and 1878 ppm) in the teenager's and child's bedroom, respectively. The moderate category was selected during the nights with open as well as closed doors. The average concentration of CO_2 during nights defined as moderate with open doors was 766 ppm and 903 ppm in the teenager's and the child's bedroom, respectively, while, during nights with closed doors, the CO_2 concentration was 968 ppm and 1185 ppm, respectively. The satisfied category of sleep was selected only by the younger girl during nights with open doors, corresponding to an average CO_2 concentration of 769 ppm.

4. Discussion

To ensure that the CO_2 concentration is at a lower level, it is necessary to supply an adequate amount of fresh air instead of used air. Bekö et al. [38] summarised that the most influential variables on the air change rate (ACR) are room volume, number of individuals sleeping in the bedroom, average window and door opening habits, location of the measured room (ground or higher floor), and year of building construction. The authors underlined that more door opening results in a higher air change rate. Our study was performed in two similar bedrooms occupied by girls aged 13 and 9 years old. Due to a low IAQ outside the building, the window opening was very limited; thus, we focussed on door opening and compared it with subjective evaluation of sleep satisfaction. Research by Bekö et al. [38] revealed that, when the bedroom doors are ajar or fully open instead of closed, the ACR increases from 0.48 to 0.55 (14%) and 0.71 (48%). Although we did not calculate the ACH, the observed decrease in CO_2 concentration while keeping the bedroom doors open during the night was 55% and 64% in the teenager's and the child's bedroom, respectively.

Generally, adults appear to be additional immune to the consequences of sleep deprivation, whereas children, teenagers and particularly young children tend to be a vulnerable subpopulation as they spend the majority of their time indoors at home while their respiratory and other systems are under development [8]. When an individual does not get enough sleep to feel awake and alert, they begin to experience symptoms of sleep deprivation such as yawning, irritability, fatigue, inability to concentrate, moodiness, forgetfulness, lack of motivation, depression, and poor perspective on life [3]. In our study, the average concentration during the night was similar in both bedrooms, at 1077 ppm (teenager) and 1103 ppm (child), which is comparable with the average CO_2 concentration in unoccupied bedrooms (999 ppm and 1236 ppm in South and North bedrooms, respectively) as reported by Bouvier et al. [25]. In occupied bedrooms, the authors in [25] reported higher (1585 ppm and 1760 ppm in South and North bedrooms, respectively) than our average concentration of CO_2, which was 766 ppm and 860 ppm with open doors and 1402 ppm and 1337 ppm with closed doors, in the teenager's and the child's bedroom, respectively (Figure 3). However, it is similar to the average CO_2 concentration of 716 ppm measured in 13 bedrooms located in Athens' residences [39].

During the night with the doors closed, in the teenager's and the child's bedroom, 48% and 43% of CO_2 concentration measurements were higher than 950 ppm above outdoor CO_2 concentration (398 ppm), which corresponds to category IV–low indoor environment quality (IEQ_{IV}), and, when open, it was < 1% in both bedrooms. The PN-EN 16798-1 standard [20] underlines that a lower level will not provide any health risk but may decrease comfort. At the same time, the categories are related to the levels of expectations. A normal level would be medium, but, for occupants with special needs (children, elderly, persons with disabilities, etc.), a higher level might be selected. During the research with opened doors, a high IEQ prevailed during 51% of the night in the teenager's bedroom, and during 26% of the night in the younger girl's bedroom. In comparison, the authors in [25] reported that time spent by an occupant in a CO_2 concentration over 1500 ppm in a mechanically ventilated detached house was approximately 30% of an entire day; thus, the IEQ in researched bedrooms could be satisfying with the door open; however, only the younger girl evaluated sleep satisfaction as high

(16%). The subjective assessment of sleep satisfaction points to a lower IEQ in the case of the older girl's bedroom, which can be supported by a greater CO_2 concentration range (Figure 2) in the teenager's bedroom. Nevertheless, the sharing of categories of environment calculated based on the PN-EN 16798-1 standard [20] does not explain the lower sleep satisfaction in the older girl's assessment. The reason can be seen in Table 1, which presents average CO_2 concentrations and ranges depending on the day of the week. In the teenager's bedroom, the maximum CO_2 concentrations were >3000 ppm on four days of the week (Mondays, Fridays, Saturdays, and Sundays), while, in the younger girl's bedroom, the maximum CO_2 concentration exceeded 2000 ppm only on Saturday nights. These results indicate the important role of short-term high concentrations in sleep satisfaction assessment.

Increasing the effectiveness of the building ventilation system is to decrease the number of exceedances of CO_2 concentration in naturally ventilated buildings. A cheap and simple solution is the application of ventilation grills to the window frames [40]. We would recommend simple passive grilles, where the airflow is due to the pressure drop between the indoor and the outdoor environment in addition to the typical pressure drop of the grille itself. Humidity-sensitive grills might not be very effective in this case because of lower humidity values (29.7–50.4%) in the bedrooms during the heating season (Supplementary Materials Table S1).

5. Conclusions

We investigated the variability of CO_2 concentration in naturally ventilated bedrooms occupied by a teenage girl and a female child according to their sleep satisfaction. Despite the fact that the average CO_2 concentration was 1077 ppm in the teenager's bedroom and 1103 ppm in the child's bedroom, the concentration ranges were much wider, from 402 to 3320 ppm and 458 to 2176 ppm, respectively. The average concentration during the night with the doors closed was 1402 ppm and 1336 ppm, respectively; however, the simple action of door opening decreased the CO_2 concentration by 55% and 64% in the older and younger girl's bedroom, respectively.

In the studied period of heating season (from 16 September 2018 to 2 March 2019) during the night (9:00 p.m. to 7:00 a.m.), the time spent by the teenage and younger girls at a concentration over 1348 ppm (IV—low IEQ category) was generally 21% and 24%, respectively. With the doors open, the low IEQ contributed to <1% of the time in both bedrooms; however, keeping the doors closed increased the contribution of low IEQ to 48% and 46% in the teenager's and the girl's bedroom, respectively. This highlights the strong influence of simple door opening during children's sleep.

The evaluation of sleep satisfaction highlights that the teenage girl was more dissatisfied in the mornings than was the younger girl, indicating the important role of short-term CO_2 concentrations >3000 ppm in sleep satisfaction assessment.

There are several limitations to this work. This study was limited by small sample size and was confined to one house. The suburban site where the house was located may be different from the general population of children and teenagers in Poland with respect to socioeconomic status, education, and other risk factors for adverse health outcomes, all of which may affect the level of IEQ of the home. The relationships presented here cannot be directly linked to other groups of children, and may not be applicable to other regions due to community differences and the air quality in the area. In Poland, during winter, suburban and rural areas according to individual coal heating systems tend to be more polluted than urban areas connected to collective heating systems, which influences the air quality in the area. Finally, these results apply to residential indoor exposure and may not apply to indoor exposure experienced in settings such as multi-family buildings. Of two bedroom IEQ measures during nights, we found that the parent's approval of children's and teenager's privacy by keeping the doors closed increased airtightness in the bedrooms, favoring low ventilation rates and poor indoor environment quality (IEQ). Caution should be taken regarding the fact that the measurement results may not be able to be interpreted as a human exposure level under regularly ventilated conditions since the data were obtained after the windows were closed. However, it could be useful for predicting exposure of CO_2 in children and teenagers during the night, since it is not only in Poland where occupants tend to close the windows and doors while sleeping. Examination of these relationships in other buildings, over a longer period of time, as well as involving wilder groups of participants, could help to explain what parameters affect indoor environment and sleep quality, which has effects on both physical and mental health.

Supplementary Materials: The following are available online at http://www.mdpi.com/2076-3417/9/17/3533/s1, Table S1: General characteristic of studied bedrooms as well as the distribution of nightly average indoor temperature and humidity in each studied bedroom depending on door opening.

Author Contributions: Conceptualization, A.M.; Methodology, A.M.; Investigation, A.M.; Data curation, A.M.; Writing—original draft preparation, A.M.; Writing—review and editing, E.Z.-Z.; Visualization, A.M.; Project administration, E.Z.-Z.

Funding: This research was funded by the Faculty of Energy and Environmental Engineering, Silesian University of Technology (statutory research).

Acknowledgments: We would like to thank our colleague Walter Mucha for drawing the scheme of the building.

Conflicts of Interest: The authors declare no conflict of interest.

References

1. Landtblom, A.M.; Engström, M. The sleepy teenager-diagnostic challenges. *Front. Neurol.* **2014**, *5*, 1–5. [CrossRef]
2. Herrmann, J.; Koeppen, K.; Kessels, U. Do girls take school too seriously? Investigating gender differences in school burnout from a self-worth perspective. *Learn. Individ. Differ.* **2019**, *69*, 150–161. [CrossRef]
3. Preety, R.; Devi, R.G.; Priya, A.J. Sleep deprivation and cell phone usage among teenagers. *Drug Invent. Today* **2018**, *10*, 2073–2075.
4. Zhang, N.; Cao, B.; Zhu, Y. Indoor environment and sleep quality: A research based on online survey and field study. *Build. Environ.* **2018**, *137*, 198–207. [CrossRef]
5. Rafihi-Ferreira, R.E.; Laura, M.; Pires, N.; Ferreira, E.; Silvares, M. Behavioral intervention for sleep problems in childhood: A Brazilian randomized controlled trial. *Psicol. Reflexão Crítica* **2019**, *32*, 1–13. [CrossRef]
6. Matricciani, L.; Paquet, C.; Galland, B.; Short, M.; Olds, T. Children's sleep and health: A meta-review. *Sleep Med. Rev.* **2019**, *46*, 136–150. [CrossRef]
7. Hall, R.; Hardin, T.; Ellis, R. School indoor air quality best management practices manual. 2003. Available online: https://www.doh.wa.gov/portals/1/Documents/Pubs/333-044.pdf (accessed on 26 August 2019).
8. Stamatelopoulou, A.; Saraga, D.; Asimakopoulos, D.; Vasilakos, C.; Maggos, T. The link between residential air quality and children's health. *Fresenius Environ. Bull.* **2017**, *26*, 162–176.
9. Mainka, A.; Zajusz-Zubek, E. Indoor air quality in urban and rural preschools in Upper Silesia, Poland: Particulate patter and carbon dioxide. *Int. J. Environ. Res. Public Health* **2015**, *12*, 7697–7711. [CrossRef]
10. Mainka, A.; Zubek, E.Z.; Kaczmarek, K. PM10 composition in urban and rural nursery schools in Upper Silesia, Poland: A trace elements analysis. *Int. J. Environ. Pollut.* **2017**, *61*, 98. [CrossRef]
11. Tham, K.W. Indoor air quality and its effects on humans—A review of challenges and developments in the last 30 years. *Energy Build.* **2016**, *130*, 637–650. [CrossRef]
12. Monn, C. Exposure assessment of air pollutants: A review on spatial heterogeneity and indoor/outdoor/personal exposure to suspended particulate matter, nitrogen dioxide and ozone. *Atmos. Environ.* **2001**, *35*, 1–32. [CrossRef]
13. Pathirana, S.M.; Sheranie, M.D.M.; Halwatura, R.U. Indoor thermal comfort and carbon dioxide concentration: A comparative study of air conditioned and naturally ventilated houses in Sri Lanka. In Proceedings of the Moratuwa Engineering Research Conference (MERCon), Moratuwa, Sri Lanka, 29–31 May 2017; pp. 401–406.
14. Khalequzzaman, M.; Kamijima, M.; Sakai, K.; Chowdhury, N.A.; Hamajima, N.; Nakajima, T. Indoor air pollution and its impact on children under five years old in Bangladesh. *Indoor Air* **2007**, *17*, 297–304. [CrossRef]
15. Santamouris, M.; Synnefa, A.; Asssimakopoulos, M.; Livada, I.; Pavlou, K.; Papaglastra, M.; Gaitani, N.; Kolokotsa, D.; Assimakopoulos, V. Experimental investigation of the air flow and indoor carbon dioxide concentration in classrooms with intermittent natural ventilation. *Energy Build.* **2008**, *40*, 1833–1843. [CrossRef]

16. Persily, A.; de Jonge, L. Carbon dioxide generation rates for building occupants. *Indoor Air* **2017**, *27*, 868–879. [CrossRef]
17. Government of Canada: Health Canada Exposure Guidelines for Residential Indoor Air Quality. Available online: http://publications.gc.ca/collections/Collection/H46-2-90-156E.pdf (accessed on 26 August 2019).
18. Regulation of the Minister of Family, Labor and Social Policy Regarding the Highest Permissible Concentrations and Intensities of Harmful Factors in the Work Environment. Available online: http://www.ilo.org/dyn/natlex/natlex4.detail?p_lang=en&p_isn=99664 (accessed on 26 August 2019).
19. PN-EN13779. Ventilation for Non-Residential Buildings. Performance Requirements for Ventilation and Room-Conditioning Systems. Available online: http://www.cres.gr/greenbuilding/PDF/prend/set4/WI_25_Pre-FV_version_prEN_13779_Ventilation_for_non-resitential_buildings.pdf (accessed on 26 August 2019).
20. PN-EN 16798-1: 2019. Indoor environmental input parameters for design and assessment of energy performance of buildings addressing indoor air quality, thermal environment, lighting and acoustics. Available online: http://sklep.pkn.pl/pn-en-16798-1-2019-06e.html (accessed on 27 August 2019).
21. CEN/TR 16798-2. Technical Report: Energy Performance of Buildings-Ventilation for Buildings-Part 2: Interpretation of the Requirements in EN 16798-1-Indoor Environmental Input Parameters for Design and Assessment of Energy Performance of Buildings Addressing Indoor. Available online: https://epb.center/documents/centr-16798-2/ (accessed on 26 August 2019).
22. Bekö, G.; Lund, T.; Nors, F.; Toftum, J.; Clausen, G. Ventilation rates in the bedrooms of 500 Danish children. *Build. Environ.* **2010**, *45*, 2289–2295. [CrossRef]
23. Satish, U.; Mendell, M.J.; Shekhar, K.; Hotchi, T.; Sullivan, D. Is CO_2 an indoor pollutant? Direct effects of low-to-moderate CO_2 concentrations on human decision-making performance. *Environ. Health Perspect.* **2012**, *120*, 1671–1678. [CrossRef]
24. Norbäck, D.; Nordström, K.; Zhao, Z.; Nordstro, K.; Norba, D. Carbon dioxide (CO_2) demand-controlled ventilation in university computer classrooms and possible effects on headache, fatigue and perceived indoor environment: An intervention study. *Int. Arch. Occup. Environ. Health* **2013**, *86*, 199–209. [CrossRef]
25. Bouvier, J.L.; Bontemps, S.; Mora, L. Uncertainty and sensitivity analyses applied to a dynamic simulation of the carbon dioxide concentration in a detached house. *Int. J. Energy Environ. Eng.* **2019**, *10*, 47–65. [CrossRef]
26. Belmonte, J.F.; Barbosa, R.; Almeida, M.G. CO_2 concentrations in a multifamily building in Porto, Portugal: Occupants' exposure and differential performance of mechanical ventilation control strategies. *J. Build. Eng.* **2019**, *23*, 114–126. [CrossRef]
27. Bakó-Biró, Z.; Clements-Croome, D.J.; Kochhar, N.; Awbi, H.B.; Williams, M.J. Ventilation rates in schools and pupils' performance. *Build. Environ.* **2012**, *48*, 215–223. [CrossRef]
28. Telejko, M.; Zander-Świercz, E. Attempt to improve indoor air quality in kindergartens. *Procedia Eng.* **2016**, *161*, 1704–1709. [CrossRef]
29. Zender-Swiercz, E.; Telejko, M. Indoor air quality in kindergartens in Poland. In Proceedings of the IOP Conference Series: Materials Science and Engineering, 3rd World Multidisciplinary Civil Engineering, Architecture, Urban Planning Symposium (WMCAUS 2018), Prague, Czech Republic, 18–22 June 2018; Volume 471.
30. Krawczyk, D.A.; Rodero, A.; Gładyszewska-Fiedoruk, K.; Gajewski, A. CO_2 concentration in naturally ventilated classrooms located in different climates—Measurements and simulations. *Energy Build.* **2016**, *129*, 491–498. [CrossRef]
31. Mainka, A.; Zajusz-Zubek, E.; Kozielska, B.; Brągoszewska, E. Investigation of air pollutants in rural nursery school–a case study. In Proceedings of the E3S Web of Conferences, X-th Scientific Conference Air Protection in Theory and Practice, Zakopane, Poland, 18–21 October 2017; Volume 28, pp. 1–8.
32. Karjalainen, S. Thermal comfort and gender: A literature review. *Indoor Air* **2012**, *22*, 96–109. [CrossRef] [PubMed]
33. Seppanen, O.A.; Fisk, W.J.; Mendell, M.J. Association of ventilation rates and CO_2 concentrations with health and other responses. *Indoor Air* **1999**, *9*, 226–252. [CrossRef] [PubMed]
34. Mahyuddin, N.; Awbi, H.B. A Review of CO_2 measurement procedures in ventilation research. *Int. J. Vent.* **2012**, *10*, 353–370. [CrossRef]
35. Eusebio, L.; Derudi, M.; Capelli, L.; Nano, G.; Sironi, S. Assessment of the indoor odour impact in a naturally ventilated room. *Sensors* **2017**, *17*, 778. [CrossRef] [PubMed]

36. Nguyen, J.L.; Schwartz, J.; Dockery, D.W. The relationship between indoor and outdoor temperature, apparent temperature, relative humidity, and absolute humidity. *Indoor Air* **2014**, *24*, 103–112. [CrossRef]
37. WHO. *WHO Air Quality Guidelines for Europe*; WHO: Copenhagen, Denmark, 2000.
38. Bekö, G.; Toftum, J.; Clausen, G. Modeling ventilation rates in bedrooms based on building characteristics and occupant behavior. *Build. Environ.* **2011**, *46*, 2230–2237. [CrossRef]
39. Stamatelopoulou, A.; Asimakopoulos, D.N.; Maggos, T. Effects of PM, TVOCs and comfort parameters on indoor air quality of residences with young children. *Build. Environ.* **2019**, *150*, 233–244. [CrossRef]
40. D'Ambrosio Alfano, F.R.; Ficco, G.; Palella, B.I.; Riccio, G.; Ranesi, A. An experimental investigation on the air permeability of passive ventilation grilles. *Energy Procedia* **2015**, *78*, 2869–2874. [CrossRef]

applied sciences

Article

Cooking/Window Opening and Associated Increases of Indoor $PM_{2.5}$ and NO_2 Concentrations of Children's Houses in Kaohsiung, Taiwan

Yu-Chuan Yen [1], Chun-Yuh Yang [1], Kristina Dawn Mena [2], Yu-Ting Cheng [1] and Pei-Shih Chen [1,3,4,5,*]

[1]   Department of Public Health, College of Health Science, Kaohsiung Medical University, Kaohsiung City 807, Taiwan; ponder10003@gmail.com (Y.-C.Y.); chunyuh@kmu.edu.tw (C.-Y.Y.); llssdog@hotmail.com (Y.-T.C.)
[2]   Epidemiology, Human Genetics, and Environmental Sciences, School of Public Health, University of Texas Health Science Center at Houston, Houston, TX 77046, USA; Kristina.D.Mena@uth.tmc.edu
[3]   Institute of Environmental Engineering, College of Engineering, National Sun Yat-Sen University, Kaohsiung City 807, Taiwan
[4]   Department of Medical Research, Kaohsiung Medical University Hospital, Kaohsiung City 807, Taiwan
[5]   Research Center for Environmental Medicine, Kaohsiung Medical University, Kaohsiung City 807, Taiwan
*   Correspondence: pschen@kmu.edu.tw; Tel.: +886-7-3110811

Received: 29 August 2019; Accepted: 2 October 2019; Published: 14 October 2019

Abstract: High concentrations of air pollutants and increased morbidity and mortality rates are found in industrial areas, especially for the susceptible group, children; however, most studies use atmospheric dispersion modeling to estimate household air pollutants. Therefore, the aim of this study was to assess the indoor air quality, e.g., CO, CO_2, NO_2, SO_2, O_3, particulate matter with aerodynamic diameter less than 2.5 µm ($PM_{2.5}$), and their influence factors in children's homes in an industrial city. Children in the "general school", "traffic school", and "industrial school" were randomly and proportionally selected. Air pollutants were sampled for 24 h in the living rooms and on the balcony of their houses and questionnaires of time–microenvironment–activity-diary were recorded. The indoor CO concentration of the traffic area was significantly higher than that of the industrial area and the general area. In regard to the effects of window opening, household NO_2 and $PM_{2.5}$ concentrations during window opening periods were significantly higher than of the reference periods. For the influence of cooking, indoor CO_2, NO_2, and $PM_{2.5}$ levels during the cooking periods were significantly higher than that of the reference periods. The indoor air quality of children in industrial cities were affected by residential areas and household activities.

Keywords: indoor air quality; children's house; industrial city; window opening; cooking

---

1. Introduction

According to the Environmental White Paper of Taiwan Environmental Protection Agency (Taiwan EPA), the annual average concentrations of ambient CO, NO_2, SO_2, and O_3 in 2008 were 0.47 ppm, 16.90 ppb, 4.35 ppb, and 29.09 ppb, respectively. The Kaohsiung–Pingtung area was the worse polluted area in Taiwan and accounted for 5.93% of station-days of the Pollutant Standards Index (PSI) > 100. Especially, Kaohsiung is a heavy industrial city. In industrial areas, high concentrations of air pollutants and increased morbidity and mortality rates are found, depending on the types of industrial activities and exposure concentrations in residential areas [1,2]. Children are more susceptible to the health effects of air pollution than adults due to not having full development of their pulmonary metabolic capacity [3]. Long-term exposure of air pollution may affect children's lung development [4]. Previously, most of the studies revealed that ambient pollution such as particulate matter with aerodynamic diameter less than 10 and 2.5 µm (PM_{10} and $PM_{2.5}$), sulfur dioxide (SO_2), nitrogen dioxide (NO_2), volatile organic compounds (VOCs), etc. in industrial areas may increase the risk of respiratory symptoms, and attacks of asthma in children [1,5,6]. Therefore, indoor air quality of children's homes may be very important to children's health, especially in industrial cities, since children spend most of their time at home [7].

Indoor air quality may be affected by indoor human activities such as cooking, smoking, cleaning, etc. and the infiltration of outdoor pollutants produced from the traffic or industrial sources [8–11]. For example, SO_2, NO_x, $PM_{2.5}$, and carbon monoxide (CO), the major conventional air pollutant in steel plants, oil refineries, and vehicular exhaust emissions [12–14], may enter a house through cracks and windows [15,16]. In addition, if the indoor air is not well ventilated, the air pollutants may accumulate in the indoor environment, and then seriously affects the health of the inhabitants [17].

Previously, atmospheric dispersion modeling was used to estimate the household concentrations of indoor air pollutants in industrial areas [1,18,19]. Only a few studies actually measured individual exposure [20] and household concentrations [21–23], and these studies only focused on PM mass concentrations, elemental composition, and VOCs concentrations. However, other air pollutants e.g., CO, carbon dioxide (CO_2), NO_2, SO_2, and ozone (O_3) in households in industrial cities also need to be considered. Therefore, the main aim of this study was to assess the indoor air quality including CO, CO_2, NO_2, SO_2, O_3, and $PM_{2.5}$, temperature and relative humidity, and their influence factors (e.g., window opening and cooking) in children's homes in an industrial city—Kaohsiung City. To our knowledge, this is the first study to assess the indoor air quality including CO, CO_2, NO_2, SO_2, and O_3 in children's homes in an industrial city. In addition, the second aim was to evaluate potential determinants of indoor air pollutants levels of occupants' activities, including cooking and window opening, etc. It is also the first study to reveal the differences of air pollutants between cooking periods/window opening periods and reference periods through a time–microenvironment–activity-diary via a questionnaire in one-hour time segments.

2. Materials and Methods

2.1. Study Area

Kaohsiung City (22°38′ N, 120°17′ E), located in southern Taiwan and with the population density of 9962.6/km^2 in 2010, is the largest industrialized harbor city in Taiwan with intense traffic and heavy industries including the largest steel plant (the China Steel Corporation, which also ranked the 19th steel mill in the world in 2005), the largest oil refinery (the CPC Corporation), the largest international shipbuilding (it ranked 6th in the world in 2005) in Taiwan, and many petrochemical industries.

2.2. Study Design

In April 2010, we selected three elementary schools in Kaohsiung City. One elementary school had a general air quality monitoring station of Taiwan EPA on the roof of the 4th floor, so we called this school a "general school". Another elementary school was 0.33 km from Taiwan EPA's traffic air quality monitoring station and was regarded as a "traffic school". The "industrial school" was an elementary school located near the Xiaogang Industrial Zone in Kaohsiung City and about 0.30 km from Taiwan EPA's air monitoring station. The study population was limited to children who attended these schools. The number of students in the "general school", "traffic school", and "industrial school" were 1669, 987, and 960, respectively. After obtaining the assented of the child and the permission of the parents, we recorded the subjects who agreed to home visits for environmental sampling. Children were randomly and proportionally selected from each school to participate in this study. Finally, the home visits of 32, 16, and 12 participants in the "general school", "traffic school", and "industrial school", respectively, were completed between April 2010 and October 2010.

2.3. Air Sampling

Indoor air pollutants including CO, CO_2, NO_2, SO_2, O_3, $PM_{2.5}$, temperature, and relative humidity were measured by real-time monitoring equipment for 24 h in the living rooms. We also measured the atmospheric CO, CO_2, NO_2, SO_2, O_3, and $PM_{2.5}$ on the balcony as outdoor concentrations. All instruments were placed on the bench at a height of approximately 1 m above the ground. The PM was measured by a real-time optical scattering instrument (DUSTTRAK™ DRX Aerosol Monitor Models 8533, TSI Incorporated, Shoreview, MN, USA) and the measurements were taken every 1 s by the flow rate of 3.0 L/min with detectable concentration from 0.001 to 150 mg/m^3. The CO, CO_2, NO_2, SO_2, indoor temperature, and relative humidity were also recorded (KD-airboxx, KD Engineering, Blaine, WA, USA) every 15 s with the measuring range of 0 to 500 ppm, 0 to 10,000 ppm, 0 to 20 ppm, 0 to 20 ppm, 0 to 50 °C, and 5% to 95%, respectively. The accuracy of CO, CO_2, NO_2, and SO_2 were ±3% of reading or 2 ppm (whichever was greater), ±5% of reading or 60 ppm (whichever was greater), 0.25 ppm, and 0.25 ppm, respectively. The resolution of CO, CO_2, NO_2, and SO_2 were 0.1 ppm, 1 ppm, 0.01 ppm, and 0.01 ppm, respectively. In terms of O_3, it was detected by a real-time monitoring (Model 202 Ozone monitor™, 2B Technologies Inc, Boulder, CO, USA) every 5 min with the measuring range of 0 to 250 ppm.

All real-time monitors were manufacturer-calibrated for the study in the beginning of this study and every six months. Before every field sampling, the DUSTTRAK™ DRX Aerosol Monitor Models 8533 was calibrated using emery oil aerosol and nominally adjusted to the respirable mass of standard ISO 12103-1, A1 test dust, (Arizona Dust); and the KD-airboxx, the Model 202 Ozone monitor™ were calibrated using zero gas and span gas. In addition, the zero calibrators of instruments were carried out, and the flow rate of sampling pump also was adjusted by Gilian Gilibrator-2NIOSH Primary Standard Air Flow Calibrator (Sensidyne, St. Petersburg, FL, USA) before every household sampling.

2.4. Household Characteristics

In addition, household characteristics including the number of occupants, air-conditioning use, smoking, incense burning, etc. were also recorded in the questionnaires. In addition to household characteristics, data on potential determinants of indoor air pollutants levels of occupants' activities, including cooking and window opening, etc. were obtained through a time–microenvironment–activity-diary via a questionnaire in one-hour time segments. We also actually evaluated the effects of window opening and cooking on indoor air pollutants. The window opening periods were defined from a time–microenvironment–activity-diary and two one-hour periods before and after window opening periods were defined as the reference periods. In terms of cooking, the cooking periods were the periods recorded by participants as cooking from a time–microenvironment–activity-diary and the reference periods were defined as the one-hour periods before the cooking periods.

2.5. Ethics

This study was approved by the Institutional Review Board of the Kaohsiung Medical University Chung-Ho Memorial Hospital (the protocol number was KMU-IRB-990045). Informed written consent was obtained from each child (the phonetic version of the consent form that the children read and signed) and their legal guardians.

2.6. Statistical Analyses

Statistical analyses in this study were performed using SAS version 9.3 (SAS Institute of Taiwan Ltd, Taipei, Taiwan). Descriptive statistics were used to describe the 24-hour of average of exposure data (indoor/outdoor air pollutant concentrations, temperature, and relative humidity). The concentrations of air pollutants were not normally distributed (data not shown), therefore we analyzed our data by nonparametric statistics, also known as distribution-free statistics. A paired Student's *t*-test was used to assess the difference in the average concentration of air pollutants between indoor and outdoor, between window opening periods and reference periods, and between cooking periods and reference periods. With the objective of evaluating significant differences among the areas (general, traffic, and industry) for all air pollutants variables, data were analyzed using one-way analysis of variance (ANOVA) with Scheffe multiple comparison test. The generalized estimating equations (GEE) is a general statistical method in a longitudinal study with small samples for adjusting time interference, in which each time point is an independent event. Finally, the relationships between the 24-hour average concentrations of indoor air pollutants (dependent variable) and household characteristics (independent variable) were analyzed using GEE, adjusting for other household characteristics, and time interference. A p-value of less than 0.05 was considered significant.

3. Results

Table 1 shows the descriptive statistics of 24-h average indoor and outdoor air pollutants, temperature, and relative humidity in 60 houses. When indoor air pollutants were paired with outdoors within the same home, we found that the 24-hour average concentrations of indoor CO, CO_2, and NO_2 were significantly higher than the 24-hour average of outdoors concentrations, whereas, outdoor O_3 and $PM_{2.5}$ concentrations were significantly higher than indoor concentrations (all $p < 0.01$). The average distance between homes of subjects and their school were 0.86 km, 0.94 km, and 1.46 km in general, traffic, and industrial areas, respectively, as well as, the average distance between homes of subjects and the nearest air monitoring station were 1.07 km, 0.97 km, and 1.46 km in general, traffic, and industrial areas, respectively.

Table 1. Descriptive statistics of 24-h average indoor and outdoor air pollutants, temperature, and relative humidity in 60 houses.

		Mean	Median	Standard Deviation	Minimum	Maximum	p-Value [#]
CO (ppm)	indoor	3.47	0.83	4.29	0.00	12.27	0.004 [‡]
	outdoor	0.60	0.38	0.55	0.00	1.98	
CO_2 (ppm)	indoor	655.43	479.55	321.60	413.82	1320.00	<0.001 [‡]
	outdoor	322.22	319.92	17.23	285.83	353.90	
NO_2 (ppb)	indoor	185.30	177.97	41.52	127.28	251.41	0.008 [‡]
	outdoor	107.54	118.22	36.83	39.90	149.80	
SO_2 (ppm)	indoor	0.00	0.00	0.00	0.00	0.00	0.193
	outdoor	0.01	0.00	0.02	0.00	0.06	
O_3 (ppb)	indoor	11.04	8.50	8.93	1.06	32.29	0.006 [‡]
	outdoor	13.46	9.20	12.34	0.24	45.50	
$PM_{2.5}$ ($\mu g/m^3$)	indoor	60.00	40.00	50.00	10.00	210.00	0.001 [‡]
	outdoor	110.00	90.00	90.00	30.00	410.00	
Temperature (°C)	indoor	31.00	31.00	1.76	26.00	34.00	-
Relative humidity (%)	indoor	72.00	72.00	4.98	62.00	84.00	-

[#] Paired Student's t-test, [‡] $p < 0.01$.

In comparison with household air pollutants of three areas, Table S1 shows descriptive statistics of 24-h average concentration of indoor air pollutants in the houses of traffic, industry, and general areas. Figure 1 shows the 24-h average concentration of indoor air pollutants (A) CO, (B) CO_2, (C) NO_2, and (D) O_3 in the houses of traffic, industry, and general areas. We found the 24-hour average concentration of indoor CO concentration of the traffic area was significantly higher than that of the industrial area, and the general area with all $p < 0.01$ (Figure 1, Table S1). In addition, the 24-hour average concentration of indoor CO_2 level of the general area was significantly lower than that of the traffic area, and industrial area (all $p < 0.01$) (Figure 1, Table S1). Finally, both the 24-hour average concentration of household NO_2 and O_3 concentrations of the industrial area were significantly lower than that of the traffic area, and general area (all $p < 0.01$) (Figure 1, Table S1). Moreover, there was no statistical significant difference of the 24-hour average concentration of indoor SO_2 and $PM_{2.5}$ between the three areas (Table S1).

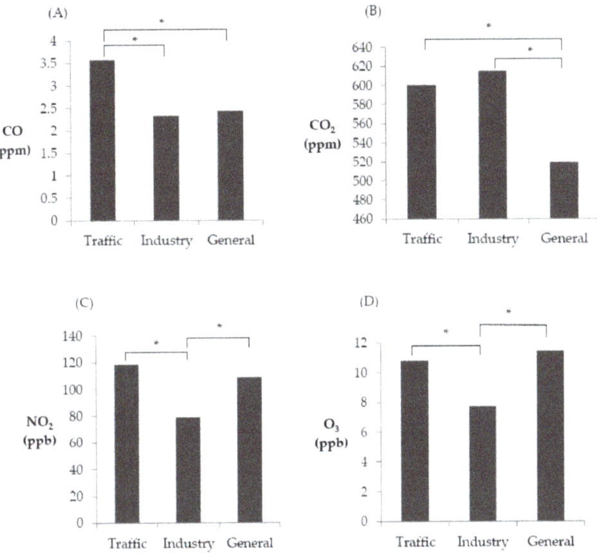

Figure 1. The 24-h average concentration of indoor air pollutants (**A**) CO, (**B**) CO_2, (**C**) NO_2, and (**D**) O_3 in the houses of traffic, industry, and general areas. * Scheffe test $p < 0.01$.

Table 2 shows the percentage of household characteristics including window opening, residents >4 people, cooking, etc. in traffic, industry, and general areas. We found compared with traffic area and industrial area, the general area had a higher percentage of window opening, cooking, and air-conditioning use; moreover, a lower percentage of residents > 4 people, smoker, incense burning, mosquito coil burning, and essential oil using.

Table 2. The percentage (%) of household characteristics in traffic, industry, and general areas.

	Area		
	Traffic	**Industry**	**General**
Window opening	68.75	66.67	87.50
Occupants (>4 people)	40.40	57.01	34.41
Cooking	87.50	83.33	95.83
Air-conditioning use	62.50	83.33	95.83
Making tea	31.25	30.00	0
Smoker	63.64	40.00	26.09
Incense burning	72.73	50.00	29.17
Mosquito coil burning	37.50	22.22	12.50
Essential oil using	31.25	33.33	25.00

The following Table 3 shows the ratios of air pollutants during window opening periods to the reference periods and the differences in air pollutants between window opening periods and reference periods. The median ratios of pollutants during window opening periods to the reference periods for NO_2 and $PM_{2.5}$ were 1.56 and 1.13, respectively with the maximum values up to 5.23 and 1.85 respectively (Table 3). The NO_2 and $PM_{2.5}$ levels during window opening periods were significantly higher than that of the reference periods, and the maximum increased values were 53.25 ppb and 44 $\mu g/m^3$, respectively. Table 4 shows the ratios of air pollutants during cooking periods to reference periods and the differences in air pollutants between window opening periods and reference periods. The median ratios of pollutants during cooking periods to the reference periods for CO, CO_2, NO_2, and $PM_{2.5}$ were 0.93, 1.06, 1.11, and 1.09, respectively. The concentrations of CO_2, NO_2, and $PM_{2.5}$ during the cooking periods were significantly higher than those of reference periods with increased concentrations of 26.17 ppm, 5.40 ppb, and 5 $\mu g/m^3$, respectively. However, the CO level during cooking periods was significantly lower than that of the reference periods with the decreased concentration of 0.25 ppm.

Table 3. The ratios of air pollutants during window opening periods to reference periods and the differences in air pollutants between window opening periods and reference periods.

	Ratios (Window Opening Periods/Reference Periods [§])				Differences (Window Opening Periods − Reference Periods [§])				*p*-Value [#]
	Median	S.D.	Min.	Max.	Median	S.D.	Min.	Max.	
CO (ppm)	0.98	1.34	0.57	4.44	0.00	1.31	−2.42	3.67	0.53
CO_2 (ppm)	1.05	0.18	0.73	1.43	29	128	−141	296	0.21
NO_2 (ppb)	1.56	1.30	0.94	5.23	18.71	16.05	−9.40	53.25	<0.01 [‡]
SO_2 (ppm)	0.00	0.92	0.00	3.27	0.00	0.02	0.00	0.08	0.21
O_3 (ppb)	1.18	0.59	0.56	2.19	0.81	4.44	−11.91	10.05	0.52
$PM_{2.5}$ ($\mu g/m^3$)	1.13	0.31	0.69	1.85	7	16.20	−6	44	0.04 [†]

[#] Paired Student's *t*-test, [†] $p < 0.05$, [‡] $p < 0.01$. [§] Reference periods were two one-hour periods before and after window opening periods.

Table 4. The ratios of air pollutants during cooking periods to reference periods and the differences in air pollutants between during cooking periods and reference periods.

	Ratios (Cooking Periods/Reference Periods [§])				Differences (Cooking Periods − Reference Periods [§])				*p*-Value [#]
	Median	S.D.	Min.	Max.	Median	S.D.	Min.	Max.	
CO (ppm)	0.93	0.22	0.46	1.51	−0.25	0.84	−3.53	0.61	<0.01 [‡]
CO_2 (ppm)	1.06	0.14	0.85	1.58	26.17	90.21	−111.67	342.5	<0.01 [‡]
NO_2 (ppb)	1.11	0.98	0.51	5.43	5.40	29.71	−71.17	101.75	<0.01 [‡]
O_3 (ppb)	1.08	0.69	0.46	4.36	0.27	8.89	−35.14	17.08	0.94
$PM_{2.5}$ ($\mu g/m^3$)	1.09	0.30	0.60	2.56	5	14	−45	56	0.04 [†]

[#] Paired Student's *t*-test, [†] $p < 0.05$, [‡] $p < 0.01$. [§] Reference periods were the one-hour period before cooking periods.

Table 5 shows the association between air pollutants concentrations (24-h average concentration of air pollutants in each house as dependent variable), and household characteristics by using the generalized estimating equations model. This study revealed that CO concentrations were positively associated with the number of occupants, cleaning, smoking, incense burning, mosquito coil burning, and negatively correlated to cooking with a statistical significance. Indoor CO_2 concentrations were positively associated with the number of occupants, air-conditioning use, smoking, incense burning, and negatively correlated to mosquito coil burning with a statistical significance. In addition, significantly higher NO_2 levels were found in the homes with smokers than homes without smokers. There were significantly positive associations between indoor SO_2 concentrations and smoking and incense burning. In terms of O_3, indoor O_3 concentrations were positively associated with the window opening and negatively correlated to the number of occupants, incense burning, and essential oil use with a statistical significance. For $PM_{2.5}$, it was positively associated with cleaning and incense burning with a statistical significance.

Table 5. Association between air pollutants concentrations (24-h average concentration of air pollutants in each house as dependent variable), and household characteristics: generalized estimating equations.

	CO (ppm)	CO_2 (ppm)	NO_2 (ppb)	SO_2 (ppm)	O_3 (ppb)	$PM_{2.5}$ ($\mu g/m^3$)
Window opening (Yes vs. No)	0.32	84.84	−0.61	0.44	24.34 ‡	−0.021
Occupants	0.52 ‡	51.62 ‡	3.02	−0.008	−3.49 ‡	0.004
Cleaning (Yes vs. No)	4.73 †	−317.49	1.39	0.43	−6.24	0.047 †
Cooking (Yes vs. No)	−3.89 †	228.02	−28.01	−0.21	1.79	0.065
Fan using (Yes vs. No)	1.42	−32.97	10.58	−0.0003	−2.07	0.002
Air- conditioning use (Yes vs. No)	−1.22	246.99 ‡	87.87	0.25	21.59	0.008
Making tea (Yes vs. No)	37.04	-	−0.45	−0.13	14.21	−0.050
Smoking (Yes vs. No)	17.21 †	1988.44 ‡	547.36 ‡	2.98 ‡	1.69	0.173
Incense burning (Yes vs. No)	18.21 †	2927.87 †	193.11	3.66 ‡	−108.9 ‡	0.416 ‡
Mosquito coil burning (Yes vs. No)	41.55 ‡	−892.64 †	673.52	2.67	2.29	-
Essential oil use (Yes vs. No)	12.76	269.25	74.89	−0.66	−89.29 ‡	−0.022

Generalized estimating equations (GEE) † $p < 0.05$, ‡ $p < 0.01$.

4. Discussion

Our results showed that the outdoor concentrations of O_3 and $PM_{2.5}$ were significantly higher than indoor concentrations. The Kaohsiung City is a city with intense traffic and heavy industries, and previous studies believed SO_2, NO_x, $PM_{2.5}$, and CO were the major conventional air pollutant in steel plants, oil refineries, and vehicular exhaust emissions [12–14,24]. In addition, outdoor O_3 might be formed by the photochemical reaction of nitrogen oxides absorbing sunlight, and VOCs [25,26]. According to the PSI database from 2010 to 2012 of Taiwan EPA, only O_3 and total suspended particulate (TSP) would exceed the standard [27]. This may be the reason why outdoor $PM_{2.5}$ and O_3 concentrations were higher than indoor concentrations. In our study, outdoor median $PM_{2.5}$ levels (90 $\mu g/m^3$) were higher than both the National Ambient Air Quality Standards of UAS and Taiwan EPA with the 24-hour standard for $PM_{2.5}$ of 35 $\mu g/m^3$. In addition, the median value of indoor $PM_{2.5}$ concentrations (40 $\mu g/m^3$) was also higher than the criteria of indoor air quality (IAQ) standards of Taiwan EPA (35 $\mu g/m^3$/24 h). In our study, indoor CO, CO_2, and NO_2 levels were significantly higher than outdoor levels. The number of occupants and human activities such as cooking, smoking, etc. might be the factors affecting indoor pollutants whereas liquefied petroleum gas (LPG), not electric stoves, was the main cooking way in Kaohsiung City [28,29]. In addition, most of the houses were just by the roads and very close to the mobile sources in Kaohsiung City, which was thought of as a traffic-intensive city with the number of cars and motorcycles of approximately 430,000 and 1,230,000, respectively, in 2010 [30]. Thus, the main combustion products of vehicular engines such as CO, NO_x, etc. entering the houses through cracks and windows might be the reason why the indoor concentrations of CO, CO_2, and NO_2 were higher than outdoor concentrations [15,16].

In comparison with traffic, industrial, and general areas, the highest household CO concentration was found in the traffic area among the three areas. According to the previous study, the greatest source of CO (more than 90%) in cities was motor vehicles [24]. The high traffic flow in the traffic area might be the reason for the observation. For CO_2, our study indicated that the lowest household CO_2 level was in the general area among the three areas. The main source of CO_2 was from human respiration [24,31]. The number of residents might be one possible reason since the number of residents > 4 people in the traffic area, the industrial area, and general area were 40.40%, 57.01%, and 34.41%, respectively. We also found both household NO_2 and O_3 concentrations of the industrial area were lowest among the three areas, which was not consistent with the observations of previous studies that ambient NO_2 was related to industrial activities [24], and outdoor O_3 might be formed by the photochemical reaction of nitrogen oxides absorbing sunlight, and VOCs [25,26]. We believed these may be related to Taiwan EPA' s policies and efforts to control air pollution from stationary sources after that the "Stationary Pollution Source Air Pollutant Emissions Standards" was passed in 1992, and the "Air Pollution Control Act Enforcement Rules" was also implemented in 2003.

In regard to the effects on the window opening, our study displayed that household NO_2 and $PM_{2.5}$ concentrations during window opening periods were significantly higher than that during reference periods. NO_X and PM were related to traffic emissions [24,32], and most of the houses in Taiwan were adjacent to roads, so window opening might increase indoor NO_2 and $PM_{2.5}$. For the influence of cooking, there were many simulated experiments exploring the air pollutant emissions of cooking-related fuel combustion [29,33–36], and they demonstrated that CO, CO_2, NO_X, and $PM_{2.5}$ would be emitted by the process of the experiments. Although CO also was produced by cooking, it was revealed that combustion of high-grade fuels (such as natural gas, and LPG which contained propane, butane, etc.), the main fuel-burning stoves use in Taiwan households usually produce much less CO than combustion of low-grade fuels [29,33]. In the previous study, Delp et al. revealed the residential cooking exhaust hoods could not completely capture the pollutants and their efficiency was highly variable [37]. Our results showed that indoor CO_2, NO_2, and $PM_{2.5}$ levels during cooking periods were significantly higher than during reference periods, but the indoor CO level during cooking periods was lower than during reference periods, possibly indicating that the emission rate of CO_2, NO_2, and $PM_{2.5}$ might be higher than the capture rate of the exhaust hood and the emission rate of CO might be lower than the pollutants capture rate of the exhaust hood.

In terms of influence factors, we found there were significantly positive correlations between the number of occupants and CO and CO_2 concentrations. Our study was consistent with the observations of the previous study that CO_2 was produced by human respiration [24,31]. In addition to the combustion, the indoor CO also was related to the status of residents; the previous studies revealed either a smoking person or person with inflammatory diseases exhaled higher CO levels than control group [38,39]. We also found smoking was significantly positively associated with household CO in our study. According to previous studies, smoking, incense burning, and mosquito coil burning were significantly positively associated with CO, CO_2, SO_2, NO_X, and PM [40–42], and these results were consistent with our observation. The cleaning behavior would increase indoor $PM_{2.5}$ and CO levels; it was consistent with the previous study that indoor $PM_{2.5}$ and PM_5 levels could be elevated by the cleaning behavior of dry dust, and vacuuming [43]. In addition, commercial cleansers and disinfectants contain VOCs [44], and El Fadel et al. found VOCs concentration was positively correlated with CO concentration [45]. We also revealed that air-conditioning use was positively associated with indoor CO_2 concentrations with a statistical significance, which was consistent with a previous observation that CO_2 levels were higher in mechanically ventilated buildings than in naturally ventilated buildings [46]. There was a significantly negative association between essential oil use and O_3 concentration. The commercially available essential oils contain many VOCs (e.g., D-limonene, α- pinene, etc.) [47], in addition, a study displayed that indoor VOCs level had increased significantly after burning essential oils [48]. O_3 was one of the indoor oxidants [49,50], and Waring et al. demonstrated that 68% of all O_3 reactions were with D-limonene, and 26% of all O_3 reactions are with α-pinene [50]. This might be the reason why the essential oil use could decrease the O_3 level. Finally, by questionnaire, it was found that window opening was significantly correlated with increased O_3 concentration, which was not consistent with the results from the time–microenvironment–activity-diary that only NO_2 and $PM_{2.5}$ levels during the window opening periods were significantly higher than that of reference periods. We believed O_3 was a major component of photochemical pollution, so it is more relevant to outdoor sources than indoor sources. Thus, compared with the households which closed the windows, the households which opened the windows had a significantly higher 24-hour average concentration of O_3. When comparing the window opening periods with the reference periods (two one-hour periods before and after window opening periods), there was no significant variation in atmospheric O_3 concentration in a short time (within three hours). For $PM_{2.5}$ and NO_X levels, there was no significant difference between households which closed and opened the windows, the possible reason might be that $PM_{2.5}$ and NO_X could come from both indoor (cooking) and outdoor (traffic) sources.

5. Conclusions

This study explored the concentration of indoor air pollutants in different areas including traffic, industrial, and general areas within an industrial city. Moreover, this study also revealed household NO_2 and $PM_{2.5}$ concentrations during window opening periods were significantly higher than that of the reference periods with increased concentrations of 18.71 ppb, and 7 $\mu g/m^3$, respectively. For the influence of cooking, indoor CO_2, NO_2, and $PM_{2.5}$ levels during the cooking periods were significantly higher than that of the reference periods with increased concentrations of 26.17 ppm, 5.40 ppb, and 5 $\mu g/m^3$, respectively.

Supplementary Materials: The following are available online at http://www.mdpi.com/2076-3417/9/20/4306/s1, Table S1: Descriptive statistics of 24-h average concentration of indoor air pollutants in the houses of traffic, industry, and general areas.

Author Contributions: Conceptualization, C.-Y.Y.; formal analysis, Y.-C.Y.; writing—original draft preparation, Y.-T.C. and Y.-T.C.; writing—review and editing, P.-S.C. and K.D.M.

Funding: This research was funded by "Wang Jhan-Yang Public Trust Fund", grant number 108-002-6.

Conflicts of Interest: The authors declare no conflict of interest.

References

1. Idavain, J.; Julge, K.; Rebane, T.; Lang, A.; Orru, H. Respiratory symptoms, asthma and levels of fractional exhaled nitric oxide in schoolchildren in the industrial areas of Estonia. *Sci. Total. Environ.* **2019**, *650*, 65–72. [CrossRef] [PubMed]
2. Martuzzi, M.; Pasetto, R.; Martin-Olmedo, P. Industrially Contaminated Sites and Health. *J. Environ. Public Health* **2014**, *2014*, 198574. [CrossRef] [PubMed]
3. Kurt, O.K.; Zhang, J.; Pinkerton, K.E. Pulmonary health effects of air pollution. *Curr. Opin. Pulm. Med.* **2016**, *22*, 138–143. [CrossRef] [PubMed]
4. Rice, M.B.; Rifas-Shiman, S.L.; Litonjua, A.A.; Oken, E.; Gillman, M.W.; Kloog, I.; Luttmann-Gibson, H.; Zanobetti, A.; Coull, B.A.; Schwartz, J.; et al. Lifetime Exposure to Ambient Pollution and Lung Function in Children. *Am. J. Respir. Crit. Care Med.* **2016**, *193*, 881–888. [CrossRef] [PubMed]
5. Deng, Q.; Lu, C.; Norbäck, D.; Bornehag, C.G.; Zhang, Y.; Liu, W.; Yuan, H.; Sundell, J. Early life exposure to ambient air pollution and childhood asthma in China. *Environ. Res.* **2015**, *143*, 83–92. [CrossRef] [PubMed]
6. Marchetti, P.; Marcon, A.; Pesce, G.; Paolo, G.; Guarda, L.; Pironi, V.; Fracasso, M.E.; Ricci, P.; de Marco, R. Children living near chipboard and wood industries are at an increased risk of hospitalization for respiratory diseases: A prospective study. *Int. J. Hyg. Environ. Health* **2014**, *217*, 95–101. [CrossRef] [PubMed]
7. Silvers, A.; Florence, B.T.; Rourke, D.L.; Lorimor, R.J. How Children Spend Their Time: A Sample Survey for Use in Exposure and Risk Assessments. *Risk Anal.* **1994**, *14*, 931–944. [CrossRef]
8. Cannistraro, G.; Cannistraro, A.; Cannistraro, M.; Galvagno, A. Analysis of air pollution in the urban center of four cities sicilian. *Int. J. Heat Technol.* **2016**, *34*, S219–S225.
9. Cannistraro, M.; Cannistraro, G.; Chao, J.; Ponterio, L. New Technique Monitoring and Transmission Environmental Data with Mobile. *Instrum. Meas. Metrol.* **2018**, *18*, 549–562. [CrossRef]
10. Cannistraro, M.; Chao, J.; Ponterio, L. Experimental Study of Air Pollution in the Urban Centre of the City of Messina. *Model. Meas. Control C* **2018**, *79*, 133–139. [CrossRef]
11. Leung, D.Y.C. Outdoor-indoor air pollution in urban environment: Challenges and opportunity. *Front. Environ. Sci.* **2015**. [CrossRef]
12. Brand, A.; McLean, K.E.; Henderson, S.B.; Fournier, M.; Liu, L.; Kosatsky, T.; Smargiassi, A. Respiratory hospital admissions in young children living near metal smelters, pulp mills and oil refineries in two Canadian provinces. *Environ. Int.* **2016**, *94*, 24–32. [CrossRef] [PubMed]

13. Tian, L.; Zeng, Q.; Dong, W.; Guo, Q.; Wu, Z.; Pan, X.; Li, G.; Liu, Y. Addressing the source contribution of PM2.5on mortality: An evaluation study of its impacts on excess mortality in China. *Environ. Res. Lett.* **2017**, *12*, 104016. [CrossRef]

14. Topacoglu, H.; Katsakoglou, S.; Ipekci, A. Effect of exhaust emissions on carbon monoxide levels in employees working at indoor car wash facilities. *Hippokratia* **2014**, *18*, 37–39. [PubMed]

15. Srithawirat, T.; Latif, M.T.; Sulaiman, F.R. *Indoor PM10 and Its Heavy Metal Composition at a Roadside Residential Environment*; Atmosfera: Phitsanulok, Thailand, 2016; Volume 29, pp. 311–322.

16. United States Environmental Protection Agency. Available online: https://www.epa.gov/report-environment/indoor-air-quality (accessed on 20 August 2019).

17. World Health Organization. Household (Indoor) Air Pollution. 2014. Available online: https://www.who.int/indoorair/en/ (accessed on 20 August 2019).

18. Han, I.; Guo, Y.; Afshar, M.; Symanski, E.; Stock, T.H. Comparison of trace elements in size-fractionated particles in two communities with contrasting socioeconomic status in Houston, TX. *Environ. Monit. Assess.* **2017**, *189*, 67. [CrossRef] [PubMed]

19. Hart, J.E.; Garshick, E.; Dockery, D.W.; Smith, T.J.; Ryan, L.; Laden, F. Long-term ambient multipollutant exposures and mortality. *Am. J. Respir. Crit. Care Med.* **2011**, *183*, 73–78. [CrossRef] [PubMed]

20. Milà, C.; Salmon, M.; Sanchez, M.; Ambrós, A.; Bhogadi, S.; Sreekanth, V.; Nieuwenhuijsen, M.; Kinra, S.; Marshall, J.D.; Tonne, C. When, Where, and What? Characterizing Personal PM2.5 Exposure in Periurban India by Integrating GPS, Wearable Camera, and Ambient and Personal Monitoring Data. *Environ. Sci. Technol.* **2018**, *52*, 13481–13490. [CrossRef]

21. Pekey, B.; Bozkurt, Z.B.; Pekey, H.; Doğan, G.; Zararsız, A.; Efe, N.; Tuncel, G.; Zararsiz, A. Indoor/outdoor concentrations and elemental composition of PM10/PM2.5 in urban/industrial areas of Kocaeli City, Turkey. *Indoor Air* **2010**, *20*, 112–125. [CrossRef]

22. Pekey, H.; Arslanba, D. The Relationship Between Indoor, Outdoor and Personal VOC Concentrations in Homes, Offices and Schools in the Metropolitan Region of Kocaeli, Turkey. *Water Air Soil Pollut.* **2008**, *191*, 113–129. [CrossRef]

23. Liu, W.; Zhang, J.; Zhang, L.; Turpin, B.; Weisel, C.; Morandi, M.; Stock, T.; Colome, S.; Korn, L. Estimating contributions of indoor and outdoor sources to indoor carbonyl concentrations in three urban areas of the United States. *Atmos. Environ.* **2006**, *40*, 2202–2214. [CrossRef]

24. Najjar, Y. Gaseous Pollutants Formation and Their Harmful Effects on Health and Environment. *Innov. Energy Policies* **2011**, *1*. [CrossRef]

25. Du, X.; Liu, J. Relationship between outdoor and indoor ozone pollution concentration. *Trans. Tianjin Univ.* **2009**, *15*, 330–335. [CrossRef]

26. Jianhui, B.; Singh, N.; Chauhan, S.; Singh, K.; Saud, T.; Saxena, M.; Soni, D.; Mandal, T.K.; Bassin, J.K.; Gupta, P.K. Study on surface O_3 chemistry and photochemistry by UV energy conservation. *Atmos. Pollut. Res.* **2010**, *1*, 118–127. [CrossRef]

27. Taiwan Environmental Protection Agency. Available online: https://taqm.epa.gov.tw/taqm/tw/PSIOver100MonthlyReport.aspx (accessed on 20 August 2019).

28. Persily, A.; De Jonge, L. Carbon dioxide generation rates for building occupants. *Indoor Air* **2017**, *27*, 868–879. [CrossRef] [PubMed]

29. Penney, D.; Benignus, V.; Kephalopoulos, S.; Kotzias, D.; Kleinman, M.; Verrier, A. *WHO Guidelines for Indoor Air Quality: Selected Pollutants*; WHO Regional Office for Europe: Denmark, 2010.

30. Taiwan Environmental Protection Agency. Available online: https://erdb.epa.gov.tw/DataRepository/Statistics/StatSceAreapop.aspx (accessed on 20 August 2019).

31. Satish, U.; Mendell, M.J.; Shekhar, K.; Hotchi, T.; Sullivan, D.; Streufert, S.; Fisk, W.J. Is CO_2 an Indoor Pollutant? Direct Effects of Low-to-Moderate CO_2 Concentrations on Human Decision-Making Performance. *Environ. Health Perspect.* **2012**, *120*, 1671–1677. [CrossRef]

32. Chao, H.-R.; Hsu, J.-W.; Ku, H.-Y.; Wang, S.-L.; Huang, H.-B.; Liou, S.-H.; Tsou, T.-C. Inflammatory Response and PM2.5 Exposure of Urban Traffic Conductors. *Aerosol. Air Qual. Res.* **2018**, *18*, 2633–2642. [CrossRef]

33. Bilsback, K.R.; Dahlke, J.; Fedak, K.M.; Good, N.; Hecobian, A.; Herckes, P.; L'Orange, C.; Mehaffy, J.; Sullivan, A.; Tryner, J.; et al. A Laboratory Assessment of 120 Air Pollutant Emissions from Biomass and Fossil Fuel Cookstoves. *Environ. Sci. Technol.* **2019**, *53*, 7114–7125. [CrossRef]

34. Shen, G.; Hays, M.D.; Smith, K.R.; Williams, C.; Faircloth, J.W.; Jetter, J.J. Evaluating the Performance of Household Liquefied Petroleum Gas Cookstoves. *Environ. Sci. Technol.* **2018**, *52*, 904–915. [CrossRef]

35. Shen, G.; Gaddam, C.K.; Ebersviller, S.M.; Wal, R.L.V.; Williams, C.; Faircloth, J.W.; Jetter, J.J.; Hays, M.D. A Laboratory Comparison of Emission Factors, Number Size Distributions, and Morphology of Ultrafine Particles from 11 Different Household Cookstove-Fuel Systems. *Environ. Sci. Technol.* **2017**, *51*, 6522–6532. [CrossRef]

36. Smith, K.R.; Uma, R.; Kishore, V.V.N.; Zhang, J.; Joshi, V.; Khalil, M.A.K. Greenhouse Implications of Household Stoves: An Analysis for India. *Annu. Rev. Energy Environ.* **2000**, *25*, 741–763. [CrossRef]

37. Delp, W.W.; Singer, B.C. Performance Assessment of U.S. Residential Cooking Exhaust Hoods. *Environ. Sci. Technol.* **2012**, *46*, 6167–6173. [CrossRef] [PubMed]

38. Deveci, S.E.; Deveci, F.; Açik, Y.; Ozan, A.T. The measurement of exhaled carbon monoxide in healthy smokers and non-smokers. *Respir Med.* **2004**, *98*, 551–556. [CrossRef] [PubMed]

39. Ryter, S.W.; Choi, A.M. Carbon monoxide in exhaled breath testing and therapeutics. *J. Breath Res.* **2013**, *7*, 017111. [CrossRef] [PubMed]

40. Alberts, W.M. Indoor air pollution: NO, NO_2, CO, and CO_2. *J. Allergy Clin. Immunol.* **1994**, *94*, 289–295. [CrossRef] [PubMed]

41. Lee, S.C.; Guo, H.; Kwok, N.H. Emissions of air pollutants from burning of incense by using large environmental chamber. In Proceedings of the Indoor Air 2002: 9th International Conference on Indoor Air Quality and Climate, Monterey, CA, USA, 30 June–5 July 2002; Volume 1.

42. Lee, S.C.; Wang, B. Characteristics of emissions of air pollutants from mosquito coils and candles burning in a large environmental chamber. *Atmos. Environ.* **2006**, *40*, 2128–2138. [CrossRef]

43. Ferro, A.R.; Kopperud, R.J.; Hildemann, L.M. Source Strengths for Indoor Human Activities that Resuspend Particulate Matter. *Environ. Sci. Technol.* **2004**, *38*, 1759–1764. [CrossRef]

44. United States Environmental Protection Agency. Available online: https://www.epa.gov/indoor-air-quality-iaq/volatile-organic-compounds-impact-indoor-air-quality (accessed on 20 August 2019).

45. El Fadel, M.; Alameddine, I.; Kazopoulo, M.; Hamdan, M.; Nasrallah, R. Carbon Monoxide and Volatile Organic Compounds as Indicators of Indoor Air Quality in Underground Parking Facilities. *Indoor Built Environ.* **2001**, *10*, 70–82. [CrossRef]

46. Sribanurekha, V.; Wijerathne, S.N.; Wijepala, L.H.S.; Jayasinghe, C. Effect of Different Ventilation Conditions on Indoor CO_2 Levels. In Proceedings of the International Conference on Disaster Resilience, At Kandalama, Sri Lanaka, 19–21 July 2011.

47. Nematollahi, N.; Kolev, S.D.; Steinemann, A. Volatile chemical emissions from essential oils. *Air Qual. Atmos. Heal.* **2018**, *11*, 949–954. [CrossRef]

48. Chao, C.J.; Wu, P.C.; Chang, H.Y.; Su, H.J. The effects of evaporating essential oils on indoor air quality. In Proceedings of the Indoor Air 2005, 10th International Conference on Indoor Air Quality and Climate, Beijing, China, 4–9 September 2005.

49. Young, C.J.; Zhou, S.; Siegel, J.A.; Kahan, T.F. Illuminating the Dark Side of Indoor Oxidants. *Environ. Sci. Process. Impacts* **2019**, *21*, 1229–1239. [CrossRef]

50. Waring, M.S.; Wells, J.R. Volatile organic compound conversion by ozone, hydroxyl radicals, and nitrate radicals in residential indoor air: Magnitudes and impacts of oxidant sources. *Atmos. Environ.* **2015**, *106*, 382–391. [CrossRef]

applied sciences

Article

Evaluation of Performance of Inexpensive Laser Based PM$_{2.5}$ Sensor Monitors for Typical Indoor and Outdoor Hotspots of South Korea

Sungroul Kim *, Sujung Park and Jeongeun Lee

Department of Environmental Health Sciences, Soonchunhyang University, Asan 31538, Korea;
psj57732398@gmail.com (S.P.); l01025263029@gmail.com (J.L.)
* Correspondence: Sungroul.kim@gmail.com; Tel.: +82-41-530-1266

Received: 27 March 2019; Accepted: 25 April 2019; Published: 12 May 2019

Featured Application: In consideration of relatively stable outcomes with the application of a correction factor for relative humidity, recently introduced inexpensive real-time monitors (IRMs), ESCORTAIR (ESCORT, Seoul, Korea) or PurpleAir (PA) (PurpleAir U.S.A.), our study supports their usage in PM$_{2.5}$ monitoring for various urban hotspots.

Abstract: Inexpensive (<$300) real-time particulate matter monitors (IRMs), using laser as a light source, have been introduced for use with a Wi-Fi function enabling networking with a smartphone. However, the information of measurement error of these inexpensive but convenient IRMs are still limited. Using ESCORTAIR (ESCORT, Seoul, Korea) and PurpleAir (PA) (PurpleAir U.S.A.), we evaluated the performance of these two devices compared with the U.S. Environmental Protection Agency (EPA) Federal Equivalent Monitoring (FEM) devices, that is, GRIMM180 (GRIMM Aerosol, Germany) for the indoor measurement of pork panfrying or secondhand tobacco smoking (SHS) and Beta-ray attenuation monitor (BAM) (MetOne, Grants Pass, OR) for outdoor measurement at the national particulate matter (PM$_{2.5}$) monitoring site near an urban traffic hotspot in Daejeon, South Korea, respectively. The PM$_{2.5}$ concentrations measured by ESCORTAIR and PA were strongly correlated to FEM (r = 0.97 and 0.97 from indoor pan frying; 0.92 and 0.86 from indoor SHS; 0.85 and 0.88 from outdoor urban traffic hotspot). The two IRMs showed that PM$_{2.5}$ mass concentrations were increased with increased outdoor relative humidity (RH) levels. However, after applying correction factors for RH, the Median (Interquartile range) of difference compared to FEM was (14.5 (6.1~23.5) %) for PA and 16.3 (8.5–28.0) % for ESCORTAIR, supporting their usage in the home or near urban hotspots.

Keywords: PM$_{2.5}$; sensor; correction; pan frying; secondhand smoke; urban traffic

---

1. Introduction

A large volume of previous epidemiological studies relied on the use of ground-based fixed national monitoring stations [1,2]. However, recently, inexpensive (<$300) particulate matter (PM) monitors (PM) have been introduced for home usage in South Korea. These devices can provide PM distribution patterns at high temporal and spatial resolution [3–5] which is a substantial improvement on establishing a pollution monitoring networking system as well as environmental epidemiologic study [6], as compared to traditional approaches that relied on relatively small number of ground-based fixed national air monitoring stations or mobile sampling techniques.

Most of these low-cost devices are classified into two groups, that is, optical particle counters (OPCs) or photometers. OPCs use the light scattered from individual particle to estimate the concentration of particles in different size ranges [7,8]. These data, along with assumptions of the particle shape and density, can be converted to estimate mass concentrations that compare favorably with reference instruments [7,8]. However, it has been reported that there may be bias when aerosol type or size is unknown [7,8].

Photometers use a light source to illuminate sensing zones that contain many particles at one time [9,10] and obtained that the mass concentration of aerosol scales linearly with the amount of light scattered by an assembly of particles captured at a discrete angle from the incident light [11]. The light scattered by the assembly of particles is measured by a photodetector at an angle specific to the photometer model, often 90° from the incident light [12]. The intensity of scattered light is directly proportional to gravimetrically measured mass concentration, although the relationship is dependent on the light scattering characteristics, density and size distribution of the particle [12].

The cost of research grade light scattering instruments (approximately, $10,000 or higher) limits their use to conduct studies at high temporal and/or spatial resolution. In Korea, a laser-based inexpensive OPC based IRM, for example, ESCORTAIR, was recently introduced. However, the reliability of ESCORTAIR have been unknown. Therefore, it may be necessary to test this new device before it is applied in a high spatial-temporal resolution exposure assessment study. As a proper calibration protocol providing a correction factor can have a dramatic impact on precision, accuracy and bias of a real-time monitor, researchers evaluated the OPC or photometers for use in the laboratory, outdoors or in the home in other countries [3,9,13–16]. A recent article reported that "one size fits all" approach to obtain $PM_{2.5}$ mass concentrations by OPC result in relatively high uncertainty in complex exposure situations. Although OPC Therefore, corresponding conversion curve approach may be most valuable when a relatively high contrast is expected in exposure levels for example, daytime home with indoor combustion sources, BBQ or secondhand smoke versus night time or day time outside with heavy traffic volume versus night time [16]. To our knowledge, no one has rigorously evaluated the performance of IRMs, operated with OPC or photometer, in Korea with comparison of the U.S. Federal Equivalent Method (US FEM) [17].

In this study, we evaluated the performance of inexpensive (less than $300) real-time PM monitors (IRMs), with high cost (about $2000–$10,000) and cross-comparisons between them and research grade PM monitors (RGMs). We used US FEMs as reference instruments (approximately $20,000 or higher) and provided a final error of mass concentration ($PM_{2.5}$) measurement after applying correction factors in this study.

2. Materials and Methods

2.1. $PM_{2.5}$ Real-Time Monitors

A laser-based light-scattering $PM_{2.5}$ sensor monitor (ESCORTAIR, ESCORT, Seoul, Korea) (weight <300 g, volume <510 cm^3) consisted of an optical particle counting (OPC) PM sensor (INNOSIPLE1), CO_2 sensor, temperature relative humidity sensor, data transfer networking module and light-emitting diode (LED) display screen (Figure 1). In the ESCORTAIR, the sensing volume is illuminated with a laser and airborne particles are counted and processed one at a time. There were various IRMs commercially available in South Korea. In this study, however, we chose ESCORTAIR as they allowed us to directly transfer data to our data server using its Wi-Fi function.

For comparison purposes, another inexpensive photometer type of PM monitor (PA, PurpleAir, Draper, UT, USA) (https://www.purpleair.com/), mounting two Plantower sensors in a monitor, was used. PA is recommended its usage by AQ-SPEC (Air quality sensor performance evaluation center, South Coast Air Quality Management District, CA, USA) or US EPA (Environmental Protection Agency, NC, USA) as an IRM. The inlet system of these IRMs did not have an impactor or a cyclone unlike that of the RGM.

The performance of the two IRMs (one OPC, that is, ESCORTAIR and one photometer, that is, PA) costed less than $300 were simultaneously compared with those of high-cost devices ($10,000 or so), that is, research-grade laser photometers including PDR-1500 (Thermo Scientific, Waltham, MA, USA) and SIDEPAK AM510 (TSI, Inc., Shoreview, MN, USA).

These research-grade monitors have a cyclone inlet or an impact inlet for the measurement of the respirable fraction of airborne particulate matters in different environments and can provide real-time data. The SIDEPAK is a portable battery-operated personal aerosol monitor with an impact inlet and light-scattering laser photometer that provides real-time aerosol mass concentration. The PDR-1500 (Personal DataRAM 1500) is a nephelometric monitor with a cyclone inlet for the measurement of the respirable fraction of the airborne particulate matters. The PDR-1500 can simultaneously collect particles on a 37 mm filter for the gravimetric analysis by passing through the sensing zone.

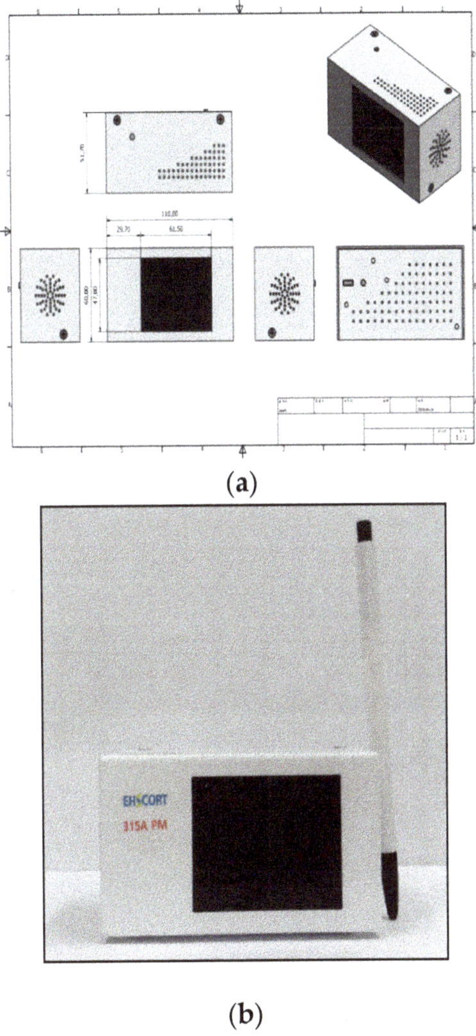

(a)

(b)

Figure 1. Layout of ESCORTAIR installed INNOSOPLE-1 PM sensor for measuring particulate matter (PM) concentrations.

2.2. Federal Equivalent Method

As mentioned above, to conduct a comparative measurement of IRMs, we used the U.S. Environmental Protection Agency (EPA) Federal Equivalent Method (FEM), that is, GRIMM 180 (GRIMM Aerosol, Technik Ainring GmbH & Co. KG, Ainring, Germany) for indoor testing and BAM-1022 (MetOne, Grants Pass, OR, USA) for outdoor field tests [17].

The Grimm Technologies, Inc. Model EDM 180 PM2.5 Monitor is a light scattering OPC monitor operated for 24 h at a volumetric flow rate of 1.2 L/min, configured with a Nafion$^®$- type air sample dryer. BAM-1022, a beta-ray attenuation mass monitor has a $PM_{2.5}$ particle size separator. Using BAM, we obtained 24 1-h average measurements at the national $PM_{2.5}$ monitoring supersite operated by National Institute of Environmental Research, Daejeon, Korea.

2.3. Flow Rate Inspection

Before each indoor and outdoor experiment, RGMs, that is, the PDR-1500 and SIDEPAK were zeroed with an in-line high efficiency particulate air (HEPA) filter and the flow of each device (1.52 L/min, 1.7 L/min and 0.5 L/min) was checked by a mass flowmeter (TSI, Inc., Shoreview, MN, USA). The temperature and relative humidity data were also downloaded from the devices if the device is equipped with sensors for these data. Flow rate of GRIMM, one of FEMs, was also checked with a similar way. The values from BAM, the other FEM, were used as reference, since it is located at the KOREA national $PM_{2.5}$ monitoring site (Daejeon) and operated with high QAQC programs [18,19] to report hourly outdoor $PM_{2.5}$ data.

2.4. Experimental Setting

We collected $PM_{2.5}$ concentrations by performing both an indoor exposure test on March 2018 and outdoor $PM_{2.5}$ monitoring at the national supersite located in Daejeon, Korea from June to July 2018 including rainy days. We used the 2 sets of each IRM or RGM for indoor and outdoor testing (serial numbers of devices: Table 1).

2.4.1. Indoor Test

The indoor test included scenarios of frying pork in a pan and exposure to secondhand smoke (SHS). Indoor pan-frying tests were conducted at inside of an empty laboratory (4 m × 10 m × 3.5 m, W D H), according to the protocol described in detail in our previous article [20]. In brief, this experiment was carried out over a 2 h measurement period per trial including first 9 min simulating the barbequing of pork belly (100 g). Standard operating protocol: a portion of pork belly (100 g) was pan-fried for 9 min: 3 min on Side A, 3 min on Side B; then 1.5 min on Side A again and finally 1.5 min on Side B again. When we fried pork belly, after the first 9 min, we opened a window (0.5 m × 0.8 m) to allow the ventilating air to naturally reduce the $PM_{2.5}$ concentration.

Measurement of the $PM_{2.5}$ levels from the exposure to secondhand smoke was conducted with a lighted cigarette burned. We opened the same window after 30 min during our secondhand smoke exposure level test. Then, we collected concentration data over next two hours.

During our indoor test, we maintained a minimum distance of 20 cm among these devices, at least 50 cm from the emission sources and 1 m above the floor. To minimize the effect of additional source contribution to our $PM_{2.5}$ measurement results, we reported our $PM_{2.5}$ results after subtracting the field background $PM_{2.5}$ concentrations measured at the baseline. We collected with the frying pan test and the secondhand exposure test three times with GRIMM 180 on separate days. The indoor test data from each device (80-s interval for PA, 60 s for the remaining devices) were calculated to the 5 min average level to be compared with the outcomes from GRIMM. The final number of data for the 2 h panfrying test was approximately 50 (2 sets of each device × 12 data point/h × 2.0 h) and that for the 2.5 h secondhand smoke exposure test was about 60 (2 sets of each device × 12 data point/h × 2.5 h) for $PM_{2.5}$, as well as temperature and relative humidity.

Table 1. Summary of $PM_{2.5}$ measurement range of concentration, measurement interval, weight and Wi-Fi availability reported by manufacturers and unit cost in South Korea by 30 June 2018.

	Device Classification [a]	Sensor Type [b]	Measurement Range	Sampling Pump Flow Rate	Precision [c]	Log Interval [c]	Unit Price ($)	Weight (g)	Wi-Fi
GRIMM (EDM180) [1] (GRIMM Aerosol, Germany) S/N #: 11R15047	FEM	OPC	0~3000,000 particles/Liter	1.2 L/min,	97% over the whole measuring range	5 s to 1 h	19,000	20,000	No
BAM-1020 [2]. (MetOne, OR) S/N #: N11181	FEM	Beta ray Attenuation	0~1000 mg/m^3	16.7 L/min	Exceeds US-EPA Class III $PM_{2.5}$ FEM standards	1 min to 1 h	23,750	24,500	No
ESCORTAIR [3] (ESCORT, Seoul, South Korea) S/N #: 6a:c6:3a:c7:83:bf 6a:c6:3a:c7:88:b1	IRM	OPC	1000 μg/m^3	NA	±10%@100~500 μg/m^3	30 s	300	400	Yes
PA [4] (PurpleAir, CA, USA) S/N #: A0:20:A6:A:AD:1B. A0:20:A6:B:83:32	IRM	Photometer	0~500 μg/m^3 as effective range	NA	±10%@100~500 μg/m^3	80 s	300	450	Yes
PDR-1500 [5] (Thermo Scientific, MA, USA) S/N #: CM17422007, CM17422017	RGM	Photometer	0.001~400 mg/m^3	Adjustable 0 to 3.5 L/min	±2% of reading or ±0.005 mg/m^3	1 s to 1 h	9000	1200	No
SIDEPAK [6] (TSI, MN, USA) S/N #: 11104037, 11008055	RGM	Photometer	0.001~100 mg/m^3	Adjustable 0 to 1.8 L/min	±0.001 mg/m^3 over 24 h as zero stability	1 s to 60 s	6000	460	No

[a] Federal equivalent method (FEM), Research Grade Monitor (RGM) and inexpensive real-time monitor (IRM); [b] Optical particle count (OPC); [c] Information from manufacture. 1. GRIMM: https://www.grimm-aerosol.com/fileadmin/files/grimm-aerosol/General_Downloads/The_Catalog_2018_web.pdf. 2. BAM: https://metone.com/wp-content/uploads/2017/08/bam-1020-9803_touch_screen_manual_rev_k.pdf. 3. ESCORTAIR: This study. 4. PA: https://www.purpleair.com/sensors. 5. PDR: https://www.newstarenvironmental.com/air-toxic-monitors/personal-dataram-pdr1500-aerosol-monitor.html?_vsrefdom=adwords&gclid=CjwKCAiA2fijBRAjEiwAuewS_cYiCFBPpx0d0bTaRKtmxe-1Kt22JVs352pQKq9e63XyqT_plbAZsRoCn9UQAvD_BwE. 6. SIDEPAK https://www.tsi.com/getmedia/84b5be22-c339-49bc-ab97-e3c4baee16c1/SidePak%20AM520_US_5001737_Web_1.

2.4.2. Outdoor Test

Using the same configuration of monitoring devices, we also measured the ambient $PM_{2.5}$ concentration at the outdoor Roof-top of one of the national $PM_{2.5}$ Supersites located in Daejeon, South Korea, operating BAM (MetOne, Grants Pass, OR, USA). Main body of BAM was installed at inside of an experiment laboratory of the Supersite while the inlet of BAM was located at the Roof-top. The outdoor temperature and RH during were measured by the sensors in ESCORTAIR, PA and PDR-1500 and the values were crosschecked. Measurements from IRMs were collected every minute, except for PA, which provided a response every 80 s. To compare the hourly concentration values provided by the Supersite, we calculated the hourly mean values using measured values acquired at hourly intervals from each IRM device. Final sample size for the outdoor data was 240 (2 sets of each device × 24 data points/day × 5 days) for $PM_{2.5}$, as well as temperature and relative humidity.

2.5. Statistical Analyses

The Spearman correlation tests were used to evaluate the associations among the outcomes of devices measured at indoor or outdoor environments, considering that variables were not normally distributed.

Using outdoor measurement data, we evaluate the association of device response with various relative humidity level. We also evaluated the associations of the daily mean concentrations from IRMs with those obtained with FEM methods using multivariate linear regression models. We used the values obtained with the FEM as dependent variables and those obtained with real-time devices as independent variables to obtain a correction factor. The hourly mean temperature and relative humidity data for each sampling date, which was previously compared with the nearest national meteorological monitoring sites, were used to adjust the effects of relative humidity on the association between IRMs and FEM outcomes. The corresponding slopes and coefficients of determinant (R2) for each RT monitor were also provided. The final measurement error (%) was calculated based on the U.S. EPA performance evaluation program for PM instruments;

$$Difference_i = \frac{Measurement_i - FEM_i}{FEM_i} \qquad (1)$$

EPA specifies that the percent bias goal for acceptable measurement uncertainty should be within ±20% [17]. Here, we reported median (IQR) value of the differences per device because the distribution of differences was not normal. We also provided mean of the differences just to compare with value in the guideline [21]. All analyses were conducted with SAS (Version 9.4) and R software (Version 2.15.3, R Development Core Team).

3. Results

3.1. $PM_{2.5}$ Concentration

In our indoor test, the median (IQR) $PM_{2.5}$ concentration over the pan-frying test was 86.8 μg/m^3 (17.8–254.4 μg/m^3) by the real-time ESCORTAIR, 104.9 μg/m^3 (43.9–228.2 μg/m^3) and 236.2 μg/m^3 (49.3–648.7 μg/m^3) by the real-time PA or PDR-1500 devices and 153.2 μg/m^3 (46.2–409.7 μg/m^3) by using GRIMM (Table 2). The median (IQR) concentrations for secondhand smoke (SHS) test were 20.9 (17.4– 156.6) μg/m^3, 31.2 (14.4–194.3) μg/m^3 and 28.4 (12.8–314.0) μg/m^3 for ESCORTAIR, PA and PDR, respectively, whereas GRIMM provided 23.5 (15.9–107.1) μg/m^3.

Simultaneous outdoor $PM_{2.5}$ monitoring results are also provided in Table 2. The median (IQR) of the hourly average values of ESCORTAIR and PA were 13.7 (7.3~21.2) and 19.7 (9.3~35.8) μg/m^3, which were an overestimation of the values obtained by BAM, one of U.S. EPA FEMs. During indoor testing, the median temperature and RH were approximately 20~22 °C and 37%. During outdoor testing, the median values (IQR) were 30.7 (25.6~40.7) °C and 56.4 (34.8~71.4) %, respectively (Table 2).

Table 2. $PM_{2.5}$ mass concentration ($\mu g/m^3$) (median, IQR) measured by real-time sensor devices and the federal equivalent method as well as the temperature and relative humidity throughout the sampling period.

	Indoor—Pan-Frying (n = 50)	Indoor—SHS (n = 60)	Outdoor—Urban Traffic Hotspot (n = 240)
GRIMM	153.2 (46.2–409.7)	23.5 (15.9–107.1)	NA
BAM	NA	NA	9.0 (4.0–22.0)
ESCORTAIR	86.8 (17.8–254.4)	20.9 (17.4–156.6)	13.7 (7.3–21.2)
PA	104.9 (43.9–228.2)	31.2 (14.4–194.3)	19.7 (9.3–35.8)
PDR-1500	236.2 (49.3–648.7)	28.4 (12.8–314.0)	13.8 (6.8–34.8)
SIDEPAK	261.3 (71.5–800.0)	50.0 (21.0–652.0)	29.1 (15.6–59.9)
Temp. (°C)	21.7 (21.1–21.7)	20.1 (19.7–20.5)	30.7 (25.6–40.7)
RH (%)	37.0 (35.0–39.0)	37.0 (35.0–38.0)	56.4 (34.8–71.4)

3.2. Correlations among Devices and the Fem

High-level correlations were obtained between the IRMs (ESCORT, PA) and FEM. The Spearman correlation coefficients were 0.97 (P = 0.0001) or 0.92 (P = 0.0001) between ESCORTAIR and GRIMM for indoor pan-frying or the SHS test, respectively. The correlation coefficients between PA and GRIMM were similar (0.97 (P = 0.0001) or 0.86 (P = 0.0001)) for the indoor test (Table 3). Our outdoor test also showed high correlation between ESCORTAIR and BAM (0.84 (P = 0.0001)). The correlation coefficient for PA was 0.88 (P = 0.0001).

Similarly, the measurements by RGMs (PDR-1500, SIDEPAK) and FEM were highly correlated: 0.97–0.98 for the Pan-frying test, 0.88~0.96 for the SHS test and 0.84–0.91 for the outdoor test. Between IRMs, that is, ESCORTAIR and PA, the association of measurements were strong to each other (r = 0.93: Indoor pan frying; 0.85: Indoor-SHS; 0.93: Outdoor-urban traffic). In addition, they showed similar correlation patterns to PDR-1500 or SIDEPAK (Table 3).

Table 3. Scatter plots and Spearman correlation coefficients between real-time $PM_{2.5}$ monitoring devices and FEM.

	Indoor-Pan-Frying					Indoor-SHS					Outdoor Urban Traffic Hotspot				
	FEM	E	PA	P	S	FEM	E	PA	P	S	FEM	E	PA	P	S
FEM	1					1					1				
E	0.97	1				0.92	1				0.85	1			
PA	0.97	0.93	1			0.86	0.85	1			0.88	0.93	1		
PDR	0.98	0.95	0.99	1		0.96	0.93	0.94	1		0.84	0.93	0.99	1	
S	0.98	0.99	0.96	0.98	1	0.88	0.86	0.88	0.93	1	0.91	0.91	0.99	0.99	1

FEM: Federal Equivalent Method: In this study Indoor: GRIMM Optical particle courting, OPC), Outdoor: BAM (Beta ray attenuation monitor). E: ESCORTAIR, PA: PurpleAir, PDR: PDR-1500, S: SIDEPAK.

3.3. Effects of Ambient Humidity for Outdoor Measurement

In this study, as seen in Supplementary Figure S1, we observed that the $PM_{2.5}$ concentration of IRMs were significantly increased with the increase in relative humidity level. The slope obtained from a simple regression line of FEM on IMRs, that is, (BAM = Slope * ESCORTAIR + intercept) at a relative humidity above 80%, was smaller than the slope at < 20.0%, 20.1~40.0%, 40.1~60.0%, 60.1~80.0%. The degree of decreasing trend was larger with IRMs, compared to two RGMs (Figure 2).

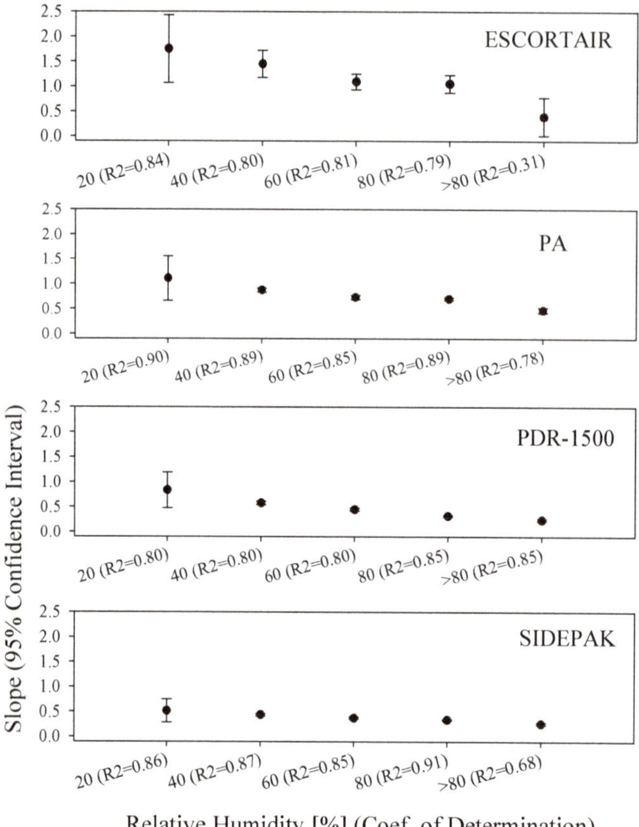

Figure 2. Change of the slope with 95% confidence interval obtained from the regression line of BAM (Beta ray attenuation monitor) on IRMs (Inexpensive real-time particulate matter monitors) or RGMs (research grade monitors) by the degree of relative humidity.

3.4. Correction Factor

Since we found the effect of RH on the measurements of ESCORTAIR or PA, we developed our own correction models for those IRM devices by performing stepwise linear regression. The correction factors were obtained from the regression models (1.11 for ESCORTAIR and 1.92 for PA, $p < 0.01$), (1.00 for ESCORTAIR and 0.87 for PA, $p < 0.01$) and (1.15 for ESCORTAIR and 0.70 for PA, $p < 0.01$) for the measurement of $PM_{2.5}$ resulting from indoor pan-frying, SHS or the urban traffic hotspot of South Korea, respectively.

After adjusting for temperature and relative humidity, the results were unchanged for the indoor tests (1.10 for ESCORTAIR and 1.90 for PA, $p < 0.01$), (0.97 for ESCORTAIR and 0.81 for PA, $p < 0.01$) (Table 3). For the outdoor measurement, the change of slopes (0.72 for ESCORTAIR and 0.77 for PA, $p < 0.01$) was relatively small but the R2 values were changed (Table 4).

Table 4. Slopes obtained from stepwise linear calibration models with adjusted R^2 (Dependent variable: US EPA FEM, Independent variable: IRM).

	Indoor—Pan-Frying				Indoor—SHS				Outdoor—Urban Traffic Hotspot			
	Single		Multivariate *		Single		Multivariate *		Single		Multivariate *	
	β	R^2	β	R^2	β	R^2	β	R^2	β	R^2	β	R^2
ESCORTAIR	1.11	0.98	1.10	0.98	1.00	0.92	0.97	0.92	1.15	0.70	1.14	0.81
PA	1.92	0.94	1.90	0.94	0.87	0.89	0.81	0.90	0.70	0.83	0.71	0.87
PDR-1500	0.33	0.98	0.33	0.98	0.54	0.91	0.49	0.92	0.33	0.72	0.36	0.80
SIDEPAK 1	0.34	0.98	0.32	0.99	0.28	0.90	0.31	0.92	0.35	0.84	0.36	0.89

* results obtained after adjusting for temperature and relative humidity.

3.5. Bias after Application of Correction Factors

We then determined the extent to which the original $PM_{2.5}$ data measured by ESCORTAIR were improved after applying the correction factor obtained from the model by performing comparative analyses using the outcomes of multivariate regression models (Table 4). Using the corrected data, the final coefficient of determination (R2) between FEM (y) and ESCORTAIR (x) was 0.81. The coefficient for PA was 0.87. We found the difference (median (IQR)) with the calibrated data, compared to FEM, to be 16.3. (8.5~28.0)% for ESCORTAIR and 14.5 (6.1 to 23.5)% for PA for outdoor environments (Figure 3). The bias (mean of the difference) was 13.1% for ESCORTAIR and 7.8% for PA for outdoor. The bias for indoor data was at least similar or lower than the bias level obtained from the outdoor test.

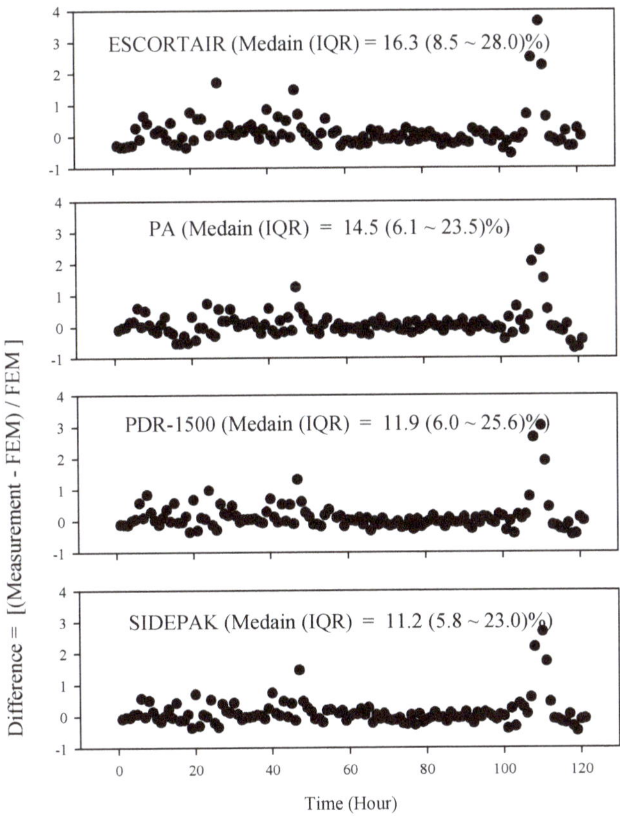

Figure 3. Distribution of difference of outdoor measurements against the FEM; BAM for outdoors.

4. Discussion

In this study, we compared $PM_{2.5}$ concentrations measured with IRMs and RGMs to those measured with U.S. EPA FEMs. A relationship between the real-time $PM_{2.5}$ concentration and FEM were acceptable (R = 0.97 to 0.98: Indoor pan frying; 0.86 to 0.96: Indoor-SHS; 0.84 to 0.91: Outdoor-urban traffic, respectively).

This study conducted indoor testing with common $PM_{2.5}$ sources of Korea, that is, frying pork in a pan or smoking. The indoor test was conducted in an indoor laboratory in which the room temperature was maintained at 20~22 °C and the relative humidity at 37% because we assumed that most low-cost PM sensing devices would be established inside susceptible populations' homes where the indoor temperature and humidity level would be relatively stable, compared to the outdoor environment.

In addition, in this study, we extended our comparison test by performing them outside with a federal equivalent method. We discovered that, on days with a high level of outdoor RH (80% or higher), our IRMs overestimated the $PM_{2.5}$ level. Thus, finally, we got the correction model providing a correction factor for ESCORTAIR to adjust for the effect of temperature and relative humidity as such a process has been conducted in a regulatory monitoring site (South Coast Air Quality Monitoring District, SCAQMD) for the field use of these kinds of inexpensive sensing devices including PA [22,23].

Several previous studies provided a correction factor for SIDEPAK monitors: 0.77 in Northern California, U.S.A. (ambient air), 0.43 or 0.52 in Italy (ambient air in urban or rural areas) and 0.42 in Italy (indoor-outdoor mixed environment) [24–26], which were comparable to our results (0.36). Our correction factor (0.36) for PDR-1500 was smaller than results reported by Wang et al. (2016) [27] (0.71 compared to PDR-1500 using its own filtering method) but very similar to the values obtained by Ramachandran et al. (2000) (0.33) and Wallace et al. (2011) (0.38), who conducted their studies on atmospheric environments [28,29].

A lack of quantitative information on the speciation of particles, traffic volume, type of vehicle or the particular sampling time or season limits further exploration of the basis for the differences in the correction factors between these studies and ours. Nevertheless, our correction model for ESCORTAIR, with consideration of the RH level, was derived in a similar way to that with which we obtained the factor for SIDEPAK or PDR-1500. Thus, we consider no significant systematic errors to have been involved in the calculation process. A good linear relationship has been obtained between the $PM_{2.5}$ mass concentration of FEM and the responses of low-cost PM sensors as reported previously in other country [8,9,25,30]. We demonstrated the urban hotspot specific correction factor for a light-scattering sensor in Korean urban environment of interest to enable our findings and methodology to be extended and replicated by researchers who are interested in the utility of low-cost sensing device, such as ESCORTAIR, in South Korea.

OPCs are reportedly good at estimating mass if they have numerous bins, such as GRIMM [31]. However, estimating the mass concentration from a limited number of bins may be subject to a measurement error during the conversion process with the factory-provided internal conversion algorithm. Therefore, we used an additional correction factor for ESCORTAIR after comparison to GRIMM for proper usage under Korean circumstances.

It is well established that the response of monitors based on light scattering varies with aerosol size distribution, composition and optical properties and need a proper calibration process [13,32,33]. No single calibration model (or correction model) can enable accurate performance for all particle sources in microenvironments. This challenge applies to both research and consumer monitors. Although gravimetric measurements may be used to determine a source- or environment-specific calibration for a research study, the approach is not practical for routine monitoring in homes. A key objective of continuous monitoring—to activate controls—can be achieved if the monitor reliably and clearly responds to sources that account for the majority of particles in the home even if responses are not quantitative.

We conducted this study by assuming that indoor $PM_{2.5}$ emission sources exist, that is, from frying pork in a pan and smoking, with consideration of Korean life style [20] and relatively high smoking prevalence [34]. We determined that our linear model obtained from the indoor test for each single-aerosol type showed excellent performance (R2 = 0.98 or 0.92 for ESCORTAIR; 0.94 or 0.90 for PA), compared to FEM responses. However, the responses of ESCORTAIR as well as PA were relatively less precise but good (R2 = 0.81, 0.87) for monitoring in urban traffic hotspots suggesting that IRMs need a site-specific calibration with a reference method before they are used [13,32,33].

The shape of the response curve can be related to the type of OEM sensors integrated with the monitor and relative humidity. Because we did not have information about the internal conversion factors (count to mass concentration for OPC type or light intensity to mass concentration for photometer), using GRIMM or BAM, known as US FEM, we tried to obtain our own correction factors for usage of ESCORTAIR or PA in urban indoor or outdoor settings.

The limitations of our study should be noted. First, the sample size in this study was relatively small. However, in our tests, we determined a $PM_{2.5}$ range of 10 to 3000 $\mu g/m^3$ including both indoor and outdoor measurement. This range ensures that the concentration distributions would not be systematically biased. Our $PM_{2.5}$ concentrations might not be representative of each sampling season or area as a result of spatial-temporal variations. Additional experiments are needed to understand the stability of our correction factor in different seasons and/or in other locations, that is, in industrial or rural areas of Korea. Extending the sampling periods for each season and location would ensure that our results are more representative. Furthermore, in future studies, measurements of the wind direction and speed are expected to provide improved correction factors between the IRM and FEM methods. As we mentioned in the method section, we checked flow rate for our RGMs or FEMs prior to our experiment. For IRMs, we have considered measuring flow rate but due to its open wide inlet and very low flow rate, we could not connect it to our mass flowmeter properly. Instead, especially, before our outdoor test, we operated 5 ESCORTAIRs and 5 PAs simultaneously and checked measurement errors between devices. Then, we selected 2 of them which provide best outcomes, compared to FEM. A preparation of QC/QA test program for massive products of IRMs are recommended. And for ESCORTAIR, like PA, application of weather proof design is suggested. In addition, future studies may be necessarily conducted to obtain site-specific correction factors including at coal power plants or in rural areas.

In this study, the performances of the IRM with RGM and that of the FEM operated with a high QCQA program were compared with one-hour monitoring intervals at national $PM_{2.5}$ monitoring site. This makes this study unique compared to previous studies, which were mostly conducted with one-day interval gravimetric methods at ordinary sampling sites.

Despite the growing public interest in reducing personal exposure levels to $PM_{2.5}$ in Korea, IRM monitoring still faces challenges in terms of providing real-time concentration information. Although the number of national $PM_{2.5}$ monitoring sites in Korea is increasing, additional IRMs in hotspots or communities are required because they can detect continuous spatial—temporal variations and identify nearby exposure sources on a real-time basis in a micro-environment of hotspots.

This study found that the measurement of $PM_{2.5}$ concentrations with recently developed laser-based IRM under- or over-estimates $PM_{2.5}$ concentrations obtained from FEM while its bias could be approximately 11 to 16% even at urban outdoor hotspots with traffic sources with high relative humidity levels. Therefore, the application of a correction factor is strongly suggested for inexpensive laser-based monitoring devices.

5. Conclusions

Our study determined that on days with a high level of outdoor RH (80% or higher), our IRMs overestimated the outdoor $PM_{2.5}$ level and showed the necessity of a correction factor for IRMs to adjust for the effect of temperature and relative humidity. $PM_{2.5}$ concentrations measured with IRMs need to be subjected to quality control and quality assurance evaluation before these monitors are used for the quantification of $PM_{2.5}$ levels in urban indoor or outdoor atmospheric environments in Korea. In consideration of relatively stable outcomes with the application of correction factors for recently developed new IRMs—ESCORTAIR or PA—our study supports their usage in networking monitoring for various urban hotspots.

Supplementary Materials: The following are available online at http://www.mdpi.com/2076-3417/9/9/1947/s1. Figure S1. Scatter plots of $PM_{2.5}$ concentrations of IRMs with those of FEM, i.e., BAM, by the level of relative humidity. The data points refer to each paired hourly mean value (n = 240, except ESCORT which had 120 points due to malfunction of a device with records of high internal temperature values during our outdoor test).

Author Contributions: S.K. designed this study, wrote manuscript and conducted interpretation of the quantitative aspects of data analysis. S.P. performed modeling simulation and J.L. provided editorial efforts. S.K. supervised the whole study.

Funding: This research was funded by Environmental Health Research Center Project (2016001360002) by Korea Environmental Industry & Technology Institute, Ministry of Environment, South Korea

Acknowledgments: The authors deeply appreciate the technical comments for sensor evaluation from. Andrea Clements and Timothy Buckley at the national exposure research laboratory, U.S. EPA. The authors also appreciate the assistance of Sungmin Jung and the staff (Minhee Lee, Taekyung Hwang, Jieun Jeong) at the Korea PM2.5 supersite of Daejeon, South Korea for their support with data collection and the help of Juhee Kim in data screening. The content is solely the responsibility of the authors and does not necessarily represent the official views of the US EPA or Korea NIER (National Institute for Environmental Research).

Conflicts of Interest: The authors declare no conflict of interest.

References

1. Orellano, P.; Quaranta, N.; Reynoso, J.; Balbi, B.; Vasquez, J. Effect of outdoor air pollution on asthma exacerbations in children and adults: Systematic review and multilevel meta analysis. *PLoS ONE* **2017**, *12*, e0174050. [CrossRef]

2. Prieto Parra, L.; Yohannessen, K.; Brea, C.; Vidal, D.; Ubilla, C.A.; Ruiz Rudolph, P. Air pollution, $PM_{2.5}$ composition, source factors, and respiratory symptoms in asthmatic and nonasthmatic children in Santiago, Chile. *Environ. Int.* **2017**, *101*, 190–200. [CrossRef] [PubMed]

3. Gillooly, S.E.; Zhou, Y.; Vallarino, J.; Chu, M.T.; Michanowicz, D.R.; Levy, J.I.; Adamkiewicz, G. Development of an in~home, real-time air pollutant sensor platform and implications for community use. *Environ. Pollut.* **2019**, *244*, 440–450. [CrossRef] [PubMed]

4. Genikomsakis, K.N.; Galatoulas, N.; Dallas, P.I.; Ibarra, L.M.C.; Margaritis, D.; Ioakimidis, C.S. Development and On Field Testing of Low Cost Portable System for Monitoring $PM_{2.5}$ Concentrations. *Sensors* **2018**, *18*, 1033. [CrossRef] [PubMed]

5. Shao, W.; Zhang, H.; Zhou, H. Fine Particle Sensor Based on Multi Angle Light Scattering and Data Fusion. *Sensors* **2017**, *17*, 1033. [CrossRef] [PubMed]

6. Aneja, V.P.; Pillai, P.R.; Isherwood, A.; Morgan, P.; Aneja, S.P. Particulate matter pollution in the coal producing regions of the Appalachian Mountains: Integrated ground based measurements and satellite analysis. *J. Air Waste Manag. Assoc.* **2017**, *67*, 421–430. [CrossRef] [PubMed]

7. Burkart, J.; Steiner, G.; Reischl, G.; Moshammer, H.; Neuberger, M.; Hitzenberger, R. Characterizing the performance of two optical particle counters (Grimm OPC1.108 and OPC1.109) under urban aerosol conditions. *J. Aerosol. Sci.* **2010**, *41*, 953–962. [CrossRef] [PubMed]

8. Sousan, S.; Koehler, K.; Hallett, L.; Peters, T.M. Evaluation of the Alphasense Optical Particle Counter (OPC N2) and the Grimm Portable Aerosol Spectrometer (PAS 1.108). *Aerosol. Sci. Technol.* **2016**, *50*, 1352–1365. [CrossRef]

9. Dacunto, P.J.; Klepeis, N.E.; Cheng, K.C.; Acevedo Bolton, V.; Jiang, R.T.; Repace, J.L.; Ott, W.R.; Hildemann, L.M. Determining $PM_{2.5}$ calibration curves for a low cost particle monitor: Common indoor residential aerosols. *Environ. Sci. Process. Impacts* **2015**, *17*, 1959–1966. [CrossRef] [PubMed]

10. Lanki, T.; Alm, S.; Ruuskanen, J.; Janssen, N.A.; Jantunen, M.; Pekkanen, J. Photometrically measured continuous personal PM(2.5) exposure: Levels and correlation to a gravimetric method. *J. Expo. Anal. Environ. Epidemiol.* **2002**, *12*, 172–178. [CrossRef] [PubMed]

11. O'Shaughnessy, P.T.; Slagley, J.M. Photometer response determination based on aerosol physical characteristics. *AIHA J.* **2002**, *63*, 578–585. [CrossRef]

12. Baron, P.A. Aerosol Photometers for Respirable Dust Measurements. In *NIOSH Manual of Analytical Methods*; 1998. Available online: https://www.cdc.gov/niosh/docs/2003--154/pdfs/chapter~{}g.pdf (accessed on 25 March 2019).

13. Kelly, K.E.; Whitaker, J.; Petty, A.; Widmer, C.; Dybwad, A.; Sleeth, D.; Martin, R.; Butterfield, A. Ambient and laboratory evaluation of a low cost particulate matter sensor. *Environ. Pollut.* **2017**, *221*, 491–500. [CrossRef] [PubMed]

14. Semple, S.; Ibrahim, A.E.; Apsley, A.; Steiner, M.; Turner, S. Using a new, low cost air quality sensor to quantify second hand smoke (SHS) levels in homes. *TOB Control.* **2015**, *24*, 153–158. [CrossRef] [PubMed]

15. Zhou, X.; Aurell, J.; Mitchell, W.; Tabor, D.; Gullett, B. A small, lightweight multipollutant sensor system for ground mobile and aerial emission sampling from open area sources. *Atmos. Environ.* **2017**, *154*, 31–41. [CrossRef] [PubMed]

16. Franken, R.; Maggos, T.; Stamatelopoulou, A.; Loh, M.; Kuijpers, E.; Bartzis, J.; Steinle, S.; Cherrie, J.W.; Pronk, A. Comparison of methods for converting Dylos particle number concentrations to $PM_{2.5}$ mass concentrations. *Indoor Air* **2019**, *29*, 450–459. [CrossRef]

17. Williams, R.; Vasu Kilaru, E.; Snyder, A.; Kaufman, T.; Dye, A.; Rutter, A.; Russell, A.; Hafner, H. *Air Sensor Guidebook*; EPA/600/R-14/159 (NTIS PB2015-100610); U.S. Environmental Protection Agency: Washington, DC, USA, 2014.

18. National Institute of Environmental Research (NIER). *2015 Annual Report of Intensive air Quality Monitoring Station, NIER GP2016–160*; NIER: Incheon, Korea, 2016.

19. Yu, G.; Park, S.; Ghim, Y.; Shin, H.; Lim, C.; Ban, S.; Yu, J.; Kang, H.; Seo, Y.; Kang, K.; et al. Difference in Chemical Composition of $PM_{2.5}$ and Investigation of its Causing Factors between 2013 and 2015 in Air Pollution Intensive Monitoring Stations. *J. Korean Soc. Atmos. Environ.* **2018**, *34*, 16–37. [CrossRef]

20. Lee, S.; Yu, S.; Kim, S. Evaluation of Potential Average Daily Doses (ADDs) of $PM_{2.5}$ for Homemakers Conducting Pan Frying Inside Ordinary Homes under Four Ventilation Conditions. *Int J. Environ. Res. Public Health* **2017**, *14*, 78. [CrossRef]

21. Rosner, B. *Hypothesis Testing, Fundamentals of Biostatistics*; Duxbury: Parific Grove, CA, USA, 2000; Chapter 7.

22. Kim, B.M.; Teffera, S.; Zeldin, M.D. Characterization of $PM_{2.5}$ and PM_{10} in the South Coast Air Basin of southern California: Part 1 Spatial variations. *J. Air Waste Manag. Assoc.* **2000**, *50*, 2034–2044. [CrossRef]

23. Kim, B.M.; Teffera, S.; Zeldin, M.D. Characterization of $PM_{2.5}$ and PM_{10} in the South Coast Air Basin of southern California: Part. 2 Temporal variations. *J. Air Waste Manag. Assoc.* **2000**, *50*, 2045–2059. [CrossRef] [PubMed]

24. Borgini, A.; Tittarelli, A.; Ricci, C.; Bertoldi, M.; De Saeger, E.; Crosignani, P. Personal exposure to $PM_{2.5}$ among high school students in Milan and background measurements: The EuroLifeNet study. *Atmos. Environ.* **2011**, *45*, 4147–4151. [CrossRef]

25. Jiang, R.T.; Acevedo Bolton, V.; Cheng, K.C.; Klepeis, N.E.; Ott, W.R.; Hildemann, L.M. Determination of response of real-time SidePak AM510 monitor to secondhand smoke, other common indoor aerosols, and outdoor aerosol. *J. Environ. Monit.* **2011**, *13*, 1695–1702. [CrossRef]

26. Karagulian, F.; Belis, C.A.; Lagler, F.; Barbiere, M.; Gerboles, M. Evaluation of a portable nephelometer against the Tapered Element Oscillating Microbalance method for monitoring PM(2.5). *J. Environ. Monit* **2012**, *14*, 2145–2153. [CrossRef] [PubMed]

27. Wang, Z.; Calderon, L.; Patton, A.P.; Sorensen Allacci, M.; Senick, J.; Wener, R.; Andrews, C.J.; Mainelis, G. Comparison of real-time instruments and gravimetric method when measuring particulate matter in a residential building. *J. Air Waste Manag. Assoc.* **2016**, *66*, 1109–1120. [CrossRef]

28. Ramachandran, G.; Adgate, J.L.; Hill, N.; Sexton, K.; Pratt, G.C.; Bock, D. Comparison of short term variations (15 min averages) in outdoor and indoor $PM_{2.5}$ concentrations. *J. Air Waste Manag. Assoc.* **2000**, *50*, 1157–1166. [CrossRef]

29. Wallace, L.A.; Wheeler, A.J.; Kearney, J.; Van Ryswyk, K.; You, H.; Kulka, R.H.; Rasmussen, P.E.; Brook, J.R.; Xu, X. Validation of continuous particle monitors for personal, indoor, and outdoor exposures. *J. Expo. Sci Environ. Epidemiol.* **2011**, *21*, 49–64. [CrossRef] [PubMed]

30. Shi, J.; Chen, F.; Cai, Y.; Fan, S.; Cai, J.; Chen, R.; Kan, H.; Lu, Y.; Zhao, Z. Validation of a light scattering $PM_{2.5}$ sensor monitor based on the long term gravimetric measurements in field tests. *PLoS ONE* **2017**, *12*, e0185700. [CrossRef]

31. Peters, T.M.; Ott, D.; O'Shaughnessy, P.T. Comparison of the Grimm 1.108 and 1.109 portable aerosol spectrometer to the TSI 3321 aerodynamic particle sizer for dry particles. *Ann. Occup. Hyg.* **2006**, *50*, 843–850. [PubMed]
32. Njalsson, T.; Novosselov, I. Design and Optimization of a Compact Low Cost Optical Particle Sizer. *J. Aerosol. Sci.* **2018**, *119*, 1–12. [CrossRef] [PubMed]
33. Northcross, A.L.; Edwards, R.J.; Johnson, M.A.; Wang, Z.M.; Zhu, K.; Allen, T.; Smith, K.R. A low cost particle counter as a realtime fine particle mass monitor. *Environ. Sci. Process. Impacts* **2013**, *15*, 433–439. [CrossRef]
34. Kim, S.; Jung, A. Optimum cutoff value of urinary cotinine distinguishing South Korean adult smokers from nonsmokers using data from the KNHANES (2008–2010). *Nicotine Tob Res.* **2013**, *15*, 1608–1616. [CrossRef]

Article

Real-Time Monitoring of Indoor Air Quality with Internet of Things-Based E-Nose

Mehmet Taştan * and Hayrettin Gökozan

Department of Electronic, Turgutlu Vocational School, Manisa Celal Bayar University, 45400 Manisa, Turkey
* Correspondence: mehmet.tastan@cbu.edu.tr; Tel.: +90-236-313-55-02

Received: 22 June 2019; Accepted: 15 August 2019; Published: 20 August 2019

Abstract: Today, air pollution is the biggest environmental health problem in the world. Air pollution leads to adverse effects on human health, climate and ecosystems. Air is contaminated by toxic gases released by industry, vehicle emissions and the increased concentration of harmful gases and particulate matter in the atmosphere. Air pollution can cause many serious health problems such as respiratory, cardiovascular and skin diseases in humans. Nowadays, where air pollution has become the largest environmental health risk, the interest in monitoring air quality is increasing. Recently, mobile technologies, especially the Internet of Things, data and machine learning technologies have a positive impact on the way we manage our health. With the production of IoT-based portable air quality measuring devices and their widespread use, people can monitor the air quality in their living areas instantly. In this study, e-nose, a real-time mobile air quality monitoring system with various air parameters such as CO_2, CO, PM_{10}, NO_2 temperature and humidity, is proposed. The proposed e-nose is produced with an open source, low cost, easy installation and do-it-yourself approach. The air quality data measured by the GP2Y1010AU, MH-Z14, MICS-4514 and DHT22 sensor array can be monitored via the 32-bit ESP32 Wi-Fi controller and the mobile interface developed by the Blynk IoT platform, and the received data are recorded in a cloud server. Following evaluation of results obtained from the indoor measurements, it was shown that a decrease of indoor air quality was influenced by the number of people in the house and natural emissions due to activities such as sleeping, cleaning and cooking. However, it is observed that even daily manual natural ventilation has a significant improving effect on air quality.

Keywords: internet of things; e-nose; indoor air quality; smart home; ESP32

---

1. Introduction

In recent years, air pollution has been a major environmental problem and a global concern that has exceeded recommended national limits. Air pollution has negative effects on human health and ecosystems, as well as affecting the world's climate [1]. Air pollution can be classified as internal or external air pollution, depending on where the activities take place [2]. Outdoor air pollution occurs in an open environment, covering the entire atmosphere. Fossil fuels used to meet the energy needs of factories, industries and vehicles are the main activities contributing to agricultural and mining outdoor air pollution. The external air pollutants are mainly composed of nitrogen oxides (NOx), nitrogen dioxide (NO_2), sulfur dioxide (SO_2), ozone (O_3), carbon monoxide (CO), hydrocarbons and particulate matter (PM) of different particle sizes. Indoor air pollution, found in offices, hospitals, schools, libraries, entertainment areas, gymnasiums, public transport vehicles, etc., is classified as the pollution of the air of indoor areas [3]. Major indoor air pollutants include NOx, SO_2, O_3, CO, carbon dioxide (CO_2), volatile and semi-volatile organic compounds VOCs, PM, radon and microorganisms. The air quality index includes an internationally adopted parameter for assessing air quality.

Ground-level ozone reflects five pollution standards including PM, carbonic oxide, SO_2 and NO_2 [4]. Indoor air quality (IAQ) is very important for many people who spend most of their lives in closed spaces, such as the elderly, disabled, infants and patients [5]. Indoor air pollution occurs due to household activities and products used in these activities. Home cleaning products and paint materials emit toxic chemicals into the air and cause air pollution. Exposure to air pollution, especially indoor air pollution, is one of the largest environmental health risk factors and is directly related to millions of premature deaths worldwide each year [6]. Air pollution comes second on the list of deaths caused by non-contagious health reasons [7].

In the United States, indoor and outdoor air quality is regulated by the Environmental Protection Agency (EPA). The EPA has shown that indoor pollutant levels may be up to 100 times higher than the level of external pollutants and poor air quality is one of the five most important environmental risks threatening public health [8]. PM, a complex mixture of solid and liquid particles of organic and inorganic substances suspended in the air, is considered the most common pollutant. Air pollution from PM ($PM_{2.5}$ and PM_{10}) poses a significant health threat to people living in cities. The health-damaging PM, which can penetrate deep into the lungs, contributes to the risk of developing cardiovascular diseases as well as lung cancer [9]. In addition to health problems affecting the low IAQ nervous system, its effects are associated with other long-term diseases [10]. Simple interventions, such as natural ventilation provided by homeowners, have significant positive effects on IAQ. However, consideration of the household's thermal comfort and climate conditions in natural ventilation is crucial [11]. Furthermore, power losses due to manually regulated natural ventilation should be considered [12]. Therefore, indoor real-time monitoring of IAQ is very important to detect unhealthy conditions [13].

Today, cities face interesting challenges and problems to meet socio-economic development and quality of life goals, and the concept of "smart cities" responds to these challenges. The smart city is directly related to a strategy to reduce the problems caused by rapid urbanization [14]. The IoT is a network where physical objects are linked to each other or to larger systems. This network collects large amounts of data from different devices we use in our daily life and converts it into utilizable information [15]. The rapidly emerging IoT concept supports many different areas and applications including health, education, agriculture, industry and environmental monitoring. Real-time environmental monitoring and analysis is an important area of research for IoT [16]. IoT provides remarkable features to smart cities which improve environmental quality control, innovative real-life solutions and services. The smart home is an indispensable element of smart cities. In the future, smart homes will be fully managed through smartphone applications and will include IoT wearable devices [17] that are supported by microsensors. To date, several have been conducted in the literature with the aim of establishing real-time monitoring solutions for air quality analysis.

Numerous IoT-based air quality monitoring systems using micro-sensors for data collection have been proposed, including open source technologies for data processing and transmission [18]. According to an IAQ analysis [19], it has been reported that indoor $PM_{2.5}$ pollution mainly comes from outdoor sources. In another study, three typical activities such as a having fireplace in the house, cooking with kitchen appliances and toasting were investigated, and indoor pollutant levels such as $PM_{2.5}$, PM_{10} and VOC were found to be above the limit values [20]. Monitoring of the PM concentration for 63 participants between the ages of 18–65 years in the Perth metropolitan area indicates that the PM concentration increases the number of heartbeats by 4–6 beats per minute (bpm) and has a significant effect on systolic blood pressure (SBP) [21]. An IoT-based IAQ measurement system was created using the WEMOS D1 mini microcontroller and the PMS5003 PM sensor. This system allows the household individual to intervene for ambient assisted living (AAL) [22].

In this study, a low-cost, portable, IoT-based and real-time monitoring system that can measure ambient air quality with a range of sensors is proposed. For the proposed system, an Android interface has been designed primarily by the Blynk platform. Then, with the embedded architecture ESP32 module, the controller unit of the system is created. Using this IoT controller with an internal Wi-Fi module, all measured air quality data are displayed in the mobile app and these data are stored on a cloud server. The mobile interface provides users with numerical and graphical data on contaminating gas concentrations, temperature and humidity. The proposed e-nose measurement system sends a notification to the users via the mobile application if any gas concentration levels reach health-threatening values. In this way, households can take measures to reduce gas concentrations when necessary. The data can be viewed graphically through the mobile interface which allows users to observe the effects of activities such as sleep, cleaning and cooking on the gas concentration. In addition, the IoT-based e-nose air quality measurement system has a low-cost (about $100), easy-to-install and open-source feature produced by a DIY approach.

2. Related Work

Due to the fast advancement in IoT and sensor techniques, interest in air quality measurements continues to grow. In [23] an environmental monitoring system was propose that included the sensors Raspberry Pi and MICS-4514, which enabled the assessment both indoors and outdoors at a university campus of multiple environmental parameters such as temperature, humidity, light, noise level, CO, NO_2. A proposed system for detecting environmental contamination [24] used various environmental detectors to identify temperature, humidity, ambient light, gas sensors and PM. All data measured by the system using the sensors STM32f4xx and Sharp GP2Y1010 for PM detection, and TGS5342 for CO, are stored in the internal storage and on an Internet server over the Wi-Fi network. Using the ATmega328AVR controller, DHT22 temperature sensor, Sharp GP2Y1010AU0F dust sensor and UVM-30A UV sensors, an Integrated Environmental Monitoring System (IEMS) [25] was proposed to detect the microenvironment. In another study [26], the Nano Environmental Monitoring System (nEMoS) was proposed using an IoT-based indoor environmental quality (IEQ) assessment system created using an Arduino Uno module and low-cost sensors such as DHT22. To determine the quality of packaged products, an IoT-based measuring system was intended, including low-cost pressure, temperature, humidity, gas sensors (BME680, DHT22 and MQ5), Arduino and XBee wireless module [27]. In the study, in which iAQ, an air quality surveillance system based on IoT architecture, was proposed for AAL [28], the wireless sensor nodes (WSN), Gateway and Android user interface provided environmental data such as temperature, humidity and CO_2 to users through the mobile application. In the IoT-based system [29] developed for real-time IAQ surveillance, multiple gases such as CO_2, NO_2, ethanol, methane and propane were detected using ESP8266 as the controller and sensors MICS-6814 and MICS-6814 as the detection unit. This system, based on open source technology, has a mobile phone application that transmits real-time notifications to users. The solution is based on the IoT concept and is entirely wireless in a study that presented an IAQ surveillance solution that can measure temperature, humidity, PM_{10}, CO_2 and light intensity in real-time [30]. The Arduino UNO emitter utilizes the ESP8266 controller and the "ThingSpeak" open-source IoT platform to record wireless data to enable wireless Internet access. In this study, a real-time surveillance system based on IoT architecture for PM surveillance was presented [31]. The iDust system was created using open source technologies and low-cost sensors. System measurement data, consisting of a WEMOS D1 Wi-Fi controller and PMS 5003 dust sensor, are transferred to users via an IoT-based implementation.

3. Materials and Methods

Low air quality poses a significant health threat for individuals who spend most of their time indoors. Some pollutants such as tobacco smoke, CO, NO_2, formaldehyde, asbestos fibres, microorganisms and allergens are known to be closely related to health problems. Temperature and humidity monitoring are part of everyday life, but in the vast majority of buildings, real-time air quality monitoring is not performed. In this study, air quality measurement system with an IoT-based e-nose has been proposed for real-time, low-cost and easy-to-install air quality monitoring. With the proposed e-nose system the ambient temperature and humidity values are measured in real-time in addition to the polluting gases such as CO_2, CO, PM_{10} and NO_2. Information on the presence of these monitored gases in excessive quantities is transmitted to users via the mobile application as a notification. This is a completely wireless solution developed using the ESP32 module, which integrates the IEEE 802.11 b/g/n network protocol into the IoT architecture.

The architecture of the proposed e-nose system is given in Figure 1. A real-time air quality monitoring system provides information about the concentration of pollutants in the environment. It provides precise and detailed information about the air quality of the living environment and helps to plan interventions that lead to improved air quality. The e-nose air quality monitoring system in Figure 1 consists of two parts: The first part is the detection and communication unit consisting of an ESP32 microcontroller-based sensor array with built-in Wi-Fi; and the second is the Android/iOS-based mobile user interface.

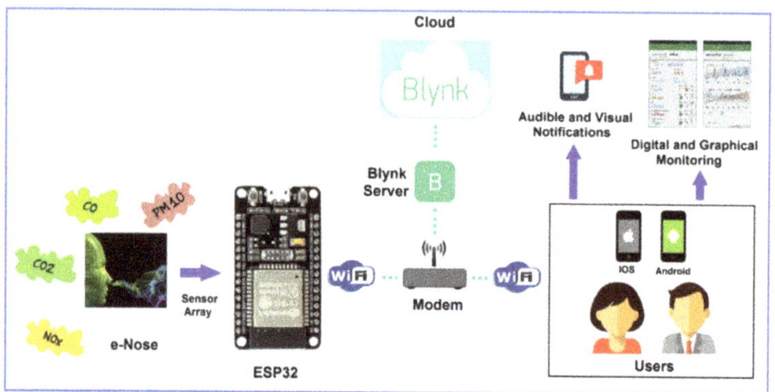

Figure 1. The proposed e-nose system architecture.

The ESP32 module with a built-in Wi-Fi module is used in the e-nose system created for monitoring air quality. The low-cost and high-performance 32-bit controller is frequently preferred in IoT applications. An ESP32 has a dual-core structure and has many internal modules such as Wi-Fi, Bluetooth, RF, IR, CAN, Ethernet module, temperature sensor, hall effect sensor and touch sensor needed for smart home applications. In the ESP32 module structure, the Harvard Tensilica Xtensa LX6 32-bit Dual Core features a processor capable of operating at up to 240 MHz. The detection unit includes the sensors GP2Y1010AU, MH-Z14, MICS-4514 and DHT22, which measure air quality parameters such as CO_2, CO, PM_{10}, NO_2, temperature and humidity.

The GP2Y1010AU is a dust sensor with an analogue output system. An infrared light-emitting diode (IRED) and a phototransistor are arranged across from each other. The IR beams reflected from dust entering the air chamber of the sensor is detected by the phototransistor and generates a corresponding voltage [32].

The MH-Z14A CO_2 sensor module uses the non-dispersive infrared (NDIR) principle. It measures between 0–5000 ppm, 5 ppm resolution with an accuracy of ±50 ppm. The sensor module sends the CO_2 concentration in three different output modes: Serial output (RS-232), analogue output and pulse width modulation (PWM) [33].

The MICS-4514 is mainly used for measuring emissions from automobile exhausts but is also used for measuring concentrations of gases such as NO_2, CO and hydrocarbons. The sensor has a built-in heating element and a micro-sensing diaphragm on the upper side. The MICS-4514 includes two sensor chips with independent heaters and delicate layers. One sensor chip detects oxidizing gases (OX) and the other sensor detects reducing gases (RED) [34].

The DHT22 consists of two parts: A thermistor temperature sensor and a capacitive humidity sensor. The DHT22 is an advanced sensor unit that provides a calibrated digital signal output. It is equipped with an 8-bit microcontroller and has a short response time. It has a relative error of ±0.5 °C in temperature measurement and ±2% rH in humidity measurement [35]. The electronic features of the sensors used in the IoT-based e-nose system are given in Table 1.

Table 1. Electronic features of the sensors in the IoT-based e-nose measuring system.

ID	Equipment Name	Types	Electronic Features
1	CO_2 gas sensor	MH-Z14 [36,37]	Detection range 0–10000 ppm; operating voltage: 4–6 V; accuracy: ± 50 ppm ±5%; resolution: 5 ppm; output Voltage: 0.4–2 V; operating temperature: 0–50 °C
2	NO_2, CO gas sensor	MICS-4514 [38–40]	Detection range 1–1000 ppm (CO); 0.05–5 ppm (NO_2); operating voltage: 4.9–5.1 V; operating temperature: −35–85 °C; heating current: 58 mA
3	Dust sensor	GP2Y1010AU [41–43]	Operating voltage: 5 V; output voltage: 0.9 (no dust)–3.4 V; operating current: max 20 mA; operating temperature: −10 to 65 °C, accuracy ±15%
4	Temperature and Humidity sensor	DHT22 [44–46]	Temperature range: −40 °C to 80 °C; humidity range: 0% to 100%; operating voltage: 3.5–5.5 V; operating current: 60 uA; output: serial; resolution: 0.1 °C and ±1 rH%; accuracy: ±0.5 °C and ±1 rH%; resolution: Temperature and humidity are 16-bit.

The images of the IoT-based e-nose system used in air quality measurements and formed from different sensors are displayed in Figure 2. The sensors used in the system are mounted in a $17 \times 12 \times 8$ cm size sealed box per the measurement specifications. In addition to the sensors used for measurement, the box includes a printed circuit board, a fan for airflow to the dust sensor, and a power supply for the energy of the system.

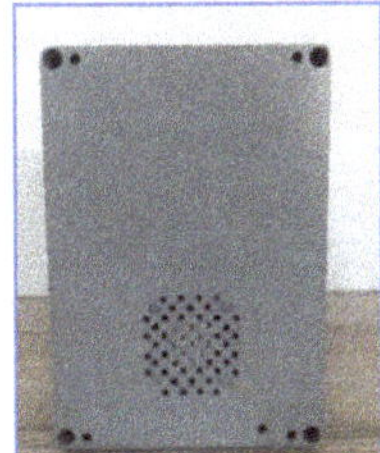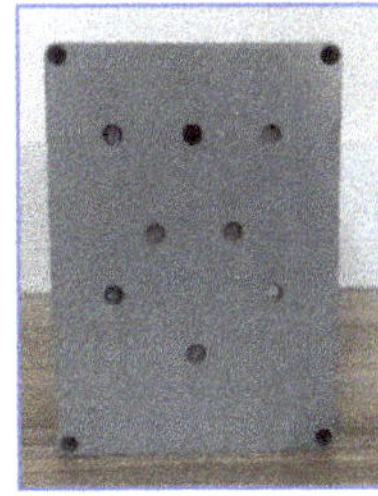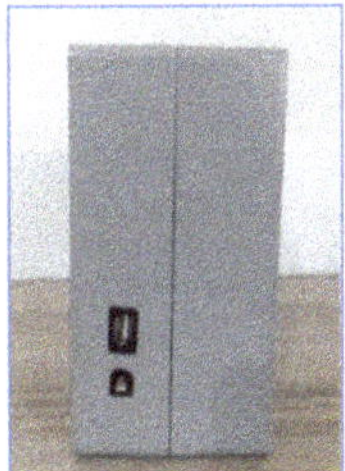

Figure 2. Images of the IoT-based e-nose system.

The measurement system can work with a 12 V DC adapter, and additionally has a mobile use feature when paired with a power bank that can be connected to the 5 V DC USB port. This feature provides a great advantage for short-term measurements. When supplied from a 5 V power supply, the measuring device draws a current of 160 mA. With a 5000 mAh power bank, it has a measuring time of approximately 30 h. Using a wireless internet connection, it provides the opportunity to measure air quality from many common living areas such as parks, gardens, highways, industrial areas, sports fields, public transportation, cafes, restaurants, schools, hospitals.

The climate parameters and gas concentrations are a median of 12 measurements taken at 5 s intervals. This minimizes the effect of erroneous measurements brought about by faulty sensors. Users are notified by exceeding the designated gas concentration and climate parameter thresholds. Five-minute averages are calculated and delivered to users as notifications, preventing false warnings that may mislead users. In this e-nose system which measures air quality, no memory element is used for data recording. The received data are recorded directly to the Blynk cloud server via the mobile interface. The data are sent to the registered e-mail address of the user when requested. The status of the internet connection is checked before each data transmission. When not connected, data packets that cannot be temporarily transmitted are stored to ensure data integrity, and then, when the internet connection is re-established, these packets are sent to the cloud server using past time tags. In this way, data loss can be prevented by providing continuity in the data flow.

The front panel images of the developed Android-based mobile user interface are given in Figure 3. Control over mobile devices in IoT applications has become very common. There are many free options available for Android and iOS devices. Blynk is one of these applications and it is an IoT platform developed for iOS and Android applications that enables management of different controllers such as Raspberry Pi, ESP8266, ESP32, chipKIT, Intel, LeMarker, Onion Omega, SparkFun and STM32. Using the Blynk cloud server service, digital data such as temperature, humidity, current, voltage measurements and control systems are stored and can be easily accessed at any time. The Blynk graphical components (widgets) allow real-time clock and calendar (RTCC) features to be used.

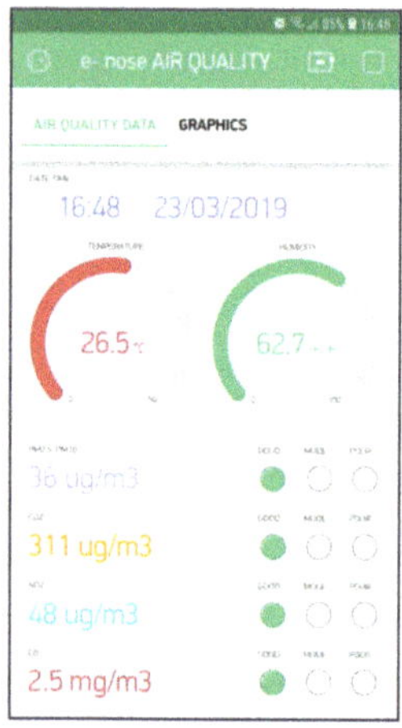
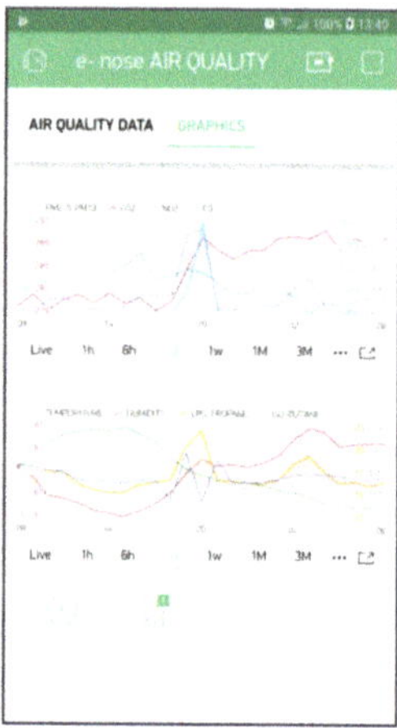

Figure 3. Mobile user interface developed for the e-nose system.

Table 2 shows the cost table of the IoT-based e-nose measurement system. The total cost of the software and hardware components required to install the system is approximately $100. Compared to most commercially accessible non-IoT-based home air quality meters, this cost is very economical and convenient.

Table 2. Cost table for IoT-based e-nose measurement system.

Component	Cost ($)
ESP-32 Controller	5
GP2Y1010AU Sensor	5
MH-Z14 Sensor	25
MICS-4514 Sensor	20
DHT22 Sensor	3
5V Fan	3
5V Power Sup.	2
PCB	3
Plastic Box	5
Cable, Socket	5
Power Supply/Bank	15
Arduino IDE	free
Blynk IoT Platform	5
Total Cost	**100 $**

4. Results and Discussion

For air quality analysis with IoT-based e-nose, data were collected in a residence for 4 days at one-minute intervals. The housing area where air quality measurements were taken is 150 m^2.

The room in which the measurements were made is 25 m^2 and the measurements were taken from a height of 1.5 m. The five-person house is heated by a central heating system. Although there were changes in the number of inhabitants during the day, there were always at least two people in the household. There was no air cleaning system in the house and the ventilation of the environment was done by manually opening the windows.

The doors in the household were not closed during the measurement and thus the air was homogeneous throughout the house. Ventilation was carried out once a day for a duration of one hour. Ventilation started at 11:30 on the first day and at 07:30 on the other days.

Figure 4 displays 4-day measurement graphs showing the relationships between humidity–CO_2, CO–CO_2, CO_2–NO_2, CO–NO_2, PM_{10}–CO and PM_{10}–NO_2.

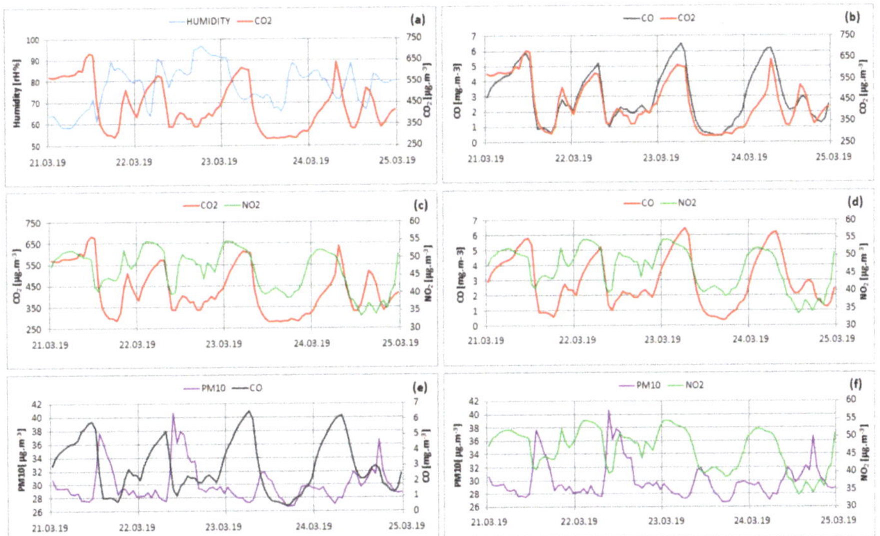

Figure 4. Relationship between gases during four-day measurement, (**a**) Humidity- CO_2, (**b**) CO-CO_2, (**c**) CO_2- NO_2, (**d**) CO-NO_2, (**e**) PM_{10}-CO, (**f**) CO-NO_2.

Figure 5 displays PM_{10}, CO_2, CO and NO_2 measurements over 4 days. When the graph is examined, it shows that the indoor PM_{10}, CO, NO_2 and CO_2 PM and gas concentrations vary depending on the number of household inhabitants and individual activities. It is seen that natural ventilation in the morning hours caused a decrease in CO, NO_2 and CO_2 gas concentrations. In addition, an increase in PM_{10} is observed in the morning as a result of increased activity in the household. These increased rates at the PM_{10} level are 27.1%, 32.2%, 14.4% and 12.5%, respectively.

Figure 6 displays the daily changes of temperature, humidity, PM_{10}, CO, CO_2 and NO_2 values. One-day time, period 1 (00:00–07:30), period 2 (07:30–16:30) and period 3 (16:30–00:00) are divided into three parts; period 1 sleeping of household inhabitants, period 2 during which the daily activities (cooking, cleaning, etc.) of the individuals in the household are performed, and the period 3 of eating and resting of the household inhabitants. The PM_{10} value seen in Figure 6a is in the order of 28.18 µg m^{-3}, 29.93 µg m^{-3} and 28.40 µg m^{-3} for periods 1-3, respectively. It can be seen that PM_{10} has the lowest value in period 1 when the household is asleep. The CO_2 values in the periods are as follows: period 1, 565 µg m^{-3}; period 2, 317 µg m^{-3}; and period 3 has a mean value of 297 µg m^{-3}. The value of CO_2 reaches the highest value in period 1 when the household inhabitants are asleep.

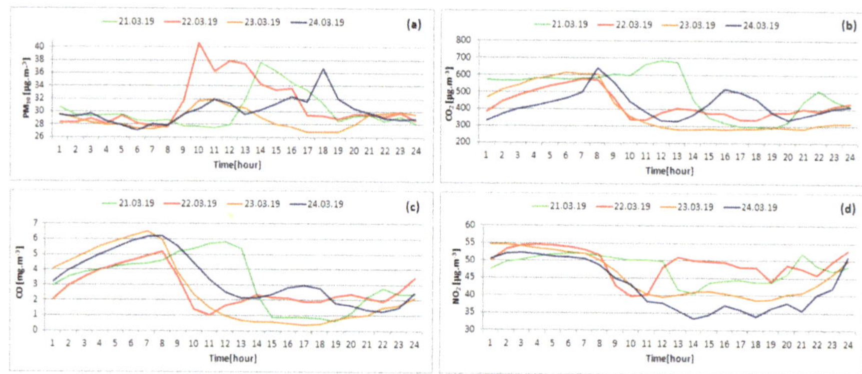

Figure 5. Daily changes of PM and gases, (**a**) PM_{10}, (**b**) CO_2, (**c**) CO, (**d**) NO_2.

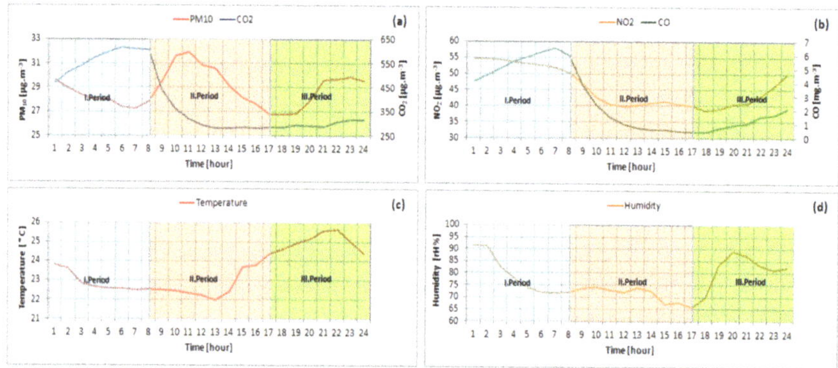

Figure 6. Hourly changes of climate parameters and gases values of day 3, (**a**) PM_{10} - CO_2, (**b**) NO_2-CO, (**c**) temperature, (**d**) humidity.

When Figure 6b is examined, it is seen that NO_2 concentration has average values of 53.15 µg m^{-3} for period 1, 41.59 µg m^{-3} for period 2 and 42.22 µg m^{-3} for period 3. It is seen that the NO_2 concentration value reached the highest value in the period 1, similar to CO_2. The CO concentration has an average of 5.48 mg m^{-3} in the period 1, 1.38 mg m^{-3} in the period 2 and 1.09 mg m^{-3} in the period 3. The CO value, and also CO_2 and NO_2 values, are observed to be highest in the period 1 while the household is asleep. Also, natural ventilation by opening the windows in the morning results in a significant reduction of CO, CO_2 and NO_2 concentrations.

As a result of natural ventilation, the indoor concentration of CO is seen to fall from 6.51 mg m^{-3} to 0.4 mg m^{-3}, NO_2 concentration from 54.7 µg m^{-3} to 38.6 µg m^{-3} and CO_2 concentration from 608 µg m^{-3} to 282 µg m^{-3}.

As a result of one hour of natural ventilation, the increased gas concentrations in the closed and unventilated environment resulted in a significant decrease of 93.8% in CO, 29.4% in NO_2 and 53.6% in CO_2.

Figure 7 shows the minimum, maximum and average values of daily gas concentrations. When the values given in Figure 7 are examined, it is seen that the largest variance is observed in the CO concentration. The CO values increased by nine times on day 1 and 16 times on day 3. Similarly, CO_2 concentration increased by 1.36 times on day 2. NO_2, with a lower change, increased by half on day 4 and PM_{10} concentration increased by half on day 2.

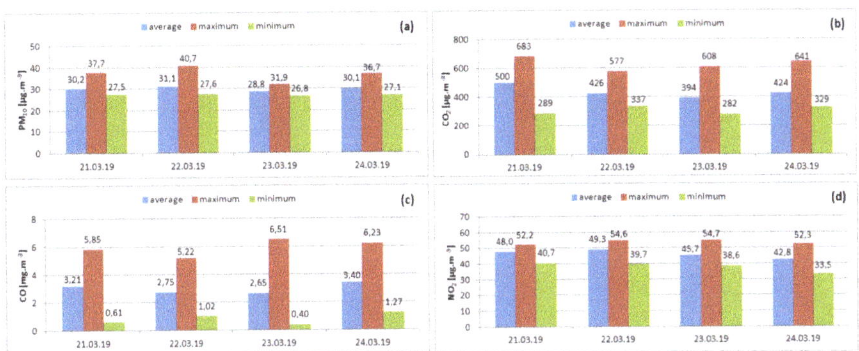

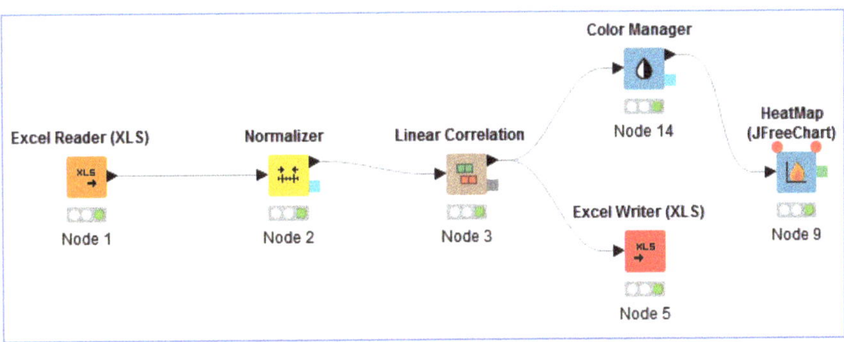

Figure 7. Minimum, maximum and average values of daily gas concentrations and PM, (**a**) PM_{10}, (**b**) CO_2, (**c**) CO, (**d**) NO_2

The daily average concentration values of indoor CO gas are as follows: 3.21, 2.75, 2.65 and 3.40 mg m^{-3}, daily concentrations of NO_2, 47.9, 49.2, 45.6 and 42.8 µg m^{-3}, CO_2, with mean values of 500, 426, 394 and 424 µg m^{-3} and PM_{10} 30.2, 31.1, 28.8 and 30.1 µg m^{-3}.

The correlation relationships between air quality dates have been estimated by the Konstanz Information Miner (KNIME) "Linear Correlation" algorithms. Figure 8 shows the linear correlation workflow of KNIME.

Figure 8. Workflow of Linear Correlation in KNIME application.

Tables 3–5 demonstrate the period 3 linear correlation matrices on 22.03.2019. Table 3 demonstrates the period 1 linear correlation matrix. There was a positive correlation between CO–CO_2, NO_2–humidity and NO_2–temperature at values of 0.943, 0.871, 0.755, respectively, during the period when the household was asleep. In addition, there were 0.979, 0.911 positive and 0.941 negative correlations between time–CO, time–CO_2 and time–NO_2, respectively. There was a strong negative correlation between climate parameters and values of CO and CO_2 and a strong positive correlation between climate parameters and values of NO_2. On the other hand, the correlation between PM10 and gas concentrations and climate parameters were minuscule. The findings indicate that in period 1 there was an elevated positive and negative correlation between gas concentrations and climatic parameters, which represents sleep hours.

Table 3. Linear correlation matrix for period 1.

	Time	CO	NO$_2$	CO$_2$	Temp	Humidity	PM$_{10}$		1
Time	1	0.979	−0.941	0.911	−0.866	−0.944	−0.251		
CO		1	−0.919	0.943	−0.907	−0.969	−0.254		
NO$_2$			1	−0.841	0.755	0.871	0.245		0
CO$_2$				1	−0.886	−0.918	−0.213		
Temp.					1	0.941	0.248		
Humidity						1	0.258		
PM$_{10}$							1		−1

Table 4 demonstrates the period 2 linear correlation matrix. There was a positive correlation between CO–CO$_2$. CO–NO$_2$ and NO$_2$–CO$_2$ at values of 0.966, 0.881, 0.861, respectively, and a negative correlation of 0.915 for temperature–humidity during this period, in which households are undertaking domestic activities (cooking, cleaning, washing, etc.). There were strong negative correlations between time, CO, CO$_2$, humidity and NO$_2$ at −0.844, −0.752, −0.735 and −0.669 during this period. In addition, this period's correlations between PM$_{10}$ value and gas levels and climate parameters were very low.

Table 4. Linear correlation matrix for period 2.

	Time	CO	NO$_2$	CO$_2$	Temp	Humidity	PM$_{10}$		1
Time	1	−0.844	−0.669	−0.752	0.640	−0.735	−0.242		
CO		1	0.881	0.966	−0.266	0.415	0.014		
NO$_2$			1	0.861	−0.098	0.254	−0.092		0
CO$_2$				1	−0.186	0.314	−0.020		
Temp.					1	−0.915	−0.327		
Humidity						1	0.317		
PM$_{10}$							1		−1

The linear correlation matrix provided in Table 5 belongs to period 3, in which families carry out activities such as eating and resting. There was a positive correlation between CO–NO$_2$ and CO–CO$_2$ at 0.833, 0.671, respectively, during this period. Furthermore, there was also a positive correlation between time–CO, time–NO$_2$, time–CO$_2$ and time–PM$_{10}$ at 0.968, 0.846, 0.622 and 0.407, respectively.

Table 5. Linear correlation matrix for period 3.

	Time	CO	NO$_2$	CO$_2$	Temp	Humidity	PM$_{10}$		1
Time	1	0.968	0.846	0.622	0.234	0.304	0.407		
CO		1	0.833	0.671	0.213	0.296	0.379		
NO$_2$			1	0.474	0.213	0.328	0.347		0
CO$_2$				1	0.051	0.001	0.192		
Temp.					1	0.629	0.230		
Humidity						1	0.169		
PM$_{10}$							1		−1

We can conclude that the cost of the proposed e-nose measurement system is very low compared to commercial products sold on the market. Moreover, the majority of domestic air quality meters available for commercial use measure only a restricted amount of air quality and climate parameters. In this system, CO, CO$_2$, NO$_2$, PM$_{10}$, and six parameters of temperature and humidity can be evaluated. Another significant advantage is that our measuring system is IoT-based and provides users with real-time data transfer through its mobile interface. The major drawback of the proposed e-nose system is that no comparison and calibration is conducted with any conventional measuring system. The measuring precision of the system is equivalent to the measuring precision indicated by manufacturers. To rectify this, in our future studies the measuring system will be calibrated first, then the sensor node created and longer-term readings made. Thus, the change in IAQ through individual activities will be disclosed more obviously.

5. Conclusions

Although many people spend most of their lives indoors, they have very limited information about the air quality in their environment. IAQ, which plays an especially important role in the health of children and the elderly, should be measured and necessary ventilation measures should be taken. In recent years, individual air quality measuring devices have been produced owing to the developing information communication technologies, Wi-Fi based microcontrollers and low-cost sensors.

In this study, an IoT-based personalized air quality measurement and monitoring system was proposed by using air quality sensors and DIY approach. The developed e-nose measurement system made data measurements at one-minute intervals and recorded these data to the cloud server. The measurement data can be monitored instantaneously via the Blynk mobile interface, and if the limit values are exceeded, the application sends a notification to the user to take the necessary measures.

According to the four-day measurement results, the following inferences have been obtained:

- IAQ is directly related to the number of people in the household and the activities carried out in the household.
- Activities such as cooking, sleeping, cleaning have a significant effect on CO, NO_2 and CO_2 gas concentrations.
- CO, NO_2, CO_2 gas concentration values of the house reach the highest level during the sleep period (period 1). During this period, PM_{10} concentration has the lowest value.
- PM_{10} concentration reaches the highest level in period 2, when daily routine tasks such as cleaning and house arrangement are carried out.
- When the daily maximum and minimum values were taken into consideration, there was a 16-fold daily maximum change on day 3 in CO concentration, with minimum values of 0.4 mg m^{-3} and maximum 6.51 mg m^{-3}.
- The largest change in PM_{10} concentration was achieved with a minimum of 27.6 µg m^{-3} and a maximum of 40.7 µg m m^{-3} on day 2, resulting in a 47% change.
- Although CO and CO had the lowest concentration in period 3, NO_2 had the lowest concentration in period 2.
- A rapid decrease in CO, NO_2, CO_2 gas concentrations was observed from the moment of natural ventilation, while a concurrent increase in PM_{10} was observed.
- In period 1, the highest positive correlations occurred between time–CO and CO–CO_2; the lowest negative correlations occurred between CO–humidity and time–humidity.
- In period 2, the largest positive correlations occurred between CO–CO_2 and CO–NO_2, whereas the smallest negative correlations occurred between temp–humidity and time–CO.
- The highest positive correlations between time–CO, time–NO_2 and CO–NO_2 occurred in period 3. There was no negative correlation between the weather parameters and gas concentrations during this period.

IAQ has been observed to change depending on the daily activities of the inhabitants. Low air quality is undoubtedly an important parameter that directly affects our health. Even simple measures such as opening only windows to reduce the concentration of harmful gases in the environment can significantly improve IAQ. The numerical data obtained show that the e-nose system is a feature that can contribute to a healthier living environment. In future studies, it is planned to model the effects of individual activities on IAQ by using the e-nose air quality measurement system used in this study.

Author Contributions: Conceptualization, M.T. and H.G.; methodology, H.G.; software, M.T.; validation, M.T.; formal analysis, M.T.; investigation, M.T. and H.G.; visualization, M.T. and H.G.; resources, H.G.; data curation, M.T.; writing—original draft preparation, M.T.; review and editing, H.G.

Funding: This research received no external funding.

Conflicts of Interest: The authors declare no conflict of interest.

References

1. Sharma, A.; Mitra, A.; Sharma, S.; Roy, S. Estimation of Air Quality Index from Seasonal Trends Using Deep Neural Network. In Proceedings of the International Conference on Artificial Neural Networks, Island of Rhodes, Greece, 5–7 October 2018; pp. 511–521.
2. Leung, D.Y. Outdoor-indoor air pollution in urban environment: Challenges and opportunity. *Front. Environ. Sci.* **2015**, *2*, 69. [CrossRef]
3. De Gennaro, G.; Dambruoso, P.R.; Loiotile, A.D.; Di Gilio, A.; Giungato, P.; Tutino, M.; Porcelli, F. Indoor air quality in schools. *Environ. Chem. Lett.* **2014**, *12*, 467–482. [CrossRef]
4. Chen, M.; Yang, J.; Hu, L.; Hossain, M.S.; Muhammad, G. Urban healthcare big data system based on crowd sourced and cloud-based air quality indicators. *IEEE Commun. Mag.* **2018**, *56*, 14–20. [CrossRef]
5. Marques, G. Ambient Assisted Living and Internet of Things. In *Harnessing the Internet of Everything (IoE) for Accelerated Innovation Opportunities*; IGI Global: Hershey, PA, USA, 2019; pp. 100–115.
6. Wei, W.; Ramalho, O.; Mandin, C. Indoor air quality requirements in green building certifications. *Build. Environ.* **2015**, *92*, 10–19. [CrossRef]
7. Neira, M.; Prüss-Ustün, A.; Mudu, P. Reduce air pollution to beat NCDs: From recognition to action. *Lancet* **2018**, *392*, 1178–1179. [CrossRef]
8. Seguel, J.M.; Merrill, R.; Seguel, D.; Campagna, A.C. Indoor air quality. *Am. J. Lifestyle Med.* **2017**, *11*, 284–295. [CrossRef] [PubMed]
9. Kampa, M.; Castanas, E. Human health effects of air pollution. *Environ. Pollut.* **2008**, *151*, 362–367. [CrossRef]
10. Annesi-Maesano, I.; Baiz, N.; Banerjee, S.; Rudnai, P.; Rive, S.; Sinphonie Group. Indoor air quality and sources in schools and related health effects. *J. Toxicol. Environ. Health Part B* **2013**, *16*, 491–550. [CrossRef]
11. Tong, Z.; Chen, Y.; Malkawi, A.; Liu, Z.; Freeman, R.B. Energy saving potential of natural ventilation in China: The impact of ambient air pollution. *Appl. Energy* **2016**, *179*, 660–668. [CrossRef]
12. Chen, Y.; Tong, Z.; Wu, W.; Samuelson, H.; Malkawi, A.; Norford, L. Achieving natural ventilation potential in practice: Control schemes and levels of automation. *Appl. Energy* **2019**, *235*, 1141–1152. [CrossRef]
13. Jones, A.P. Indoor air quality and health. *Atmos. Environ.* **1999**, *33*, 4535–4564. [CrossRef]
14. Gökozan, H.; Taştan, M.; Sarı, A. *Smart Cities and Management Strategies*; 2017 Socio-Economic Strategies; Lambert Academic Publishing: Riga, Latvia, 2017; pp. 115–123, ISBN 978-3-330-06982-4.
15. Gubbi, J.; Buyya, R.; Marusic, S.; Palaniswami, M. Internet of Things (IoT): A vision, architectural elements, and future directions. *Future Gener. Comput. Syst.* **2013**, *29*, 1645–1660. [CrossRef]
16. Rathore, P.; Rao, A.S.; Rajasegarar, S.; Vanz, E.; Gubbi, J.; Palaniswami, M. Real-time urban microclimate analysis using internet of things. *IEEE Internet Things J.* **2018**, *5*, 500–511. [CrossRef]
17. Taştan, M. IoT Based Wearable Smart Health Monitoring System. *Celal Bayar Univ. J. Sci.* **2018**, *14*, 343–350. [CrossRef]
18. Pitarma, R.; Marques, G.; Caetano, F. Monitoring indoor air quality to improve occupational health. In *New Advances in Information Systems and Technologies*; Springer: Cham, Switzerland, 2016; pp. 13–21.
19. Yang, Y.; Liu, L.; Xu, C.; Li, N.; Liu, Z.; Wang, Q.; Xu, D. Source apportionment and influencing factor analysis of residential indoor PM2. 5 in Beijing. *Int. J. Environ. Res. Public Health* **2018**, *15*, 686. [CrossRef] [PubMed]
20. Canha, N.; Lage, J.; Galinha, C.; Coentro, S.; Alves, C.; Almeida, S. Impact of Biomass Home Heating, Cooking Styles, and Bread Toasting on the Indoor Air Quality at Portuguese Dwellings: A Case Study. *Atmosphere* **2018**, *9*, 214. [CrossRef]
21. Rumchev, K.; Soares, M.; Zhao, Y.; Reid, C.; Huxley, R. The association between indoor air quality and adult blood pressure levels in a high-income setting. *Int. J. Environ. Res. Public Health* **2018**, *15*, 2026. [CrossRef]
22. Wang, S.; Bolling, K.; Mao, W.; Reichstadt, J.; Jeste, D.; Kim, H.C.; Nebeker, C. Technology to Support Aging in Place: Older Adults' Perspectives. *Healthc. Multidiscip. Digit. Publ. Inst.* **2019**, *7*, 60. [CrossRef]
23. Alvarez-Campana, M.; López, G.; Vázquez, E.; Villagrá, V.; Berrocal, J. Smart CEI moncloa: An iot-based platform for people flow and environmental monitoring on a Smart University Campus. *Sensors* **2017**, *17*, 2856. [CrossRef]
24. Cho, H. An Air Quality and Event Detection System with Life Logging for Monitoring Household Environments. In *Smart Sensors at the IoT Frontier*; Springer: Cham, Switzerland, 2017; pp. 251–270.
25. Wong, M.; Yip, T.; Mok, E. Development of a personal integrated environmental monitoring system. *Sensors* **2014**, *14*, 22065–22081. [CrossRef]

26. Salamone, F.; Danza, L.; Meroni, I.; Pollastro, M. A Low-Cost Environmental Monitoring System: How to Prevent Systematic Errors in the Design Phase through the Combined Use of Additive Manufacturing and Thermographic Techniques. *Sensors* **2017**, *17*, 828. [CrossRef] [PubMed]

27. Popa, A.; Hnatiuc, M.; Paun, M.; Geman, O.; Hemanth, D.J.; Dorcea, D.; Ghita, S. An Intelligent IoT-Based Food Quality Monitoring Approach Using Low-Cost Sensors. *Symmetry* **2019**, *11*, 374. [CrossRef]

28. Marques, G.; Pitarma, R. An indoor monitoring system for ambient assisted living based on internet of things architecture. *Int. J. Environ. Res. Public Health* **2016**, *13*, 1152. [CrossRef] [PubMed]

29. Marques, G.; Pitarma, R. A cost-effective air quality supervision solution for enhanced living environments through the internet of things. *Electronics* **2019**, *8*, 170. [CrossRef]

30. Marques, G.; Pitarma, R. Monitoring health factors in indoor living environments using internet of things. In Proceedings of the World Conference on Information Systems and Technologies, Madeira, Portugal, 11–13 April 2017; pp. 785–794.

31. Marques, G.; Roque Ferreira, C.; Pitarma, R. A system based on the Internet of Things for real-time particle monitoring in buildings. *Int. J. Environ. Res. Public Health* **2018**, *15*, 821. [CrossRef] [PubMed]

32. GP2Y1010AU0F: Compact Optical Dust Sensor. Available online: http://www.socle-tech.com/doc/IC%20Channel%20Product/sharp%20products.pdf (accessed on 11 June 2019).

33. Zhengzhou Winsen Electronics Technology CO. Ltd. MH-Z14 CO2 Module Datasheet. 2013. Available online: http://www.futurlec.com/Datasheet/Sensor/MH-Z14.pdf (accessed on 17 June 2019).

34. MICS-4514 Datasheet. Available online: http://files.manylabs.org/datasheets/MICS-4514.pdf (accessed on 11 June 2019).

35. Aosong Electronics CO. Ltd. DHT22 Temperature and Humidity Sensor Datasheet. Available online: https://www.sparkfun.com/datasheets/Sensors/Temperature/DHT22.pdf (accessed on 13 June 2019).

36. Hong, C.S.; Ghani, A.S.A.; Khairuddin, I.M. Development of an Electronic Kit for detecting asthma in Human Respiratory System. In Proceedings of the IOP Conference Series: Materials Science and Engineering, Yogyakarta, Indonesia, 7–8 December 2017; IOP Publishing: Bristol, UK, 2018; Volume 319, p. 012040.

37. Bimaridi, A.; Putra, K.D.; Djunaedy, E.; Kirom, M.R. Assasment of Outside Air Supply for Split AC system–Part A: Affordable Instrument. *Procedia Eng.* **2017**, *170*, 248–254. [CrossRef]

38. Suárez, J.I.; Arroyo, P.; Lozano, J.; Herrero, J.L.; Padilla, M. Bluetooth gas sensing module combined with smartphones for air quality monitoring. *Chemosphere* **2018**, *205*, 618–626. [CrossRef]

39. McKercher, G.R.; Vanos, J.K. Low-cost mobile air pollution monitoring in urban environments: A pilot study in Lubbock, Texas. *Environ. Technol.* **2018**, *39*, 1505–1514. [CrossRef]

40. Nguyen, T.N.T.; Ha, D.V.; Do, T.N.N.; Nguyen, V.H.; Ngo, X.T.; Phan, V.H.; Bui, Q.H. Air pollution monitoring network using low-cost sensors, a case study in Hanoi, Vietnam. In Proceedings of the IOP Conference Series: Earth and Environmental Science, Pan Pacific Hanoi, Vietnam, 23–24 January 2019; IOP Publishing: Bristol, UK, 2019; Volume 266, p. 012017.

41. Liu, H.Y.; Schneider, P.; Haugen, R.; Vogt, M. Performance assessment of a low-cost PM2.5 sensor for a near four-month period in Oslo, Norway. *Atmosphere* **2019**, *10*, 41. [CrossRef]

42. Budde, M.; El Masri, R.; Riedel, T.; Beigl, M. Enabling low-cost particulate matter measurement for participatory sensing scenarios. In Proceedings of the 12th International Conference on Mobile and Ubiquitous Multimedia, Lulea, Sweden, 2–5 December 2013; p. 19.

43. Carminati, M.; Ferrari, G.; Sampietro, M. Emerging miniaturized technologies for airborne particulate matter pervasive monitoring. *Measurement* **2017**, *101*, 250–256. [CrossRef]

44. Muangprathub, J.; Boonnam, N.; Kajornkasirat, S.; Lekbangpong, N.; Wanichsombat, A.; Nillaor, P. IoT and agriculture data analysis for smart farm. *Comput. Electron. Agric.* **2019**, *156*, 467–474. [CrossRef]

45. Xie, J.; Gao, P.; Wang, W.; Xu, X.; Hu, G. Design of Wireless Sensor Network Bidirectional Nodes for Intelligent Monitoring System of Micro-irrigation in Litchi Orchards. *IFAC-PapersOnLine* **2018**, *51*, 449–454.

46. Wijaya, D.R.; Sarno, R.; Zulaika, E.; Sabila, S.I. Development of mobile electronic nose for beef quality monitoring. *Procedia Comput. Sci.* **2017**, *124*, 728–735. [CrossRef]

Article

An Internet of Things-Based Environmental Quality Management System to Supervise the Indoor Laboratory Conditions

Gonçalo Marques and Rui Pitarma *

Unit for Inland Development, Polytechnic Institute of Guarda, Avenida Doutor Francisco Sá Carneiro N° 50, 6300-559 Guarda, Portugal; goncalosantosmarques@gmail.com
* Correspondence: rpitarma@ipg.pt; Tel.: +351-2712-20111

Received: 19 December 2018; Accepted: 24 January 2019; Published: 28 January 2019

Abstract: Indoor air quality (IAQ) is not only a determinant of occupational health but also influences all indoor human behaviours. In most university establishments, laboratories are also used as classrooms. On one hand, indoor environment quality (IEQ) conditions supervision in laboratories is relevant for experimental activities. On the other hand, it is also crucial to provide a healthy and productive workplace for learning activities. The proliferation of cost-effective sensors and microcontrollers along with the Internet of Things (IoT) architectures enhancements, enables the development of automatic solutions to supervise the Laboratory Environmental Conditions (LEC). This paper aims to present a real-time IEQ-laboratory data collection system-based IoT architecture named iAQ Plus (*iAQ+*). The *iAQ+* incorporates an integrated Web management system along with a smartphone application to provide a historical analysis of the LEC. The *iAQ+* collects IAQ index, temperature, relative humidity and barometric pressure. The results obtained are promising, representing a meaningful contribution for IEQ supervision solutions based on IoT. *iAQ+* supports push notifications to alert people in a timely way for enhanced living environments and occupational health, as well as a work mode feature, so the user can configure setpoints for laboratory mode and schoolroom mode. Using the *iAQ+*, it is possible to provide an integrated management of data information of the spatio-temporal variations of LEC parameters which are particularly significant not only for enhanced living environments but also for laboratory experiments.

Keywords: IAQ; enhanced living environments; IEQ; IoT; smart cities; LEC

---

1. Introduction

Indoor environment quality (IEQ) in buildings incorporates indoor air quality (IAQ), acoustics, thermal comfort, and lighting [1]. It is well known that poor IEQ has a negative effect on occupational health, particularly on children and old people.

Poor IAQ is a significant problem which affects particularly the underprivileged people in the world, as they remain most exposed, presenting itself as a critical problem for global health such as tobacco use or the problem of sexually transmitted diseases [2]. The Environmental Protection Agency (EPA) is responsible for indoor and outdoor air quality supervision in the United States. The EPA assessment recognises that IAQ pollutants concentration can be up to 100 times greater when compared with the outdoor pollutants concentration levels and established poor air quality, which are in the top five environmental risks to global health [3].

Indoor living environments include several types of spaces and workplaces such as offices, hospitals, public service centres, schools, libraries, leisure spaces and also the cabins of vehicles [4]. In particular, schools are an essential place to monitor. Typically, the large number of occupants, the time spent indoors, and the higher density of occupants justifies the need to develop automatic supervision systems to provide a healthful and productive workplace for students, teachers and school staff [5].

Buildings are responsible for 40% of the global energy consumption and 30% of the CO_2 emissions. Thermal comfort assurance is related to a substantial percentage of the referred energy consumption. The adoption of personalised conditioning systems is apparently a reliable approach to improve user acceptability with the environment thermal conditions since thermal comfort is a complicated subject with several interrelated aspects that need to be understood [6].

Currently, IEQ in buildings is based on random sampling. However, these procedures are only providing information relating to a specific sampling and are devoid of details of spatio-temporal variations, which are particularly relevant in laboratory tests such as thermography experiments.

The fundamental concept of the Internet of Things (IoT) is the pervasive presence of several types of devices with communication and collaboration skills between them to reach a mutual purpose [7]. The IoT will not only improve everyday life behaviours and activities in different areas such as smart homes, assisted living environments and smart-health, but also in present innovative data and computational resources to develop novel software solutions [8]. One of the fields where the IoT plays a significant role is IAQ monitoring [9]. The rapid increasing of Information and Communication Technologies (ICTs) and IoT proliferation offer significant opportunities in the development of healthcare information systems. However, challenges still exist in achieving safety, security and privacy for healthcare applications [10,11].

Ambient Assisted Living (AAL) is a research area which studies the conceptualisation of an ecosystem of diverse kinds of sensors, computers, mobile gadgets, wireless networks and software development for enhanced personal healthcare supervision and telemedicine solutions [12]. At 2050, a significant proliferation of diseases will lead to high healthcare costs, lack of caregivers, dependency and notable social impact, because 20% of the world population will be aged 60 or over [13]. Most people (87%) prefer to stay in their houses and pay a significant charge for nursing care [14]. However, there are numerous difficulties in AAL solutions conceptualisation and development such as data structure, user-interface, human–computer communication, ergonomics, usability and accessibility [15]. Furthermore, the AAL solutions adopted by the aged population is influenced by social and ethical obstacles. Therefore, privacy and confidentiality are fundamental to AAL solutions' development. Actually, technology should not substitute human care but must be seen as a useful supplement. IoT and AAL research can be conducted side by side in order to present multiple advantages to various healthcare activities such as not only the identification, authentication and tracking of objects and patients but also in automated data acquisition, consulting and storage [16].

The "smart city" approach has recently been proposed as an important method to include current urban production circumstances in a mutual framework especially to focus the ICTs influence in the last 20 years on city performance [17]. The smart city concept is associated with a recent approach to decrease the obstacles caused by the growth of urban population and rapid urbanisation [18]. The interoperability of heterogeneous technologies and devices is a significant challenge in smart cities; IoT can offer the interoperability needed to create unified urban-scale ICT [19].

IoT incorporates notable features to develop innovative real-life solutions and services for smart cities [20], particularly for environmental quality supervision. The smart home is an indispensable element in smart cities [21]. In the future, smart homes should incorporate IoT wearable gadgets managed by smartphone applications and powered by miniaturised built-in sensors [22]. Smart homes follow the AAL paradigm, which enables access to health-care services for patients and medical staff as well as other AAL applications [23]. The use of the paradigm of cognitive dynamic systems associated with IoT architectures in intelligent houses is proposed by Reference [24]. According to this study, the smart home can benefit from several building blocks of the cognitive dynamic system (CDS) along with the incorporation of IoT.

This paper aims to present an environmental quality solution based on IoT to supervise Laboratory Environmental Conditions (LEC) named *iAQ+*. This low-cost wireless solution for

IEQ supervision incorporates mobile computing technologies for data consulting, easy installation, significant notifications for enhanced living conditions, and laboratory activities.

2. Related Work

Several studies have shown the IAQ importance in schoolrooms. The IAQ of public primary Portuguese schools was analysed in 73 primary classrooms in Porto [25]. This study indicates that IAQ deficiencies can persist in classrooms with pollutant sources and defective ventilation. Therefore, pollutant source regulation procedures are the most effective methods for the prevention of unfavourable health impacts on children in scholar institutions. The effectiveness of manual airing strategies on the IAQ of naturally ventilated Italian schools was presented by Reference [26]. A study conducted in New York State primary and secondary schools organised by 501 teachers concluded that many classroom characteristics are probably associated with bad IAQ and over 40% of the teachers described a minimum of one health symptom associated with the building construction [27]. A review of the association between IAQ and its consequences on respiratory wellbeing in Malaysian students was proposed by Reference [28]. Despite the relatively small-scale epidemiologic evidence, Malaysian research proposes effective and relatively consistent evidence among IAQ and children's respiratory health. Another study that characterises the levels of several indoor air contaminants at scholar institutions in Hong Kong, correlates the calculated concentrations with proper standards, and recommend methods to decrease the exposure of students to unwanted pollutants was proposed by Reference [29].

On the one hand, in most university establishments, laboratories have a large number of polluting sources. On the other side, these spaces are very often used as classrooms. Therefore, university laboratories need to be monitored for two purposes: as a classroom (IAQ parameters, including thermal comfort) and/or to ensure different conditions for sampling and performing experiments with reliable quality and data. Most people consider thermal comfort with higher importance when compared with visual and acoustic comfort and IAQ [30]. The comfort temperature might be as low as 17 °C and as high as 30 °C. Thermal comfort is influenced by several factors such as air temperature, radiant temperature, air velocity, humidity, clothing insulation and metabolic heat. The first four factors can be measured and the last two are personalised factors [31]. Although, for laboratory experiments, the recommendation is 23 °C (±5 °C) and <70 % RH for temperature and relative humidity respectively. A study on the thermal comfort in a Portuguese school was presented by Reference [32].

Multiple studies on environmental quality supervision are accessible in the literature. This section presents several prominent low-cost solutions which include not only open-source but also mobile computing technologies.

A wireless sensor network (WSN) low-cost solution for proper greenhouse pepper cultivation which incorporates proper supervision methods such as remote administration for drip irrigation and equipment control was proposed by Reference [33].

A real-time WSN architecture for environmental supervision which provides acoustic levels, temperature, relative humidity and particulate matter concentration data collection for smart cities was proposed by Reference [34].

A WSN approach for temperature distribution supervision in large-scale indoor environments was proposed by Reference [35]. This methodology aims to enhance the quality of the data transmitted by wireless signals, classify the temperature distribution model and improve the allocation of supplied air measure flow levels to various supply air terminals which fulfill the area taking into account the temperature distribution model.

Numerous IoT solutions for IAQ supervision which merge open-source technologies not only for processing and data transmission and sensors for data collection but also to provide data consulting from distinct areas at the same time using Web and smartphone applications in real time are proposed by References [36–43].

The *iAQ+* solution proposes a valuable instrument for enhanced living environments in smart cities. The advantages for well-being, comfort and productivity of healthy IAQ levels can be enhanced by reducing the pollutant concentration when the ventilation is still unchanged [44]. Consequently, the authors propose a wireless system which incorporates an ESP8266 module which performs the IEEE 802.11 b/g/n networking protocol. The ESP8266 support built-in Wi-Fi technology, therefore, this module is used not only as a processing unit but also for data transmission. For data consulting, this solution uses a smartphone application developed using SWIFT language for the iOS operating system (Figure 1).

Taking into account the IAQ influence on health, the development of a low-cost, open-source supervision system is a trending idea. Therefore, several monitoring solutions have been created [45–52]. Regarding the quality and important contribution of the referred solutions, a review of these studies is presented in Table 1.

Table 1. A summary comparison of real-time IAQ monitoring studies.

Authors	MCU	Sensors	Architecture	Low Cost	Open-Source	Connectivity	Data Access	Easy Installation	Notifications
P. Srivatsa and A. Pandhare [45]	Raspberry Pi	CO_2	WSN/IoT	√	√	Wi-Fi	Web	×	×
F. Salamone et al. [46]	Arduino UNO	CO_2	WSN	√	√	ZigBee	×	×	×
S. Bhattacharya et al. [47]	Waspmote	CO, CO_2, PM, Temperature, Relative Humidity	WSN	×	√	ZigBee	Desktop	×	×
F. Salamone et al. [48]	Arduino UNO	Temperature, Relative Humidity, CO_2, Ligth, Air velocity	IoT	√	√	ZigBee / BLE	Mobile	×	×
Wang, S.K et al. [49]	Arduino	Temperature, Relative Humidity, CO_2	WSN	√	√	ZigBee	Desktop	×	×
Jen-Hao Liu et al. [50]	TI MSP430	CO, Temperature, Relative Humidity	WSN	√	√	ZigBee	×	×	×
J.Kang and K. Hwang [51]	TI MSP430	CO, Temperature, Relative Humidity, VOC, PM	IoT	√	×	ZigBee	×	×	×
Benammar M. et al [52]	Raspberry Pi	CO_2, CO, SO_2, NO_2, O_3, Cl_2, CO, Temperature, Relative Humidity	IoT/WSN	√	×	ZigBee/Ethernet	×	×	×

MCU: microcontroller; √: apply; ×: not apply.

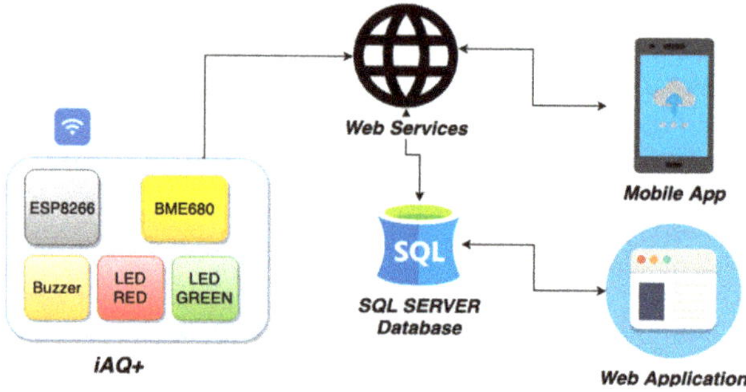

Figure 1. *iAQ+* architecture.

Proposed iAQ+ Management System

Considering the importance of developing a cost-effective system, only the Bosch BME680 sensor was chosen. This sensor was selected since it is an integrated environmental sensor produced particularly for portable applications and wearables where dimension and low power consumption are essential requirements. This sensor provides temperature, relative humidity, barometric pressure and qualitative air quality, the main parameters that should be supervised for both school and laboratory scenarios. The proposed system is composed of a hardware prototype for environmental data acquisition and a Web portal and mobile application compatible for data access (Figure 1). The *iAQ+* is based on open-source technologies and is a totality Wi-Fi solution with various benefits when related to current solutions, such as its modularity, scalability, low-cost, real-time notifications and easy installation. The data collected is stored in a SQL Server database management system and a smartphone application or a Web portal can be used for data consulting. The *iAQ+* incorporates the ESP8266 microcontroller with built-in Wi-Fi communication technology and has been tested in the context of infrared thermography experiments.

Taking into consideration the SQL Server native .NET framework integration feature, this database management system was chosen. Furthermore, the SQL Server incorporates several advantages such as data-recovery features, professional management utilities and built-in data compression and data encryption features. The SQL Server supports easy administration and audit functionalities for enhanced data management.

3. Laboratory Environmental Conditions

LEC supervision is critical from a quality perspective to provide a consistent and regulated state of control for equipment and samples for proper lab activities. LEC is influenced by the building envelope, thermal loads, obstructions and buildings climatisation system, which can produce air flow patterns of hot and cold spots. In some cases, these hot and cold spots can produce environmental conditions variations during experiment activities which lead to unwanted effects on samples as well as manipulation of the testing equipment's samples. Therefore, the LEC must be monitored in real-time not only to assure proper regulation and maintenance of indoor conditions but also to correlate the collected samples with these conditions.

On the one hand, the laboratory activities must be monitored and stored to ensure that they are stable when the test is conducted and at the data collection moment as they influence the quality of the results. LEC has a significant influence on test results and on the precision and reliability of testing records that remain concerned with environmental parameters. On the other hand, monitored results must be saved in the laboratory information management system that should provide the following requisites: 1) must incorporate authentication and access control for data safety; 2) must incorporate data recovery and data adulteration protection methods; 3) must be manipulated in an environment that satisfies the provider or laboratory specs or, in the situation of non-computerized methods, provides conditions which preserve the precision of manual reporting and transcription; 4) must be managed to guarantee the information integrity; and 5) must implement reporting system crashes support to allow prompt corrective operations.

In general, temperature, humidity and barometric pressure are assumed as the main conditions to be monitored for enhanced laboratory environments. The recommendation for LEC is 23 °C (±5 °C) and <70 % RH for temperature and relative humidity respectively. For temperature data collection, the measurements must be made with calibrated sensors that must be placed far from the equipment under analysis to anticipate every heating consequence which can lead to inaccurate ambient samples. Regarding the humidity measurements, the data collected must be done at the same altitude from the ground as the equipment under test and, preferably, in a similar place if possible. The barometric levels can be consulted from local airports. However, for critical laboratory experiences, barometer sensors are required and must be placed in the laboratory to detect indoor building environmental limitations and barometric fluctuations.

IAQ has a significant influence on LEC [53–56]. On the one hand, IAQ should be supervised in order to provide a healthy and productive workplace for the researchers. On the other hand, IAQ should be monitored to minimise the impact the laboratory experiments' samples. For instance,

IQ is extremely significant in the clinical embryology laboratory activities and is recognised as a significant parameter which influences in vitro fertilisation (IVF) success levels [57].

4. Materials and Methods

Taking into account, not only the necessity to keep a healthful and productive workplace for the students, teachers and the school staff, but also to provide a consistent and regulated state of control for laboratory activities, the *iAQ+* solution has been created by the authors. The *iAQ+* solution provides a low-cost and reliable method for IAQ supervision which incorporates easy configuration and easy installation features. This system is a low-cost, reliable system that can be easily configured and installed by the average user without supporting the cost of an installation done by certified professionals.

The main objective is to provide proper supervision of the LEC, such as air temperature, humidity and barometric pressure and a qualitative air quality index. Therefore, the authors selected a cost-effective BOSCH BME680 sensor, a four-in-one multi-functional microelectromechanical system (MEMS) environmental sensor which integrates a Volatile Organic Compounds (VOC) sensor, temperature sensor, humidity sensor and barometer. The DFRobot Gravity BME680 environmental sensor was used as it provides a Gravity I2C connector, is plug & play and is easy to connect.

The *iAQ+* is based on a microcontroller with built-in Wi-Fi compatibility, a FireBeetle ESP8266 (*DFRobot*). Figure 2 presents the hardware prototype developed by the authors. In this section, the hardware and software development materials and methods will be discussed.

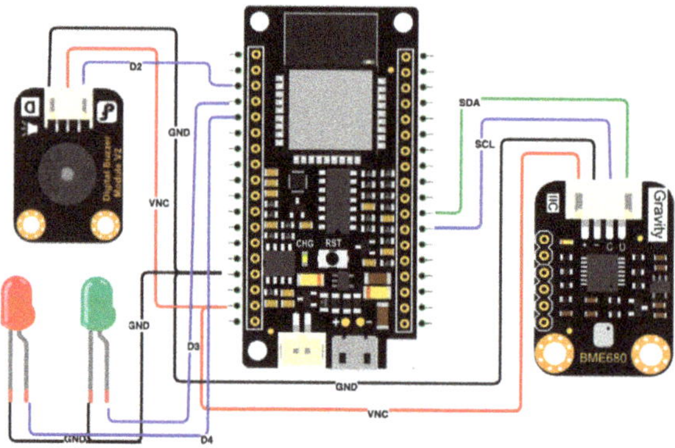

Figure 2. *iAQ+* prototype.

A brief introduction of each component used is shown below.

- **FireBeetle ESP8266** is 32-bit Tensilica L106 microcontroller module which supports IEEE802.11 b/g/n WiFi (2.4 GHz~2.5 GHz). This module support one 10-bit analogue input, 10 digital inputs which incorporate multiple interfaces such as SPI, I2C, IR, and I2S. The clock speed is 80MHz and can reach a maximum 160MHz; in addition, it includes a 50KB SRAM and 16MB flash memory. It supports a low-power-consumption mode of 46uA and the operating voltage is 3.3 V.
- **DFRobot Gravity BME680** is an I2C environmental VOC sensor, temperature sensor, humidity sensor and barometer. It supports an input voltage of 3.3–5.0 V; the operating current consumption is 5 mA without air quality sensing and 25 mA with air quality features. This sensor module size is 30 × 22 mm / 1.18 × 0.87 inches. The temperature range is from -40 °C to +85 °C with a precision of ±1.0 °C (0–65 °C). The humidity range is from 0 to 100% r.H with a precision of a ±3% r.H. (20–80% r.H., 25 °C). The atmospheric pressure measurement range is from 300 to 1100 hPa with a precision of ±0.6 hPa (300–1100 hPa, 0–65 °C).
- **DFRobot Buzzer Module** is a buzzer module that supports an input voltage of 3.3–5.0 V.

- **5V Green LED**—a 5 V green LED is used to notify the end-user of a good IEQ conditions.
- **5V Red LED**—a 5 V red LED is used to notify the end-user of poor IEQ conditions.

The BME680 sensor calculates the sum of VOCs in the surrounding air to provide qualitative air quality data. This sensor incorporates a background auto-calibration feature in order to provide reliable IAQ qualitative data. This process regards the recent measurement records to guarantee that IAQ index ~25 matches to typical good air and IAQ index ~250 states for typical polluted air. The sensor output resistance value varies according to VOCs concentrations, the higher the concentration of reducing VOCs, the lower the resistance and vice versa. The IAQ qualitative range is from 0 to 500 (the larger, the worse). Table 2 illustrates the IAQ qualitative index meaning of the selected sensor.

Table 2. BME680 qualitative IAQ index meaning.

IAQ index	Air Quality
0–50	Good air quality
51–100	Normal air quality
101–150	Little poor air quality
151–300	Poor air quality
201–300	Bad air quality
301–500	Very bad air quality

The mobile application, designated as *iAQ+Mobile,* was created using SWIFT programming language in XCode IDE (Integrated Development Environment), and the minimum requirement is the iOS 12. The *iAQ+Mobile* incorporates significant highlights as it allows not only real-time data access of the last temperature, humidity, air pressure and qualitative IAQ index information but also to receive real-time warnings to notify the occupants when the IEQ has severe deficiencies (Figure 3). The smartphone application also allows viewing the collected data in chart form.

Figure 3. *iAQ+Mobile.*

The *iAQ+* was developed by the authors to be a centralised supervision solution. Therefore, the Web application, *iAQ+Web,* was built in ASP.NET C# using the Visual Studio IDE. Using the *iAQ+Web,* the build manager can consult the IEQ data in real time. This Web portal saves the IEQ conditions history for future analysis (Figure 4).

IAQ+ Environmental Quality Monitoring

Home   Chart's Now   Config   Notifications   About

Last Environmental Quality Data

Sensor	value	Un	Time	Sensor	value	Un	Time	Sensor	value	Un	Time	Sensor	value	Un	Time
Temperature	28	°C	2018-09-24 23:20	IAQ	122	und	2018-09-24 23:20	Humidity	43	%	2018-09-24 23:20	Pressure	94632	Pa	2018-09-24 23:20
Temperature	28	°C	2018-09-24 23:20	IAQ	116	und	2018-09-24 23:20	Humidity	43	%	2018-09-24 23:20	Pressure	94625	Pa	2018-09-24 23:20
Temperature	28	°C	2018-09-24 23:19	IAQ	127	und	2018-09-24 23:19	Humidity	43	%	2018-09-24 23:19	Pressure	94632	Pa	2018-09-24 23:19
Temperature	28	°C	2018-09-24 23:19	IAQ	111	und	2018-09-24 23:19	Humidity	43	%	2018-09-24 23:19	Pressure	94625	Pa	2018-09-24 23:19
Temperature	28	°C	2018-09-24 23:19	IAQ	111	und	2018-09-24 23:19	Humidity	43	%	2018-09-24 23:19	Pressure	94632	Pa	2018-09-24 23:19
Temperature	28	°C	2018-09-24 23:18	IAQ	114	und	2018-09-24 23:18	Humidity	43	%	2018-09-24 23:18	Pressure	94634	Pa	2018-09-24 23:18
Temperature	28	°C	2018-09-24 23:18	IAQ	107	und	2018-09-24 23:18	Humidity	43	%	2018-09-24 23:18	Pressure	94632	Pa	2018-09-24 23:18
Temperature	28	°C	2018-09-24 23:17	IAQ	112	und	2018-09-24 23:17	Humidity	43	%	2018-09-24 23:17	Pressure	94634	Pa	2018-09-24 23:17
Temperature	28	°C	2018-09-24 23:17	IAQ	94	und	2018-09-24 23:17	Humidity	42	%	2018-09-24 23:17	Pressure	94632	Pa	2018-09-24 23:17
Temperature	28	°C	2018-09-24 23:16	IAQ	77	und	2018-09-24 23:16	Humidity	42	%	2018-09-24 23:16	Pressure	94632	Pa	2018-09-24 23:16
1 2				1 2				1 2				1 2			

Figure 4. *iAQ+Web*.

The *iAQ+Web* allows historical data export for enhanced reporting and analysis. The Web portal supports hourly, daily and monthly charts and tables of the monitored data for enhanced building management and audits.

To support the two main activities in the university laboratories, the laboratory experiments and teaching activities, the *iAQ+* software (Web and smartphone) supports two functional modes, the laboratory mode and the schoolroom mode. The *iAQ+* software allows the user to configure setpoints for both modes and the laboratory manager can change the status mode using the Web or the smartphone app. The smartphone app is an agile and effective way to change the status. However, the software allows the end-user to schedule the working modes (Figure 5).

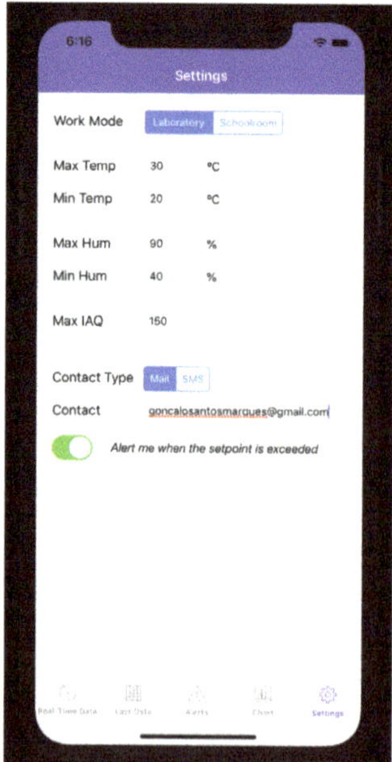

Figure 5. *iAQ+Mobile* Settings.

5. Results and Discussion

In Portugal, the majority of indoor environments have natural ventilation. The nature of dwelling, construction, heating and ventilation assumes an important influence on the air permeability variations. It is expected that over 66% of business/services constructions which support natural ventilation mechanisms are notably airtight, and the other 33% have a tendency to be leakier.

For testing purposes, one laboratory of a Portuguese university were on-site supervised using one *iAQ+* prototype. Figure 6a presents the experiment conducted by the authors of LEC supervision of thermography activities. As in most laboratories, the supervised space incorporates natural ventilation and does not have dedicated ventilation slots on the facades. The indoor air is reheated and recirculated through a couple of standard air–water fan-coils of the heating system, and the air exchange is performed through infiltrations and by opening windows manually. The outdoor air is employed to afford ventilation, to reduce the temperature or when the occupants detect the severe or disturbing odour; therefore, the IAQ is frequently deficient.

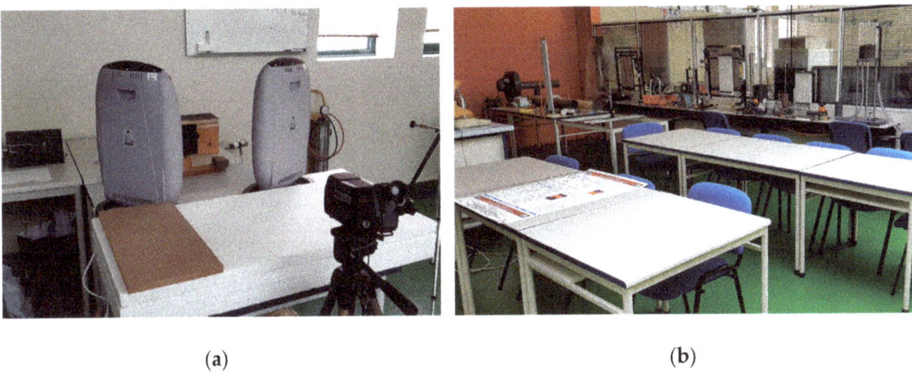

(a)   (b)

Figure 6. Laboratory layout: (**a**) experiments research area (e.g. thermography experiments supported by *iAQ+*); (**b**) classroom learning space (teaching support).

The *iAQ+* was powered by the power grid using a 230V-5V AC-DC 2A power supply. The LEC exposure data were collected in thermography experiments which showed that indoor conditions can be different from the recommended values and can affect the integrity of the collected data.

The tests conducted show the system capability not only to keep a healthful and productive workplace for the students, teachers and the school staff but also to provide a consistent and regulated state of control for laboratory activities. Figure 6b represents the laboratory space used as a schoolroom.

Figure 7 presents a sample of the graphics of the results achieved in the tests conducted. It should be noted that the graphs displayed the results obtained in the monitored rooms with induced simulations using tobacco smoke.

On one hand, monitoring environmental conditions and maintaining laboratory temperature and humidity requirements is extremely important for high-quality experimental activities. On the other hand, supervising IEQ is significant in creating a healthy and productive workplace for learning and teaching. Using the smartphone application, the user can carry all the monitored data in his pocket.

This solution not only supports alerts and setpoints configuration for real-time notifications via e-mail, SMS and push notifications but also provides an integrated dashboard for the monitored real-time collected data. Quite apart from this, the *iAQ+* incorporates a built-in visual and audio alarm, two LED's and a buzzer respectively. When the configured setpoints are met, the correspondent LED is triggered, green to inform the good IEQ status and red to notify the poor IEQ status along with the buzzer activation. In the last case, the buzzer will ring for 30 seconds with the aim to notify the occupants to act in real-time to enhance the indoor conditions.

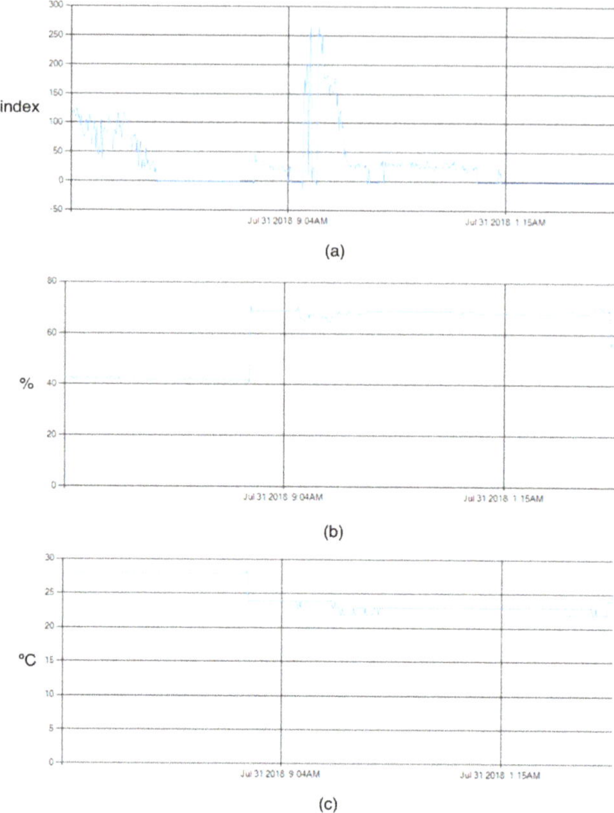

Figure 7. Results of IAQ index, air temperature and relative humidity obtained in the experiments conducted in a real laboratory environment: (**a**) IAQ index; (**b**) relative humidity (%); (**c**) air temperature (°C).

The data collected by the system is analysed before being inserted into the database. If the data exceeds the parameterised limits, the user is notified and an email or SMS is triggered (Figure 8).

The SMS notification is performed using Twilio, a cloud communication platform, as a service. This service makes it possible to programmatically send an SMS and other communication functions using its Web service APIs. The push notifications are performed using the Firebase Cloud Messaging that is a cross-platform solution for messages and notifications for Android, iOS and Web applications.

The real-time alerts promote behaviour changes. In fact, these messages alert the user to act in real-time to perform actions to increase IAQ. On the other hand, with this real-time feature, the building manager can understand when the recurrent unhealthy cases are detected and implement new adjustments to prevent them. Consecutively, *iAQ+* provides the requirements to act in real-time for enhanced living environments and laboratory conditions. The alert indicators architecture is shown in Figure 9.

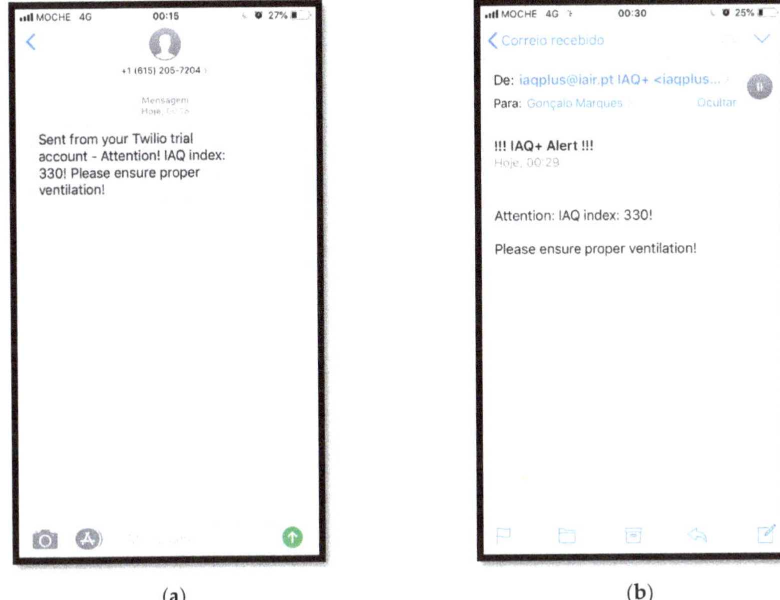

(a) (b)

Figure 8. Examples of SMS (**a**) and E-mail (**b**) notifications.

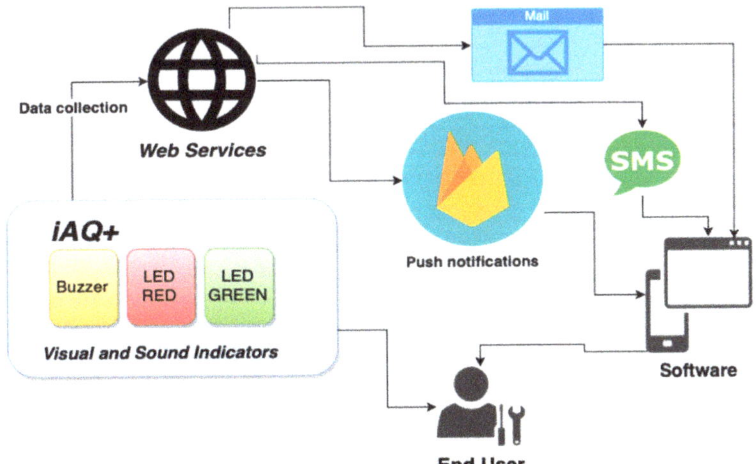

Figure 9. Alerts indicators architecture.

Firstly, *iAQ+* can help in the maintenance of LEC by providing an integrated management system that supports daily records of historical behaviours and variations as well as remote notifications when the configured setpoints are met. Secondly, this system can not only be used by the building manager to detect unhealthy situations in real-time but also as a decision-making tool to address behavioural changes to promote IEQ for enhanced productive environments.

From a quality point of view, supervising the laboratory environment is crucial to achieving a continuous LEC in the laboratory activities. Irregular maintenance of LEC can lead to unwanted effects to samples as well as to laboratory equipment's ability to produce stable results. The *iAQ+* allows the user to store the historical data of the LEC; therefore, it is possible to check the integrity of the collected data and perform data analysis and comparison taking into account the indoor conditions.

Through charts visualisation of IEQ conditions, the application provides a better acknowledgement of the supervised data when compared to the numeric table format. Consequently, the proposed solution is a significant decision-making tool to plan interventions in order to promote a healthful and productive living environment but also for IEQ analysis.

The *iAQ+* incorporates a wireless connection interface for Internet access along with an easy Wi-Fi network configuration. The *iAQ+* should be connected to a Wi-Fi hotspot for data transmission and to store the network credentials on the flash memory after successful connection. At system start-up, the *iAQ+* searches for a stored Wi-Fi network to connect to. However, if there no saved Wi-Fi networks available, the system will turn to hotspot mode and create a Wi-Fi network with an SSID "*iAQ+*". At this stage, the user must connect to the referred hotspot in order to configure the credentials of the Wi-Fi network to which the system will be connected (Figure 10).

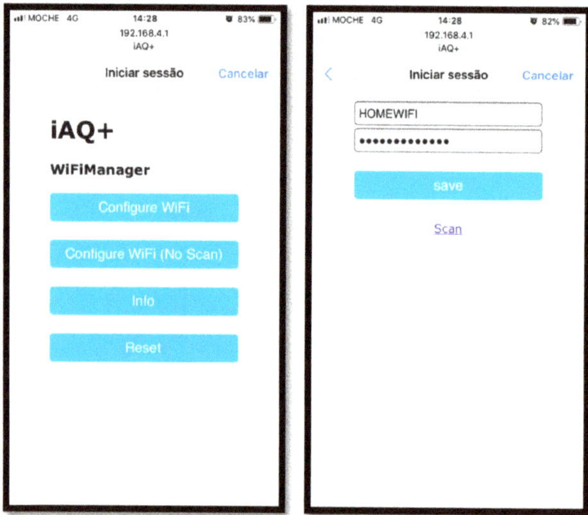

Figure 10. Wi-Fi configuration.

In this way, the *iAQ+* can be easily installed by the end-user which not only follows the original paradigm of IoT solutions but also contributes to the low-cost aim of the presented solution.

Compared to several similar solutions proposed and described in Table 1, the *iAQ+* supports push notifications to alert people in a timely way for enhanced living environments and occupational health by providing ventilation, deactivation of pollutant equipment and by activating air purification systems. When the parameters meet the setpoints for the correspondent working mode, the user is alerted to ensure a productive environment for teaching or correct laboratory environmental parameters. The *iAQ+* offers a work mode feature; the user can configure setpoints for laboratory mode and schoolroom mode. Using the *iAQ+Mobile*, the user can easily change the work mode in real-time.

On one hand, compared to other systems proposed by References [46,47,49,50] based on WSN, the *iAQ+* provides several advantages in scalability and installation in indoor living environments as it is only necessary to configure the Wi-Fi Internet connection and it is not required to configure the sensor nodes and coordinators. On the other hand, compared to the other systems proposed by References [46,50–52] which do not use mobile computing, the *iAQ+* provides smartphone and Web compatibility for data access. The solution proposed by Reference [48] supports mobile access.

However, the proposed mobile app allows only to access the last collected data and does not support historical data access. The *iAQ+Mobile* supports graphical presentation for an easy overview of the IEQ data by the end-user.

The IoT design facilitates the scalability of the proposed solution offering flexibility and expandability as the area can be monitored using only one *iAQ+* unity, but other unities can be added if needed.

Enhancements to the system hardware and software are scheduled in order to incorporate other IEQ parameters such as noise supervision and/or experiment variables. So, as future work improvements and adaptions for specific laboratory tests, such as thermographic experiments applied to wood and trees are planned. The proposed solution has multiple benefits in installation and configuration as the *iAQ+* incorporates wireless communication technology for data transmission, but also provides compatibility with both the standard domestic homes and smart homes.

Apache Kafka is an open-source stream-processing software platform for distributed high-throughput systems. This platform provides low-latency, built-in partitioning, replication and inherent fault-tolerance for real-time data handling. Apache Spark is a framework for large-scale data processing which can be used with Kafka to stream the data to solve the response time and system throughput problems. The SQL Server response time and system throughput tests were not performed. However, these evaluations are planned by the authors as long as the study of the incorporation of open-source platforms such as Apache Kafka and Apache Spark to face response time and system throughput difficulties.

It is imperative to enhance the IEQ conditions, and the authors consider that the first step is to implement real-time supervision to identify its variation and to plan interventions for enhanced living environments.

Managerial Implication

Regarding managerial implications, the results provide insights for enhanced living environments and laboratory experiments. In most university establishments, laboratories are also used as classrooms. Therefore, *iAQ+* is relevant to provide a healthful and productive living environment but also to support experimental activities. LEC has a direct impact on test results and on the accuracy and consistency of test data that are affected by environmental conditions. Thus, the proposed solution offers an integrated management system for historical data analysis in order to guarantee the integrity of test results. The two functional modes of the *iAQ+* solution, the laboratory mode and the schoolroom mode allow the user to configure setpoints for both modes. This will allow switching the functional mode in an agile way, contributing to increasing the quality of the laboratory experiments and teaching activities. The *iAQ+* easy installation feature follows the original paradigm of IoT solutions and contributes to the low cost of the proposed solution. Most importantly, the real-time alerts promote behaviour changes to support the occupants to act promptly in order to provide a healthy and productive workplace for learning and a regulated environment for laboratory activities. The proposed solution provides a mobile application and a Web portal for IAQ data analytics which facilitates data access and analysis.

6. Conclusions

This paper presented an IoT real-time supervision system architecture named *iAQ+* composed by a hardware prototype for data acquisition along with Web and mobile compatibility for data access. The proposed system provides temperature, relative humidity, barometric pressure and qualitative air quality supervision for both school and laboratory scenarios.

With the proliferation of IoT technologies, there is great potential to create automatic IEQ supervision solutions for enhanced living environments and occupational health.

The results achieved are promising, indicating an important contribution to LEC supervision solutions based on IoT. Using *iAQ+*, the monitored data can be particularly valuable to analyse and store the laboratory activities' conditions to ensure that they are stable in the course of the experiments conducted, as they influence the quality of the results.

Compared to existing systems, the *iAQ+* supports push notifications to alert people in a timely way for enhanced living environments and occupational health; in addition, by supporting a work mode feature, the user can configure setpoints for laboratory mode and schoolroom mode. This system offers flexibility and expandability as the user can start with only one *iAQ+* unity and add more unities if needed. However, quality assurance (QA) and quality control (QC) testing of the proposed study and the SQL Server response time and system throughput analysis was not done by the authors. In the future, the QA/QC should be implemented for enhanced product quality traceability and these response time and systems throughput evaluations will be conducted. The authors have also planned software and hardware improvements to adapt the system to specific laboratory tests, such as thermographic experiments applied to wood and trees. In spite of the influence of indoor environments in daily human activities, systems like this will contribute to ensuring a productive environment for teaching and proper LEC.

Author Contributions: G.M. and R.P. designed the study, developed the methodology, performed the analysis, and wrote the manuscript.

Funding: This research is framed in the project "TreeM—Advanced Monitoring & Maintenance of Trees" N. ° 023831, 02/SAICT/2016, co-financed by CENTRO 2020 and FCT, Portugal 2020 and structural funds UE-FEDER.

Acknowledgments: The financial support from the Research Unit for Inland Development of the Polytechnic Institute of Guarda is acknowledged.

Conflicts of Interest: The authors declare no conflict of interest.

References

1. Vilcekova, S.; Meciarova, L.; Burdova, E.K.; Katunska, J.; Kosicanova, D.; Doroudiani, S. Indoor environmental quality of classrooms and occupants' comfort in a special education school in Slovak Republic. *Build. Environ.* **2017**, *120*, 29–40. [CrossRef]

2. Bruce, N.; Perez-Padilla, R.; Albalak, R. Indoor air pollution in developing countries: A major environmental and public health challenge. *Bull. World Health Organ.* **2000**, *78*, 1078–1092. [PubMed]

3. Seguel, J.M.; Merrill, R.; Seguel, D.; Campagna, A.C. Indoor Air Quality. *Am. J. Lifestyle Med.* **2016**, *11*, 284–295. [CrossRef] [PubMed]

4. De Gennaro, G.; Dambruoso, P.R.; Loiotile, A.D.; Di Gilio, A.; Giungato, P.; Tutino, M.; Marzocca, A.; Mazzone, A.; Palmisani, J.; Porcelli, F. Indoor air quality in schools. *Environ. Chem. Lett.* **2014**, *12*, 467–482. [CrossRef]

5. Madureira, J.; Paciência, I.; Rufo, J.; Ramos, E.; Barros, H.; Teixeira, J.P.; de Oliveira Fernandes, E. Indoor air quality in schools and its relationship with children's respiratory symptoms. *Atmos. Environ.* **2015**, *118*, 145–156. [CrossRef]

6. Rupp, R.F.; Vásquez, N.G.; Lamberts, R. A review of human thermal comfort in the built environment. *Energy Build.* **2015**, *105*, 178–205. [CrossRef]

7. Giusto, D.; Iera, A.; Morabito, G.; Atzori, L. (Eds.) *The Internet of Things: 20th Tyrrhenian Workshop on Digital Communications*; Springer: New York, NY, USA, 2010; ISBN 978-1-4419-1673-0.

8. Gubbi, J.; Buyya, R.; Marusic, S.; Palaniswami, M. Internet of Things (IoT): A vision, architectural elements, and future directions. *Future Gener. Comput. Syst.* **2013**, *29*, 1645–1660. [CrossRef]

9. Ibaseta, D.; Molleda, J.; Díez, F.; Granda, J.C. Indoor Air Quality Monitoring Sensor for the Web of Things. *Proceedings* **2018**, *2*, 1466. [CrossRef]

10. Yin, Y.; Zeng, Y.; Chen, X.; Fan, Y. The internet of things in healthcare: An overview. *J. Ind. Inf. Integr.* **2016**, *1*, 3–13. [CrossRef]

11. Bhatt, Y.; Bhatt, C. Internet of Things in HealthCare. In *Internet of Things and Big Data Technologies for Next Generation Healthcare*; Bhatt, C., Dey, N., Ashour, A.S., Eds.; Springer International Publishing: Cham, Switzerland, 2017; Volume 23, pp. 13–33. ISBN 978-3-319-49735-8.

12. Marques, G.; Pitarma, R. An indoor monitoring system for ambient assisted living based on internet of things architecture. *Int. J. Environ. Res. Public Health* **2016**, *13*, 1152. [CrossRef]

13. United Nations. *World Population Ageing, 1950–2050*; Department of Economic and Social Affairs, Population Division, United Nations: New York, NY, USA, 2002; pp. 11–13. ISBN 92-1-051092-5.

14. Centers for Disease Control and Prevention. The state of aging and health in America 2007. N. A. on an Aging Society. 2007. Available online: http://https://www.cdc.gov/aging/pdf/saha_2007.pdf (accessed on 1 December 2018).

15. Koleva, P.; Tonchev, K.; Balabanov, G.; Manolova, A.; Poulkov, V. Challenges in designing and implementation of an effective Ambient Assisted Living system. In Proceedings of the 2015 12th International Conference on Telecommunication in Modern Satellite, Cable and Broadcasting Services (TELSIKS), Niš, Serbia, 14–17 October 2015; pp. 305–308.

16. Atzori, L.; Iera, A.; Morabito, G. The Internet of Things: A survey. *Comput. Netw.* **2010**, *54*, 2787–2805. [CrossRef]

17. Caragliu, A.; Del Bo, C.; Nijkamp, P. Smart Cities in Europe. *J. Urban Technol.* **2011**, *18*, 65–82. [CrossRef]

18. Chourabi, H.; Nam, T.; Walker, S.; Gil-Garcia, J.R.; Mellouli, S.; Nahon, K.; Pardo, T.A.; Scholl, H.J. Understanding Smart Cities: An Integrative Framework. In Proceedings of the 2012 45th Hawaii International Conference on System Sciences, Maui, HI, USA, 4–7 January 2012; pp. 2289–2297.

19. Zanella, A.; Bui, N.; Castellani, A.; Vangelista, L.; Zorzi, M. Internet of Things for Smart Cities. *IEEE Internet Things J.* **2014**, *1*, 22–32. [CrossRef]

20. Hernández-Muñoz, J.M.; Vercher, J.B.; Muñoz, L.; Galache, J.A.; Presser, M.; Hernández Gómez, L.A.; Pettersson, J. Smart Cities at the Forefront of the Future Internet. In *The Future Internet*; Domingue, J., Galis, A., Gavras, A., Zahariadis, T., Lambert, D., Cleary, F., Daras, P., Krco, S., Müller, H., Li, M.-S., et al., Eds.; Springer Berlin Heidelberg: Berlin/Heidelberg, Germany, 2011; Volume 6656, pp. 447–462, ISBN 978-3-642-20897-3.

21. Skouby, K.E.; Lynggaard, P. Smart home and smart city solutions enabled by 5G, IoT, AAI and CoT services. In Proceedings of the 2014 International Conference on Contemporary Computing and Informatics (IC3I), Mysore, India, 14–17 December 2014; pp. 874–878.

22. Alaa, M.; Zaidan, A.A.; Zaidan, B.B.; Talal, M.; Kiah, M.L.M. A review of smart home applications based on Internet of Things. *J. Netw. Comput. Appl.* **2017**, *97*, 48–65. [CrossRef]

23. Theoharidou, M.; Tsalis, N.; Gritzalis, D. Smart Home Solutions: Privacy Issues. In *Handbook of Smart Homes, Health Care and Well-Being*; Van Hoof, J., Demiris, G., Wouters, E.J.M., Eds.; Springer International Publishing: Cham, Switzerland, 2017; pp. 67–81. ISBN 978-3-319-01582-8.

24. Feng, S.; Setoodeh, P.; Haykin, S. Smart Home: Cognitive Interactive People-Centric Internet of Things. *IEEE Commun. Mag.* **2017**, *55*, 34–39. [CrossRef]

25. Madureira, J.; Paciência, I.; Pereira, C.; Teixeira, J.P.; Fernandes, E.d.O. Indoor air quality in Portuguese schools: Levels and sources of pollutants. *Indoor Air* **2016**, *26*, 526–537. [CrossRef] [PubMed]

26. Stabile, L.; Dell'Isola, M.; Russi, A.; Massimo, A.; Buonanno, G. The effect of natural ventilation strategy on indoor air quality in schools. *Sci. Total Environ.* **2017**, *595*, 894–902. [CrossRef] [PubMed]

27. Kielb, C.; Lin, S.; Muscatiello, N.; Hord, W.; Rogers-Harrington, J.; Healy, J. Building-related health symptoms and classroom indoor air quality: A survey of school teachers in New York State. *Indoor Air* **2015**, *25*, 371–380. [CrossRef] [PubMed]

28. Choo, C.P.; Jalaludin, J. An overview of indoor air quality and its impact on respiratory health among Malaysian school-aged children. *Rev. Environ. Health* **2015**, *30*, 9–18. [CrossRef]

29. Lee, S.; Chang, M. Indoor and outdoor air quality investigation at schools in Hong Kong. *Chemosphere* **2000**, *41*, 109–113. [CrossRef]

30. Yang, L.; Yan, H.; Lam, J.C. Thermal comfort and building energy consumption implications—A review. *Appl. Energy* **2014**, *115*, 164–173. [CrossRef]

31. Havenith, G.; Holmér, I.; Parsons, K. Personal factors in thermal comfort assessment: Clothing properties and metabolic heat production. *Energy Build.* **2002**, *34*, 581–591. [CrossRef]

32. Pereira, L.D.; Cardoso, E.; da Silva, M.G. Indoor air quality audit and evaluation on thermal comfort in a school in Portugal. *Indoor Built Environ.* **2015**, *24*, 256–268. [CrossRef]

33. Srbinovska, M.; Gavrovski, C.; Dimcev, V.; Krkoleva, A.; Borozan, V. Environmental parameters monitoring in precision agriculture using wireless sensor networks. *J. Clean. Prod.* **2015**, *88*, 297–307. [CrossRef]

34. Sanchez-Rosario, F.; Sanchez-Rodriguez, D.; Alonso-Hernandez, J.B.; Travieso-Gonzalez, C.M.; Alonso-Gonzalez, I.; Ley-Bosch, C.; Ramirez-Casanas, C.; Quintana-Suarez, M.A. A low consumption real time environmental monitoring system for smart cities based on ZigBee wireless sensor network. In Proceedings of the 2015 International Wireless Communications and Mobile Computing Conference (IWCMC), Dubrovnik, Croatia, 24–28 August 2015; pp. 702–707.

35. Zhou, P.; Huang, G.; Zhang, L.; Tsang, K.-F. Wireless sensor network based monitoring system for a large-scale indoor space: Data process and supply air allocation optimization. *Energy Build.* **2015**, *103*, 365–374. [CrossRef]

36. Marques, G.; Pitarma, R. IAQ Evaluation Using an IoT CO2 Monitoring System for Enhanced Living Environments. In *Trends and Advances in Information Systems and Technologies*; Rocha, Á., Adeli, H., Reis, L.P., Costanzo, S., Eds.; Springer International Publishing: Cham, Switzerland, 2018; Volume 746, pp. 1169–1177. ISBN 978-3-319-77711-5.

37. Pitarma, R.; Marques, G.; Ferreira, B.R. Monitoring Indoor Air Quality for Enhanced Occupational Health. *J. Med Syst.* **2017**, *41*, 23. [CrossRef]

38. Pitarma, R.; Marques, G.; Caetano, F. Monitoring Indoor Air Quality to Improve Occupational Health. In *New Advances in Information Systems and Technologies*; Rocha, Á., Correia, A.M., Adeli, H., Reis, L.P., Mendonça Teixeira, M., Eds.; Springer International Publishing: Cham, Switzerland, 2016; Volume 445, pp. 13–21. ISBN 978-3-319-31306-1.

39. Marques, G.; Pitarma, R. Smartwatch-Based Application for Enhanced Healthy Lifestyle in Indoor Environments. In *Computational Intelligence in Information Systems*; Omar, S., Haji Suhaili, W.S., Phon-Amnuaisuk, S., Eds.; Springer International Publishing: Cham, Switzerland, 2019; Volume 888, pp. 168–177. ISBN 978-3-030-03301-9.

40. Marques, G.; Pitarma, R. Using IoT and Social Networks for Enhanced Healthy Practices in Buildings. In *Information Systems and Technologies to Support Learning*; Rocha, Á., Serrhini, M., Eds.; Springer International Publishing: Cham, Switzerland, 2019; Volume 111, pp. 424–432. ISBN 978-3-030-03576-1.

41. Marques, G.; Pitarma, R. Monitoring Health Factors in Indoor Living Environments Using Internet of Things. In *Recent Advances in Information Systems and Technologies*; Rocha, Á., Correia, A.M., Adeli, H., Reis, L.P., Costanzo, S., Eds.; Springer International Publishing: Cham, Switzerland, 2017; Volume 570, pp. 785–794. ISBN 978-3-319-56537-8.

42. Marques, G.; Roque Ferreira, C.; Pitarma, R. A System Based on the Internet of Things for Real-Time Particle Monitoring in Buildings. *Int. J. Environ. Res. Public Health* **2018**, *15*, 821. [CrossRef]

43. Akkaya, K.; Guvenc, I.; Aygun, R.; Pala, N.; Kadri, A. IoT-based occupancy monitoring techniques for energy-efficient smart buildings. In Proceedings of the 2015 IEEE Wireless Communications and Networking Conference Workshops (WCNCW), New Orleans, LA, USA, 9–12 March 2015; pp. 58–63.

44. Wargocki, P.; Wyon, D.P.; Sundell, J.; Clausen, G.; Fanger, P.O. The Effects of Outdoor Air Supply Rate in an Office on Perceived Air Quality, Sick Building Syndrome (SBS) Symptoms and Productivity. *Indoor Air* **2000**, *10*, 222–236. [CrossRef]

45. Srivatsa, P.; Pandhare, A. Indoor Air Quality: IoT Solution. In Proceedings of the National Conference NCPCI, 19 March 2016; Volume 2016, p. 19.

46. Salamone, F.; Belussi, L.; Danza, L.; Galanos, T.; Ghellere, M.; Meroni, I. Design and Development of a Nearable Wireless System to Control Indoor Air Quality and Indoor Lighting Quality. *Sensors* **2017**, *17*, 1021. [CrossRef]

47. Bhattacharya, S.; Sridevi, S.; Pitchiah, R. Indoor air quality monitoring using wireless sensor network. In Proceedings of the 2012 Sixth International Conference on Sensing Technology (ICST), Kolkata, India, 18–21 December 2012; pp. 422–427.

48. Salamone, F.; Belussi, L.; Danza, L.; Ghellere, M.; Meroni, I. Design and Development of nEMoS, an All-in-One, Low-Cost, Web-Connected and 3D-Printed Device for Environmental Analysis. *Sensors* **2015**, *15*, 13012–13027. [CrossRef]

49. Wang, S.K.; Chew, S.P.; Jusoh, M.T.; Khairunissa, A.; Leong, K.Y.; Azid, A.A. WSN based indoor air quality monitoring in classrooms. *AIP Conf. Proc.* **2017**, *1808*, 020063.

50. Liu, J.; Chen, Y.; Lin, T.; Lai, D.; Wen, T.; Sun, C.; Juang, J.; Jiang, J.-A. Developed urban air quality monitoring system based on wireless sensor networks. In Proceedings of the 2011 Fifth International Conference on Sensing Technology, Palmerston North, New Zealand, 28 November–1 December 2011; pp. 549–554.

51. Kang, J.; Hwang, K.-I. A Comprehensive Real-Time Indoor Air-Quality Level Indicator. *Sustainability* **2016**, *8*, 881. [CrossRef]
52. Benammar, M.; Abdaoui, A.; Ahmad, S.; Touati, F.; Kadri, A. A Modular IoT Platform for Real-Time Indoor Air Quality Monitoring. *Sensors* **2018**, *18*, 581. [CrossRef] [PubMed]
53. Klein, R.C.; King, C.; Kosior, A. Laboratory air quality and room ventilation rates: An update. *J. Chem. Health Saf.* **2011**, *18*, 21–24. [CrossRef]
54. Stuart, R.; Sweet, E.; Batchelder, A. Assessing general ventilation effectiveness in the laboratory. *J. Chem. Health Saf.* **2015**, *22*, 2–7. [CrossRef]
55. Yau, Y.H.; Chew, B.T.; Saifullah, A.Z.A. Studies on the indoor air quality of Pharmaceutical Laboratories in Malaysia. *Int. J. Sustain. Built Environ.* **2012**, *1*, 110–124. [CrossRef]
56. Ugranli, T.; Gungormus, E.; Sofuoglu, A.; Sofuoglu, S.C. Indoor Air Quality in Chemical Laboratories. In *Comprehensive Analytical Chemistry*; Elsevier: Amsterdam, The Netherlands, 2016; Volume 73, pp. 859–878. ISBN 978-0-444-63605-8.
57. Morbeck, D.E. Air quality in the assisted reproduction laboratory: A mini-review. *J. Assist. Reprod. Genet.* **2015**, *32*, 1019–1024. [CrossRef]

Article

A Promising Technological Approach to Improve Indoor Air Quality

Thomas Maggos [1,*], Vassilios Binas [2], Vasileios Siaperas [3], Antypas Terzopoulos [3], Panagiotis Panagopoulos [1] and George Kiriakidis [2]

[1] Environmental Research Laboratory, NCSR "Demokritos", 15310 Ag. Paraskevi, Athens, Greece; ppanag.uoi@gmail.com

[2] Institute of Electronic Structure and Laser, Foundation for Research and Technology, 70013 Heraclion, Crete, Greece; binasbill@iesl.forth.gr (V.B.); kiriakid@iesl.forth.gr (G.K.)

[3] 691 Industrial Base Factory, 19011 Avlonas, Greece; bsiaper@gmail.com (V.S.); antyterzo@gmail.com (A.T.)

* Correspondence: tmaggos@ipta.demokritos.gr; Tel.: +30-210-6503716

Received: 18 September 2019; Accepted: 5 November 2019; Published: 12 November 2019

Abstract: Indoor Air quality (IAQ) in private or public environments is progressively recognized as a critical issue for human health. For that purpose the poor IAQ needs to be mitigated and immediate drastic measures must be taken. In environmental science and especially in advanced oxidation processes and technologies (AOPs-AOTs), photocatalysis has gained considerable interest among scientists as a tool for IAQ improvement. In the current study an innovative paint material was developed which exhibits intense photocatalytic activity under direct and diffused visible light for the degradation of air pollutants, suitable for indoor use. A laboratory and a real scale study were performed using the above innovative photo-paint. The lab test was performed in a special design photo-reactor while the real scale in a military's medical building. Nitrogen Oxide (NO) and Toluene concentration was monitored between "reference" rooms (without photo paint) and "green" rooms (with photo-paint) in order to estimate the photocatalytic efficiency of the photo-paint to degrade the above pollutants. Results of the study showed a decrease up to 60% and 16% for NO and toluene respectively under lab scale tests while an improvement of air quality up to 19% and 5% under real world conditions was achieved.

Keywords: IAQ improvement; photo-paint; NO; Toluene degradation

---

1. Introduction

Indoor air quality (IAQ) is an important determinant of human health, comfort and productivity. For that purpose, high quality indoor air is desirable. Indoor air pollution can be addressed through the two approaches of prevention and removal. The latter includes the use of air cleaning technologies especially in buildings where ventilation rates are being reduced in order to save energy. Indicative air cleaning technologies which have been developed during the recent years are: filtration and adsorption, electrostatic air purification, air filtration and gas adsorption filtration [1–3], ozonation [4–7], non-thermal plasma [8,9] and photocatalytic oxidation (PCO) [10–13]. PCO is a general air cleaning technology, which is able to degrade Volatile Organic Compounds (VOCs), such as aromatics, alkanes, odor compounds etc. Air cleaning photocatalytic technology is based on the principle that radiation of suitable wave-lengths can be absorbed by semiconductors, which leads to the creation of reactive oxygen species (ROS) that can degrade air pollutants. TiO_2 is the most commonly-used semiconductor in PCO research. However, over the last years, scientists have combined TiO_2 with other materials such as activated carbon and zeolite hybrid catalysts in order to enhance the PCO degradation of air pollutants [14–16]. Furthermore, numerous TiO_2 photocatalysts with different morphological designs have been developed: nanoparticles, nanotubes, hollow fibers and mesoporous. To this end, the need to evaluate the photocatalytic performance of the above materials in a common methodological approach has been raised and extensive research efforts have been devoted to it [17,18].

However, in most of these studies, only a single compound was tested, using a photocatalytic reactor and a methodological approach, which have been developed by the same lab that produced the material. It is well known that indoor air contains numerous contaminants; thus, tests of only one compound may be misleading. Furthermore, many studies proved the generation of by-products during the photocatalytic processes, such as formaldehyde, ozone, benzaldehyde, acetaldehyde etc. It is obvious that in some cases, by-products could be more harmful than the target pollutant [19–22]. For that purpose, and although photocatalytic technology is promising, the synthesis route of a photocatlytic material should be designed carefully in order to avoid contaminants which could lead to the formation and emission in the gas phase of intermediate products as they can be more hazardous than the target pollutant. The latter is clearly demonstrated during the evaluation of two photocatalytic air-purification in a mock-up air cabin. Although two symptoms, dizziness and claustrophobia, were reported to decrease when either one of the photocatalytic air-purification devices was operated, intermediates (acetaldehyde and formaldehyde) were detected as a result of ethanol photodegradation [23,24].

The present research addresses the indoor air purification study using a photocatalytic paint, which was tested under both laboratory and real world indoor conditions (application in building walls). A modified TiO_2 was chosen as photocatalyst in order to be activated by visible light which is the dominant spectrum in indoor environments. Traditional TiO_2 is activated only under UV light. More specifically a Mn-doped TiO_2 photocatalyst (powder) was used in the synthetic route of a photocatalytic paint production. The powder named TCM-1 [25] has already been successfully tested for the oxidation of air pollutants under indoor-like illumination conditions and when mixed either with calcareous or cementitious base matrices demonstrated a unique ability to efficiently degrade volatile organic compounds (VOCs) such as BTX (Benzene, Toluene, Xylene), formaldehyde and nitrogen oxides (NOx) [26–28]. Additionally, it has been proven effective on eliminating bacteria such as *E. coli* and *Klebsiella pneumoniae* and phages such as MS2 [29–32]. Furthermore, no dangerous by-products were produced during the degradation process. In a previous work, in order to demonstrate the photocatalytic effectiveness of TCM-1, the powder has been incorporated in/on different construction matrices such as glass, plywood, wood, ceramic and concrete substances, which were, subsequently, tested indicating very promising results [33].

In the current study TCM-1 incorporated into paint production process in order to produce paint with the ability to photocatalytically improve IAQ by degraded indoor air pollutants. NO and toluene were chosen as the target pollutants since they are typical indoor air pollutants that can be emitted from various indoor sources, such as cooking, tobacco smoke, furniture, building materials and fireplaces, as well as from outdoor sources, e.g., traffic, domestic heating. These indoor pollutants can have significant health impacts and for that purpose their elimination from the indoor environment should be appropriately addressed. In order to simulate the real indoor air conditions, visible light was used as the source for the PCO reaction in the current study. The latter is different from what we usually find in the literature. More specifically, UV light is the most commonly used parameter in the experimental procedures, which usually applied to test the efficiency of a photo-material.

The experimental study of the current work was carried out using the European Committee for Standardization (CEN) Technical Specification (TS) 16980:2016 as a basic reference to perform the lab scale experiments, while the real scale experiments were based on the comparison between the concentration level of a pollutant in a reference and the "green" room, respectively. The latter was performed at the military medical center of the Cadets Training Camp in Heraklion Crete.

2. Materials and Methods

2.1. Materials

Chemicals for the preparation of the photocatalytic powder such as Titanium (IV) oxysulfate hydrate (TiOSO4·xH2O), manganese (II) acetate tetrahydrate Mn (CH3COO)2) and ammonium hydroxide (25% NH4OH) purchased from Aldrich were applied.

Chemicals for the preparation of the photo-paint were purchased from VINAVIL EGYPT (New Cairo, Egypt), DOW chemicals (Midland, Michigan, MI, United States), BYK (Wesel, Germany), DuPont (Midland, Michigan, MI, United States) and Dionyssos Marbles (Penteli, Greece) and are readily available raw materials mainly used at the coating industry.

2.2. Methods

2.2.1. Preparation of Photo-Material (powder)

In the present study, an optimum powder (0.1% Mn-doped) named TCM-1 was used for the preparation of photocatalytic paint. TCM-1 was synthesized by a co-precipitation method with 0.1% of manganese. TCM-1 was precipitated at pH ~ 7 from aqueous solution of titanium (IV) oxysulfate hydrate and manganese by the addition of ammonia. After aging the suspension overnight, the precipitate was filtered and dried under air at 373 K. The residue was crushed to a fine powder and calcined in a furnace at 973 K for 3 h. More details in the synthesis procedure and the characterization are given in previous work along with the preparation details [25].

2.2.2. Preparation of Photo-Paint

TCM-1 powder (0.1% Mn-doped) was added to a specially formulated architectural coating. The paint was consisted 10% w/w of the TCM-1 powder partly replacing the Titanium Oxide (normally used as a white pigment). The raw materials for the preparation of the photo-paint are commonly used in the coating industry. A polyvinyl acetate copolymer binder was used and the Pigment Volume Content (PVC) was adjusted, but kept below its critical value in order for the paint to have a high quality, matte finish.

The production process of the photo-paint was consisted of three discreet phases. At phase 1, the mill-base (the minimum amount of liquid that can wet the solid particles of the paint, added at phase 2) was produced. Water was used as solvent with additives, such as sodium hexametaphosphate (dispersion agent), propylene glycol (antifreezing agent), silica modified surfactant (defoaming agent), cellulose (rheology modifier), isothiazolinone based mixture (biocide agent) and a mixture of Alkanol-amines (pH adjusting agent). At phase 2, we applied high shear forces to the mill-base by using a disperser equipped with a blade dissolver disc and gradually added the solid phase of the paint [Calcium carbonate powder (20 µm), Titanium Oxide (0.2–0.5 µm), Talc Powder (6 µm) and TCM-1 powder]. The goal of this phase is to maximize dispersion by eliminating the presence of any agglomerate in the final paint. At phase 3, we added the Polyvinyl Acetate copolymer binder emulsion and a coalescent solvent mixture, necessary for the final product to have a smooth surface (filmer). The dispersion was achieved by using low shearing forces and the pigment dispersion was stabilized in order to prevent the formation of uncontrolled flocculates by using non-ionic urethane rheology modifiers.

2.3. Characterization

Powder X-ray diffraction patterns were obtained by a Rigaku D/MAX-2000H rotating anode diffractometer (CuKα radiation) equipped with a secondary pyrolytic graphite monochromator operating at 40 kV and 80 mA over the 2θ collection range of 20–80° (scan rate was 0.05° s^{-1}). The grain size (nm) of TCM-1 was calculated from the line broadening of the X-ray diffraction peak according to the following Scherrer formula:

$$D = k\lambda/\beta cos\theta \tag{1}$$

where k is the Scherrer contact (~ 0.9), λ is the wavelength of the X-ray radiation (1.54 Å for CuKα), β is the full width at half maximum (FWHM) of the diffraction peak measured at 2θ, and θ is the Bragg angle. The morphology and elemental analysis were performed using scanning electron microscopy (SEM) and energy dispersive X-ray spectroscopy (EDX), on a JSM-6390LV microscope (Jeol, Tokyo, Japan).

The photo-paint was evaluated through typical quality control measurements. Viscosity was measured using a Brookfield KU-2 viscometer (Brookfield Engineering Laboratories, INC. Middleboro, MA, USA) at a temperature of 25 °C with the help of a laboratory water bath and the measurements were conducted according to ASTM D 562-10 (2018) standard test method.. Density was measured using a 100 ml pyknometer (density cup) by Elcometer (Manchester, United Kingdom) and an analytical scale, and the measurements were conducted according to ISO 2811 -1:2016 standard. Fineness of Grind was evaluated using a Hegman gauge and the evaluation was conducted according to ASTM D 1210-05 (2014) standard test method. Lastly, pH was measured using a Hanna HI 83141V pH meter (Hanna Instruments, Greece) and the measurements were conducted according to ISO 787-7:2009 standard.

2.4. Photocatalytic Evaluation_Lab-Scale

The experimental methodology that was applied and the required scientific equipment were based on CEN Technical Specification (TS) 16980:2016. The photocatalytic effect of the optimized material studied in a continuous flux photocatalytic reactor (Figure 1), which consists of a) a gas transfer and mixing unit in order to adjust the concentration and humidity levels; b) the photocatalytic reactor main body made of special plastic so as radiation intensity and wavelength of the radiation is not affected; (c) the sample irradiation system (OMICRON FS LED Rodgau-Dudenhofen, Germany) consisting of an LED device connected via software to a computer in order to achieve the optimum efficiency and accuracy in measuring the radiation; d) a NOx and VOCs analyzer installed on line with the reactor for continuous monitoring of the pollutant concentration.

Furthermore, in order to ensure optimum mixing of atmospheric pollutants in the reactor, a fan is installed inside the chamber, while its intensity is adjusted externally to ensure the stability of the experimental conditions throughout the experiments. The NO concentration is set to (0.50 ± 0.05) ppmv, while the relative humidity to 40 ± 5%. The illumination provided an average irradiance to the test specimen surface within the range of wavelengths that are mostly adsorbed by the photocatalyst, equal to (10.0 ± 5%) W/m^2.

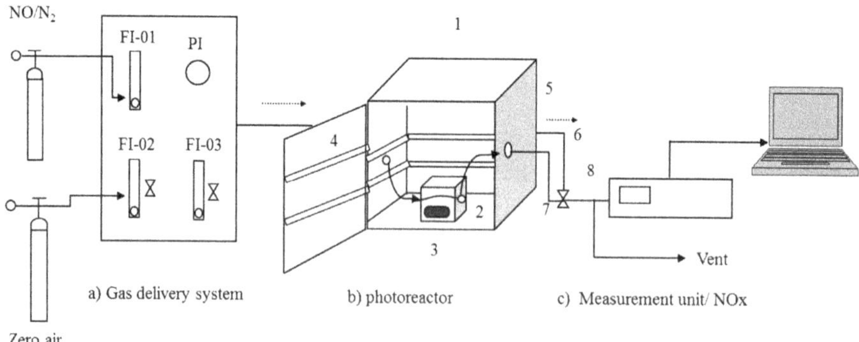

Figure 1. Photocatalytic reactor.

2.5. Photocatalytic Evaluation_Real-Scale

A building of the Hellenic Army in Crete was used to test the de-polluting efficiency of the photo-paint under real scale application. In general, the photocatalytic efficiency of a material while applied in a real scale environment and more specifically in indoor building environment could be estimated through two approaches:

a) A first approach could be the installation of air quality and environmental monitoring systems (passive inorganic (NO) and organic (BTX) samplers, temperature and humidity recorders) in the buildings prior to the application of the photo-paint. To that end, a reliable record on the concentration levels of air pollutants in the case study building (without the photo paint) for at least 12 months should be obtained. Accordingly, the photo-paint should be applied in the buildings and environmental parameters should be monitored through passive sampling techniques for another 12 months. The potential changes in the indoor air quality due to the photocatalytic action of the photo-paint will be recorded and quantified to illustrate the capability of the photo-paint to improve IAQ. A restriction of this approach could be the variations in outdoor air quality and meteorological conditions during the different sampling periods. In order to eliminate the effect of the above restrictions, the experiments has to take place the same season (e.g., winter) and the outdoor concentration should be considered on the final results.

b) A second approach to evaluate the efficiency of the photo-paints is to estimate the IAQ differences between "reference" rooms (without photo paint) and "green" rooms (with photo-paint), which are located on the same level (in a raw) and where the same activities take place. The current approach overcomes the restrictions of the previous one and for that purpose was used in the current study. More specifically a room of 120 m^2 in the ground floor was paint with the innovative photocatalytic paint ("green room") and compared with a same size and usage room ("reference room") located very close to the "green" one. The outcome of this approach was compared with the outcome of the lab tests in order to estimate the differences of the photocatalytic performance of a material when studied in a control experimental reactor and under real world environmental conditions. Passive samplers for NO and Toluene were applied. More specifically, four passive samplers/pollutants were installed in each of the rooms for 30 days and then analyzed in the lab using the well-established Saltzmann spectrophotometric method for NO, while the Toluene samplers were and desorbed by carbon disulphide and analyszed by gas chromatography (GC/FID).

3. Results

3.1. Physical and Chemical Properties

Figure 2 shows the X-Ray Difraction (XRD) pattern of TCM-1(dopant concentration 0.1 wt%), calcinated at 700 °C for 3 h 1. The peaks at 2θ values of 25.3°, 37.6°, and 48.2° correspond to the (101), (004) and (200) planes, respectively, and they are all anatase signature peaks. The grain size for the TCM-1was 38.69 nm and it was determined from the full width at half maximum (FWHM) of the (101) anatase peak according to the Scherrer's formula. In our previous work, it was proven that the doping and the role of Mn in TiO$_2$. in the case of 0.1 wt% manganese shows only 2+ oxidation state in comparison with high concentrations [34].

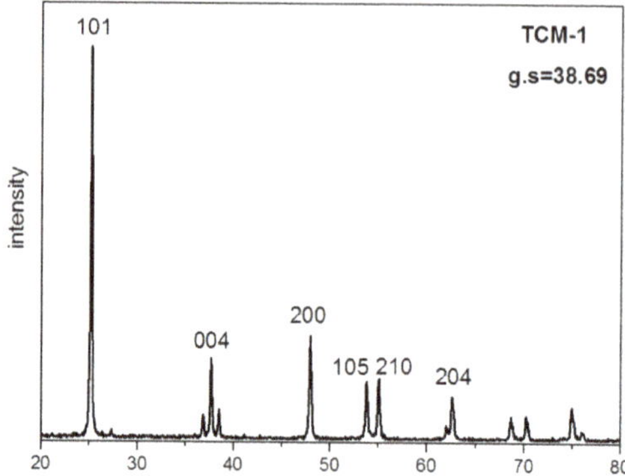

Figure 2. Powder XRD patterns of TCM-1.

Figure 3 shows the UV-Vis absorption as a function of wavelength for TCM-1 and exhibited an absorption edge in the visible light range (400–800 nm). The band gap energy was 2.87 eV for Mn doped catalyst.

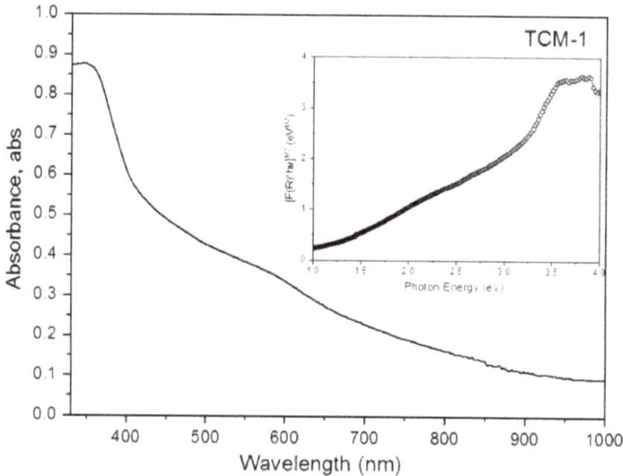

Figure 3. UV-vis absorption of TCM-1and energy gaps calculated from Kubenka–Munk plots.

Figure 4a shows the morphology of the photocatalyst were investigated with SEM, without specific morphology, while the spherical shape particles of all the samples demonstrated some degree of agglomeration and the diameter ranged from 0.1 to 40 μm. Figure 4b shows the characteristic peaks of Mn and Ti atoms.

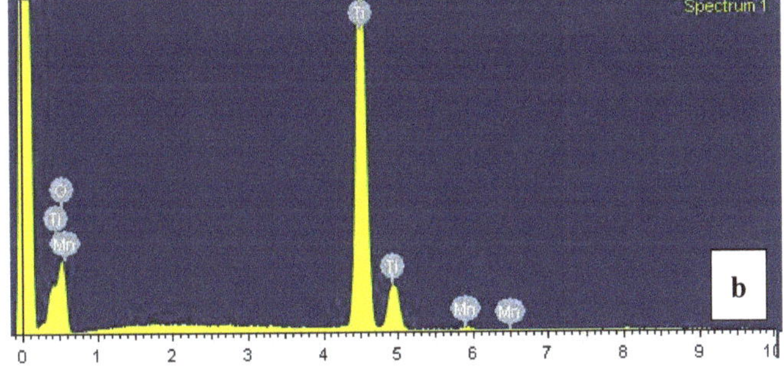

Figure 4. (a,b) SEM image of TCM-1 and EDX analysis.

As far as the photo-paint is concerned, the main physicochemical properties are listed in Table 1. The measured values are typical for an indoor emulsion architectural paint, and although TCM-1 was used at 10% w/w, TCM-1 behaved as expected from a Titanium Oxide white pigment. A paint formulated with the process described in Section 2.2.2 and with the use of Ti Pure 902+ Titanium Oxide white pigment instead of TCM-1 has identical physicochemical properties.

Table 1. Physicochemical Properties of photo-paint.

Property	Value	Test Method
Viscosity at 25°C (KU)	100–110	ASTM D 562
Density at 25°C (kg/l)	1.55	ISO 2811
Fineness and Dispersion	<40 μm	ASTM D1210
pH at 25°C	8.5–9	ISO 787-9
PVC (%)	66	Calculated
Usage rate for a 50 μm dry film thickness (m^2/kg)	5.95	Calculated

3.2. Lab and Real-Scale Photocatalytic Performance

In order to estimate the background, experiments in the absence of the photocatlatyic paint were performed. The photocatalytic experiments determine the total pollutant degradation involving both UV photolysis and photocatalysis on the photo-paint. The net photocatalytic effect is calculated by the subtraction of the background contribution from the photocatalytic experiments. More specifically, blank tests were carried out by polluting the reactor with NO and Toluene in the absence of the photocatalyst without and with irradiation, respectively; then, the same experiments were carried out in the presence of the photocatalysts.

Figure 5a,b present the elimination of NO and toluene under the irradiation of the photo-paint from Vis-light. It is obvious that just after the irradiation of the sample (Time 30), a sharp decrease in NO and a smaller but significant decrease in toluene concentration is observed, which demonstrate the immediate response of the photocatalytic system and provide the photo-efficiency of the paint.

Adsorption of NO onto the chamber's wall area and photolysis are the main sinks of NO during blank tests. Calculations have shown that both these mechanisms did not contributed to the total NO and toluene removal during the photocatalytic experiments.

The photocatalytic activity was evaluated by the calculation photocatalytic yield (% n, Equation (2)) and photodegradation rate (r, Equation (3)). The corresponding equations were used for toluene. The results in Table 2 showed the possibility of developing a very promising and highly active to air pollutants photocalatylic paint (Table 2).

$$\% \; \eta_{NO}^{total} = \frac{C_{NO}^{IN} - C_{NO}^{OUT,light}}{C_{NO}^{IN}} \times 100 \tag{2}$$

where:

C_{NO}^{IN}: the concentration of NO at reactor inlet

$C_{NO}^{OUT,light}$: the concentration of NO at reactor outlet under stable conditions with irradiation (lamp on)

The rate of photocatalytic yield of the material is calculated by the formula below (Equation (3)) and expressed in $\mu gm^{-2} \, s^{-1}$:

$$r_{NO}^{photo} = \frac{613F}{S}\left(\frac{\eta_{NO}^{total}}{\left(1 - \eta_{NO}^{total}\right)} - \frac{\eta_{NO}^{dark}}{\left(1 - \eta_{NO}^{dark}\right)} \right) \tag{3}$$

where

F: the gas flow ($m^3 \, h^{-1}$)

S: The area of the test surface (m^2)

The photodegradation rate provides a more accurate measure of the photocatalytic activity of the material in comparison with the % photocatalytic decomposition. The latter is attributed to the fact that r is taking into consideration the initial concentration of the pollutant, the sample's area and the irradiation time. It is expressed as μg of converted NO/toluene per m^2 of material per second of irradiation. The current parameter was calculated only for the lab tests, where the pollutant flow rate in the chamber was known.

The losses in the system are minimal, and as a consequence the fraction $\frac{\eta_{NO}^{dark}}{\left(1 - \eta_{NO}^{dark}\right)}$ is zero.

Table 2. Photocatalytic parameters for lab and real scale tests.

Parameter	Lab Scale		Real Scale	
	NO	Toluene	NO	Toluene
% η	59.08	16.7	18.8	5.26
rphoto ($\mu g/m^2 s$)	3.89	1.05	-	-

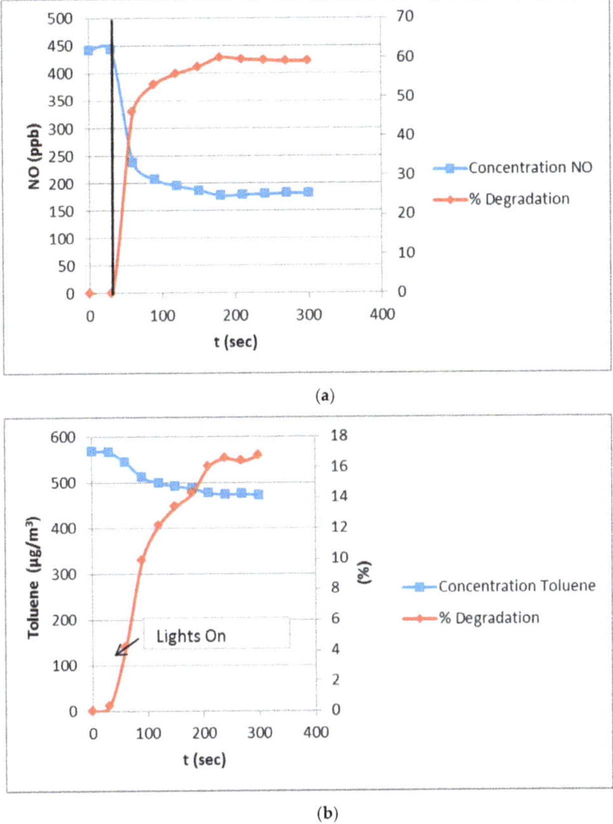

Figure 5. (**a**) Photocatalytic yields (% η) concentration profile versus time for the photo-oxidation of NO in the presence of the catalyst. (**b**) Photocatalytic yields (% η) concentration profile versus time for the photo-oxidation of Toluene in the presence of the catalyst.

The variations on NO and Toluene concentrations during the experimental procedure of photocatalysis are presented in Figure 5a and Figure 5b, respectively. At first, the pollutant was introduced into the photo-reactor and the system remained in equilibrium for 30 min. Then, irradiation was followed and a sharp reduction in NO and significant but less intense than NO elimination in toluene concentration is observed, which stabilized after 3 h and remained stable for approximately 2 h where the irradiation stopped. The results from the laboratory tests showed that for both pollutants, the photocatalytic paint gave very promising results. It is worth to note that in the case of toluene, the degradation efficiency is significantly lower than NO. However, a reduction of VOCs even at that level could have significant effects in the improvement of IAQ. Furthermore, the lower degradation efficiency in VOCs compare with NO is in line with the results of various studies [35–38]. The latter could be attributed to the low adsorption capacity of toluene molecules at the catalyst surface. The extremely hydroxylated surface of the paint material due to physically or chemically bound water, could constrain its interaction with the active radicals through surface diffusion [39,40].

3.3. Real-Scale Photocatalytic Performance

As far as the real scale application is concerned, the deppolution efficiency was calculated by the absolute difference in the concentration levels of NO and toluene, which was measured in the two rooms of the military building: the "green" and the "reference" rooms. Results are shown in Table 2. It is observed that real scale tests showed significant lower values of NO and toluene degradation than in the lab-scale application. The latter was also observed in other studies [41–45]. The control environment of the laboratory versus the more complex and polyparametric environment of a real application could be the main reason that led to significant differences in the values that characterize the photocatalytic efficiency of the paint. However, they provide an indication of the photocatalytic efficiency of the paints to degrade pollutants in situ and a basis for photo-paint applications in order to improve IAQ.

4. Discussion

The effect of photocatalytic paint in the improvement of IAQ has been studied in both laboratory and real scale tests. It is interesting to note that NO could be effectively converted by the photocatalytic paint, while toluene showed lower photocatalytic removal in the same photocatalytic process (lab and real scale). A possible explanation of the different photocatalytic performance of NO and toluene under the same application could be explained as follows: the dominant oxidants in a photocatalytic reaction are hydroxyl radicals ($OH_\bullet$) and hydroperoxyl radicals ($HO_{2\bullet}$), which are generated from the redox reactions of positive holes (h^+) and electrons (e^-) with O_2, H_2O or OH^-. The kinetic coefficients of gas phase reactions for HO_x radicals and targeted gas (NO and toluene in this case) have only minor differences [46,47]. However, NO is better adsorbed by the alkaline constituents of the paint due to its acidic property. This significantly increases their reaction potential with HO_x radicals. On the other hand, the adsorption capacity of toluene is lower than NO due to the extremely hydroxylated surface of the paint, which constrain its interaction with the active radicals. Toluene is absorbed more easily in a less hydrophilic TiO_2 surface than a more hydroxylated one [41]. It is obvious that different gaseous pollutants present significant variations in their photocatalytic activity due to the difference in their diffusion in the paint matrix. Efficient adsorption of the pollutant molecules at the catalyst surface could promote the photocatalytic reaction.

Comparing lab versus real scale experiments results, it is observed that laboratory tests showed significant higher values of both pollutants (NO and toluene) degradation than in real scale application. The translation from the laboratory results to "real" site efficiency is difficult because of the great number of parameters involved such as traffic and environmental parameters (temperature, light intensity, relative humidity wind speed). For that purpose, the more complex and polyparametric environment of a real application lead to significant differences in the values, which characterize the photocatalytic efficiency of the paint. In any case, precaution has to be taken with the interpretation of data obtained from the real scale experiments since these results are limited over time. However, they provide an indicative picture of the efficiency of the photocatalytic paints to eliminate air pollutants under real world conditions and a basis on which to improve their photocatalytic capacity for future applications. The need for large scale applications is more imperative nowadays, as many photo-materials have indicated very promising results during lab scale tests, but their capacity under real world conditions has not been proven. The latter is critical and proven by the outcome of the current work, which showed significant differences between the photo-efficiency of the same material under two different scales. Additionally, the demonstration of the effectiveness of photocatalytic materials on site should also include negative effects, such as the formation and emission of by-products during the photocatalytic reactions, as well as the durability of the photo-paint mechanical properties. The latter is of high importance, as in most of the cases, TiO_2 oxidizes not only the air pollutants on the photo-material surface, but also their organic and inorganic components. For that purpose, special attention is given in the formulation of such materials in order to avoid it. As it is not in the scope of the current manuscript to study the durability of the photo-paint mechanical properties, measurements to characterize the mechanical durability of the paint were not performed. However, it is worth mentioning that almost two years after the photo-paint application, cracks or other surface damages have not been observed. Nevertheless, beyond the visual observation, the latter should be certified by analytical measurements, which should be the topic of future study. Furthermore, better results could be obtained by using mathematical models to simulate the photocatalytic processes, validating the model using the outcomes of the lab measurements and then implement various parameters in order to assess the real life effects [48].

5. Conclusions

This article addresses the effect of a photocatalytic paint on the elimination of air pollutants, and more specifically NO and toluene, for application in indoor environments. The following conclusions can be drawn:

- The physicochemical properties (including the mechanical parameters) of the photo-paint does not seem to be affected from the introduction of the photocatalytic powder (TCM-1) in the synthesis route.
- The photocatalytic efficiency of the paint on NO removal was significantly higher than toluene. The potential of a pollutant removal depends on the intrinsic properties of gas and the chemical nature of the paint in which the TiO_2 particles are embedded. However, the removal rate of toluene was very promising for the improvement of IAQ while using the studied photo-paint.
- Lab tests showed better photocatalytic properties of the paint than the results from the real scale application due to the great number of parameters involved in the case of real scale application.
- There is a need for large scale applications to demonstrate the effectiveness of photocatalytic materials on site, including any negative effects of the application, such as the emission of by-products (e.g., carbonyl compounds, O_3 etc.).

Author Contributions: Conceptualization, T.M., V.B. and G.K.; Data curation, T.M. and P.P.; Investigation, G.K.; Methodology, T.M., V.B., V.S. and A.T.; Project administration, V.B.; Writing—original draft, T.M.; Writing—review & editing, T.M., V.B., V.S., A.T., P.P. and G.K.

Funding: This research was funded by the project "Intelligent technologies for enhancing Quality of Life through the provision of health monitoring at home and support for everyday life activities" (MIS 5002464), funded by the Operational Programme "Competitiveness, Entrepreneurship and Innovation" (NSRF 2014-2020) and co-financed by Greece and the European Union (European Regional Development Fund).

Conflicts of Interest: The authors declare no conflict of interest.

References

1. Ren, H.; Koshy, P.; Chen, W.-F.; Qi, S. Charles Christopher Sorrell Photocatalytic materials and technologies for air purification. *J. Hazard. Mater.* **2017**, *325*, 340–366. [CrossRef] [PubMed]
2. Zhang, Y.; Moa, J.; Li, Y.; Sundell, J.; Wargocki, P.; Zhang, J.; Little John, C.; Corsi, R.; Deng, Q.; Leung, M.H. Can commonly-used fan-driven air cleaning technologies improve indoor air quality? A literature review. *Atmos. Environ.* **2011**, *45*, 4329–4343. [CrossRef]
3. Bolashikov, Z.D.; Melikov, A.K. Methods for air cleaning and protection of building occupants from airborne pathogens. *Build. Environ.* **2009**, *44*, 1378–1385. [CrossRef]
4. Mamaghani, A.H.; Haghighat, F.; Lee, C.-S. Photocatalytic oxidation technology for indoor environment air purification: The state-of-the-art. *Appl. Catal. B Environ.* **2017**, *203*, 247–269. [CrossRef]
5. Fan, Z.; Lioy, P.; Weschler, C.; Fiedler, N.; Kipen, H.; Zhang, J. Ozone-initiatedreactions with mixtures of volatile organic compounds under simulatedindoor conditions. *Environ. Sci. Technol.* **2003**, *37*, 1811–1821. [CrossRef] [PubMed]
6. Boeniger, M.F. Use of ozone generating devices to improve indoor airquality. *Am. Ind. Hyg. Assoc. J.* **1995**, *56*, 590–598. [CrossRef] [PubMed]
7. Zhong, L.; Haghighat, F. Ozonation air purification technology in HVACapplications. *ASHRAE Trans.* **2014**, *120 (Pt 1)*, 8.
8. Bahri, M.; Haghighat, F.; Rohani, S.; Kazemian, H. Impact of design parameterson the performance of non-thermal plasma air purification system. *Chem. Eng. J.* **2016**, *302*, 204–212. [CrossRef]

9. Bahri, M.; Haghighat, F. Plasma-based indoor air cleaning technologies: Thestate of the art-review. *CLEAN-Soil Air Water* **2014**, *42*, 1667–1680. [CrossRef]

10. Zhong, L.; Haghighat, F.; Lee, C.-S. Ultraviolet photocatalytic oxidation for indoor environment applications: Experimental validation of the model. *Build. Environ.* **2013**, *62*, 155–166. [CrossRef]

11. Farhanian, D.; Haghighat, F. Photocatalytic oxidation air cleaner: Identification and quantification of by-products. *Build. Environ.* **2014**, *72*, 34–43. [CrossRef]

12. Mo, J.; Zhang, Y.; Xu, Q. Effect of water vapor on the by-products anddecomposition rate of ppb-level toluene by photocatalytic oxidation. *Appl. Catal. B Environ.* **2013**, *132–133*, 212–218. [CrossRef]

13. Debono, O.; Thévenet, F.; Gravejat, P.; Héquet, V.; Raillard, C.; Le, L.; Locoge, N. Gas phase photocatalytic oxidation of decane at ppb levels: Removal kinetics, reaction intermediates and carbon mass balance. *J. Photochem. Photobiol. A Chem.* **2013**, *258*, 17–29. [CrossRef]

14. Ao, C.H.; Lee, S.C.; Mak, C.L.; Chan, L.Y. Photodegradation of volatile organic compounds (VOCs) and NO for indoor air purification using TiO2: Promotion versus inhibition effect of NO. *Appl. Catal. B Environ.* **2003**, *42*, 119–129. [CrossRef]

15. Ao, C.H.; Lee, S.C.; Zou, S.C.; Mak, C.L. Inhibition effect of SO_2 on NOx and VOCs during the photodegradation of synchronous indoor air pollutants at parts perbillion (ppb) level by TiO2. *Appl. Catal. B Environ.* **2004**, *49*, 187–193. [CrossRef]

16. Selishchev, D.S.; Kolinko, P.A.; Kozlov, D.V. Adsorbent as an essential participant in photocatalytic processes of water and air purification: Computer simulation study. *Appl. Catal. A Gen.* **2010**, *377*, 140–149. [CrossRef]

17. Verbruggen, S.W. TiO_2 photocatalysis for the degradation of pollutants ingas phase: From morphological design to plasmonic enhancement. *J. Photochem. Photobiol. C Photochem.* **2015**, *24*, 64–82. [CrossRef]

18. Ismail, A.A.; Bahnemann, D.W. Mesoporous titania photocatalysts: Preparation, characterization and reaction mechanisms. *J. Mater. Chem.* **2011**, *21*, 11686–11707. [CrossRef]

19. Hodgson, A.T.; Destaillats, H.; Sullivan, D.P.; Fisk, W.J. Performance of ultraviolet photocatalytic oxidation for indoor air cleaning applications. *Indoor Air* **2007**, *17*, 305–316. [CrossRef] [PubMed]

20. Muggli, D.S.; McCue, J.T.; Falconer, J.L. Mechanism of the photocatalytic oxidation of ethanol on TiO2. *J. Catal.* **1998**, *173*, 470–483. [CrossRef]

21. Mo, J.H.; Zhang, Y.P.; Xu, Q.J.; Zhu, Y.F.; Lamson, J.J.; Zhao, R.Y. Determination and risk assessment of by-products resulting from photocatalytic oxidation of toluene. *Appl. Catal. B Environ.* **2009**, *89*, 570–576. [CrossRef]

22. Kovalevskiy, N.S.; Lyulyukina, M.N.; Selishcheva, D.S.; Kozlova, D.V. Analysis of air photocatalytic purification using a total hazard index: Effect of the composite TiO2/zeolite photocatalyst. *J. Hazard. Mater.* **2018**, *358*, 302–309. [CrossRef] [PubMed]

23. Denny, F.; Permma, E.; Scott, J.; Wang, J.; Pui, D.Y.H.; Amal, R. Integrated Photocatalytic Filtration Array for Indoor Air Quality Control. *Environ. Sci. Technol.* **2010**, *44*, 5558–5563. [CrossRef] [PubMed]

24. Yu, H.; Zhang, K.; Rossi, C. Experimental Study of the Photocatalytic Degradation of Formaldehyde in Indoor Air using a Nano-particulate Titanium Dioxide Photocatalyst Indoor. *Built Environ.* **2007**, *16*, 529–537. [CrossRef]

25. Binas, V.D.; Sambani, K.; Maggos, T.; Katsanaki, A.; Kiriakidis, G. Synthesis and photocatalytic activity of Mn-doped TiO_2 nanostructured powders under UV and visible light. *Appl. Catal. B Environ.* **2012**, *113–114*, 79–86. [CrossRef]

26. Cacho, C.; Geiss, O.; Barrero-Moreno, J.; Binas, V.D.; Kiriakidis, G.; Botalico, L.; Kotzias, D. Studies on photo-induced NO removal by Mn-doped TiO_2 under indoor-like illumination conditions. *J. Photochem. Photobiol. A Chem.* **2011**, *222*, 304–306. [CrossRef]

27. Karafas, E.S.; Romanias, M.N.; Stefanopoulos, V.; Binas, V.; Zachopoulos, A.; Kiriakidis GPapagiannakopoulos, P. Effect of metal doped and co-doped TiO_2 photocatalysts oriented to degrade indoor/outdoor pollutants for air quality improvement. A kinetic and product study using acetaldehyde as probe molecule. *J. Photochem. Photobiol. A-Chem.* **2019**, *371*, 255–263. [CrossRef]

28. Binas, V.; Stefanopoulos, V.; Kiriakidis, G.; Papagiannakopoulos, P. Photocatalytic oxidation of gaseous benzene, toluene and xylene under UV and visible irradiation over Mn-doped TiO2 nanoparticles. *J. Mater.* **2019**, *5*, 56–65. [CrossRef]

29. Venieri, D.; Tournas, F.; Gounaki, I.; Binas, V.; Zachopoulos, A.; Kiriakidis, G.; Mantzavinos, D. Inactivation of Staphylococcus aureus in water by means of solar photocatalysis using metal doped TiO_2 semiconductors. *J. Chem. Technol. Biotechnol.* **2017**, *92*, 43–51. [CrossRef]

30. Venieri, D.; Gounaki, I.; Bikouvaraki, M.; Binas, V.; Zachopoulos, A.; Kiriakidis, G.; Mantzavinos, D. Solar photocatalysis as disinfection technique: Inactivation of Klebsiella pneumoniae in sewage and investigation of changes in antibiotic resistance profile. *J. Environ. Manag.* **2017**, *195*, 140–147. [CrossRef] [PubMed]

31. Venieri, D.; Gounaki, I.; Binas, V.; Zachopoulos, A.; Kiriakidis, G.; Mantzavinos, D. Inactivation of MS2 coliphage in sewage by solar photocatalysis using metal-doped TiO$_2$. *Appl. Catal. B Environ.* **2015**, *178*, 54–64. [CrossRef]

32. Venieri, D.; Fraggedaki, A.; Kostadima, M.; Chatzisymeon, E.; Binas, V.; Zachopoulos, A.; Kiriakidis, G.; Mantzavinos, D. Solar light and metal-doped TiO$_2$ to eliminate water-transmitted bacterial pathogens: Photocatalyst characterization and disinfection performance. *Appl. Catal. B Environ.* **2014**, *154–155*, 93–101. [CrossRef]

33. Binas, V.; Papadaki, D.; Maggos, T.; Katsanaki, A.; Kiriakidis, G. Study of innovative photocatalytic cement based coatings: The effect of supporting materials. *Constr. Build. Mater.* **2018**, *168*, 923–930. [CrossRef]

34. Binas, V.; Venieri, D.; Kotzias, D.; Kiriakidis, G. Modified TiO$_2$ based photocatalysts for improved air and health quality. *J. Mater.* **2017**, *3*, 3–16.

35. Ramirez, A.M.; Demeestere, K.; De Belie, N.; Mäntylä, T.; Levänen, E. Titanium dioxide coated cementitious materials for air purifying purposes: Preparation, characterization and toluene removal potential. *Build. Environ.* **2010**, *45*, 832–838. [CrossRef]

36. Tsoukleris, D.S.; Maggos, T.; Vassilakos, C.; Falaras, P. Photocatalytic degradation of volatile organics on TiO$_2$ embedded glass spherules. *Catal. Today* **2007**, *129*, 96–101. [CrossRef]

37. Kannangara, Y.Y.; Wijesena, R.; Rajapakse, R.G.M.; Nalin de Silva, K.M. Heterogeneous photocatalytic degradation of toluene in static environment employing thin films of nitrogen-doped nano-titanium dioxide. *Int. Nano Lett.* **2018**, *8*, 31–39. [CrossRef]

38. Pei, C.C.; Leung, W.W.-F. Photocatalytic oxidation of nitrogen monoxide and o-xylene by TiO$_2$/ZnO/Bi$_2$O$_3$ nanofibers: Optimization, kinetic modeling and mechanisms. *Appl. Catal. B Environ.* **2015**, *174–175*, 515–525. [CrossRef]

39. Chen, J.; Kou, S.-c.; Poon, C.-s. Photocatalytic cement-based materials: Comparison of nitrogen oxides and toluene removal potentials and evaluation of self-cleaning performance. *Build. Environ.* **2011**, *46*, 1827–1833. [CrossRef]

40. Ardizzone, S.; Bianchi, C.L.; Cappelletti, G.; Naldoni, A.; Pirola, C. Photocatalytic degradation of toluene in the gas phase: Relationship between surface species and catalyst features. *Environ. Sci. Technol.* **2008**, *42*, 6671–6676. [CrossRef] [PubMed]

41. Boonen, E.; Beeldens, A. Photocatalytic roads: From lab tests to real scale applications. *Eur. Transp. Res. Rev.* **2013**, *5*, 79–89. [CrossRef]

42. Beeldens, A. Air purification by pavement blocks: Final results of the research at the BRRC. In Proceedings of the Transport Research Arena Europe–TRA, Ljubljana, Slovenia, 21–24 April 2008.

43. Gignoux, L.; Christory, J.P.; Petit, J.F. Concrete roadways and air quality–Assessment of trials in Vanves in the heart of the Paris region. In Proceedings of the 12th International Symposium on Concrete Roads, Sevilla, Spain, 13–15 October 2010.

44. Maggos, T.; Plassais, A.; Bartzis, J.G.; Vasilakos, C.; Moussiopoulos, N.; Bonafous, L. Photocatalytic degradation of NOx in a pilot street canyon configuration using TiO$_2$-mortar panels'. *Environ. Monit. Assess.* **2007**, *136*, 35–44. [CrossRef] [PubMed]

45. Maggos, T.; Bartzis, J.G.; Liakou, M. Gobin Photocatalytic degradation of NOx gases using TiO$_2$-containing paint: A real scale study. *J. Hazard. Mater.* **2007**, *146*, 668–673. [CrossRef] [PubMed]

46. Atkinson, R.; Baulch, D.L.; Cox, R.A.; Crowley, J.N.; Hampson, R.F.; Hynes, R.G.; Jenkin, M.E.; Rossi, M.J.; Troe, J. Evaluated kinetic and photochemical data for atmospheric chemistry: Volume I-gas phase reactions of O$_x$, HO$_x$, NO$_x$ and SO$_x$ species. *Atmos. Chem. Phys.* **2004**, *4*, 1461–1738. [CrossRef]

47. Finlayson-Pitts, B.J.; Pitts, J.N. *Chemistry of the Upper and Lower Atmosphere*; Academic Press: San Diego, CA, USA, 2000.

48. Hunger, M.; Hüsken, G.; Brouwers, H.J.H. Photocatalytic degradation of air pollutants–from modeling to large scale application. *Cem. Concr. Res.* **2010**, *40*, 313–320. [CrossRef]

 *applied sciences*

Article

Study of Passive Adjustment Performance of Tubular Space in Subway Station Building Complexes

Junjie Li [1], Shuai Lu [2,*], Qingguo Wang [3], Shuo Tian [1] and Yichun Jin [1]

[1] School of Architecture and Design, Beijing Jiaotong University, Beijing 100044, China;
    lijunjie@bjtu.edu.cn (J.L.); 17121721@bjtu.edu.cn (S.T.); 18121731@bjtu.edu.cn (Y.J.)
[2] School of Architecture and Urban Planning, Shenzhen University, Shenzhen 518060, China
[3] China Design and Research Group, Beijing 100042, China; wangqg@cadg.cn
*   Correspondence: Lyushuai@szu.edu.cn; Tel.: +86-10-18500234716; Fax: +86-10-62785691

Received: 11 January 2019; Accepted: 20 February 2019; Published: 26 February 2019

Abstract: The stereo integration of subway transportation with urban functions has promoted the transformation of urban space via extensive two-dimensional plans to intensive three-dimensional development. As sustainable development aspect, it has posed new challenges for the design of architectural space to be better environmental quality and low energy consumption. Therefore, subway station building complexes with high-performance designs should be a primary focus. Tubular space is a very common spatial form in subway station building complexes; it is an important space carrier for transmitting airflow and natural light. As such, it embodies the advantages of effectively utilizing natural resources, improving the indoor thermal and light environments, refining the air quality, and reducing energy consumption. This research took tubular space, which has a passive regulation function in subway station building complexes as its research object. It firstly established a scientific and logical method for verifying the value of tubular space by searching causal relationships among the parameterized building space information factors, occupancy satisfaction elements, physical environment comfort aspects, and climate conditions. Secondly, based on the actual field investigation, a database of physical environment performance data and users' subjective satisfaction information was collected. Through the fieldwork results and analysis, the research thirdly concluded that the potential passive utilization of tubular space in subway station building complexes can be divided into two aspects: improvement in comfort level itself and utilization of climate between natural or artificial. Finally, three typical integrated design method for tubular spaces exhibiting high levels of performance and low amounts of energy consumption in subway station building complexes was put forward. This interdisciplinary research provides a design basis for subway station building complexes seeking to achieve high levels of performance and low amounts of energy consumption.

Keywords: passive space design; tubular space; physical building environment; fieldwork test; subway station building complex

---

Highlights:

(1) Construction of a multi-criteria analysis framework to analyze the passive adjustment performance of tubular space in subway station building complexes;

(2) Establishment of a database of physical environment performance and occupants' subjective satisfaction, based on actual field investigations;

(3) Development of an integrated design idea for tubular spaces in subway station building complexes that displays a high level of performance and low amount of energy consumption as the target orientation; and

(4) Proposal of three typical design concepts for compound tubular space.

1. Introduction

1.1. Research Background

With the rapid evolution of urban construction, the Transit-Oriented Development (TOD) mode has gradually formed a new organizational model for urban public spaces [1–3]. With the expansion of city subway, subway station building complexes have also entered a period of reinvention [4,5]. The stereo integration of subway transportation with urban functions has promoted the transformation of urban spaces from extensive two-dimensional plans, to intensive three-dimensional development [6]. Due to China's rapidly-advancing urbanization, the demand for sustainable development is becoming more and more urgent [7,8], and the issues of improving occupant comfort and reducing environmental load must be optimized [9,10]. The significant flow rate of people mainly in pass-through mode has led to lower environmental quality in above- and underground spaces at the junctions in subway station, and this may directly affect occupant comfort [11] and health [12]. In addition, large and complex public buildings tend to occupy a significant proportion of a city's energy consumption, threatening the sustainable development of human living environments [13,14].

1.2. Passive Design of Tubular Space in Subway Station Building Complexes

Passive design, which affects the sustainability of architecture from the prototype stage onward, is an important aspect of green building design [15]. Passive building design does not rely on active system equipment, but it does depend on a strong capacity for climate adaptability and self-adjustment, which creates a harmonious indoor coexistence of people and the outdoor environment [16,17]. Passive architecture describes buildings that are designed to cope with climate factors by providing enduring and natural comfortable indoor conditions [18,19]. The term "passive" conveys the idea of self-defense or self-protection of users in architectural design, with respect to the local natural environment [20,21]. A quality passive design avoids the possibility of high levels of energy consumption, saving up to 50% over traditional methods [22]. Therefore, the architectural prototype generally determines the degree of sustainability of the building.

Tubular space includes horizontal and vertical corridors in buildings, usually in slender shape, such as ventilation shafts, patios and lighting tubes, and tunnel corridors for connection. Tubular space occupies an important proportion of buildings in subway station building complexes, and its passive regulation has not been deeply investigated [23]. Subway station building complexes are affected by the characteristics of the mode of space utilization, wherein it is very common to use tubular space in ground-level and underground spaces (as shown in Figure 1), including patios and lighting tubes to improve natural lighting efficiency. Other uses include ventilation shafts for improving the indoor thermal environment and air quality, a station's traffic tubes for connecting ground-level and underground stations, and tunnels for traffic transmission. Tubular space can be seen as a "communications device" that transmits people, mass, and energy to different spaces [24]. This space type is a passive adjustment strategy located between the external and interior environments of the building. It uses natural energy sources (such as wind, solar energy, and rainwater) and the natural environment to regulate microclimates and improve the indoor atmosphere. In subway station building complexes, tubular space has the potential to play an important role in passive adjustment performance, especially with regards to natural lighting and ventilation, passive cooling, etc., to optimize comfort and user satisfaction with the indoor space, and greatly reduce the energy consumption of the building's operating phase [25].

Figure 1. Examples of tubular spaces in subway station building complexes.

In their preliminary analyses of spatial design and climatic contradiction factors, some scholars have considered the particularity of using underground spaces over ground-level spaces, with respect to climate [26,27]. From the perspective of the physical environment of underground spaces, the degree of thermal and light comfort are of great importance. The comfort level of subway transit spaces and vehicle interiors has been verified by the coupling of actual tests with digital simulations [28]. After six years of actual tests of underground civil air defense space, Yong Li argued for a suitable acceptable thermal temperature range for underground areas [29]. In recent years, scholars have paid more attention to occupant health and placed a greater emphasis on ventilation and air quality, by conducting typological studies, and control and defense research on pollution and particulate matter (PM). Min Jeong Kim and others have proposed ventilation systems that can improve the platform PM10 levels and reduce ventilation energy, as compared to manual systems [30]. Practices based on this theory can be found as early as in ancient Rome, where ancient architects used tubular space to create more comfortable living environments. For instance, the underground corridor is a very good example of a kind of air cooling system in use at this time [31]. In terms of modern urban architecture, Hikarie Shibuya, a Japanese transportation complex designed as an integration concept that considers the passive utilization relationship between subway station buildings and above-ground structures, solved the problem of subway station lighting and ventilation [32]. In the retrofitting of the Les Halles area of Paris, the utilization of tubular space was adopted in underground spaces to provide natural lighting [33].

1.3. Objective of this Study

This study addressed a variety of forms of tubular space in city subway station building complexes, screening those spaces for passive adjustment potential in order to study the effect on the comfort level of the indoor environment and overall energy consumption. Through a fieldwork evaluation of building performance and occupant satisfaction in actual built projects, this research conducted a quantitative performance evaluation of passive architectural design strategies for tubular spaces. Through an analysis of the current situation and excavation of the spatial potential, this work determined the passive adjustment performance effects for subway station building complexes in terms of sustainable development, therefore providing a basis for improving architectural design methods to show higher levels of performance and lower amounts of energy consumption. This research pursued the following three objectives: (1) provide a basis for design by analyzing the types of passive function and specific variables for the technical strategies employed by subway station building complexes; (2) test the passive adjustment effects of subway station building complexes and improve the authenticity and objectivity of the designs via actual and effective environmental monitoring evaluation; and (3) explore typical further-optimized strategy models for the compound tubular space systems of subway station building complexes, and provide guidance for design optimization.

2. Methodology

This research was based on the dual perspectives of architecture and the built environment. With regards to architectural design, this work produced a space prototype and deconstructed the factors affecting the passive adjustment performance of architecture. According to a comparison of the factors that influence the quality of indoor buildings and actual built environments, a comprehensive evaluation was made of the passive adjustment performance of tubular spaces in subway station building complexes [34]. First, the factors that affect the passive adjustment performance were analyzed. Then, according to the analytic factors and taking the urban Beijing subway station building complexes as an example, a long-term physical environment test was carried out. The subject was a subway station building complex with a typical amount of tubular space. This research focused on the physical environment, as tube as the passive function of potential space and its influence on surrounding functional areas. It included actual measurement results such as the thermal conditions, air ventilation, lighting environments, indoor air quality, occupant satisfaction and comfort, and other subjective feedback. Through this objective investigation of the physical environment and subjective feedback of the occupants' degree of comfort, the problems with objective space were able to be studied and analyzed, and the potential for spatial optimization put forth from the perspective of passive adjustment performance. Finally, the database established through this research assisted in highlighting the design goals for tubular space in subway station building complexes. A model for three typical kinds of composite tubular spaces was constructed with the goal of achieving high levels of performance and low amounts of energy consumption. Therefore, this research method was divided into the following four steps. (as shown in Figure 2)

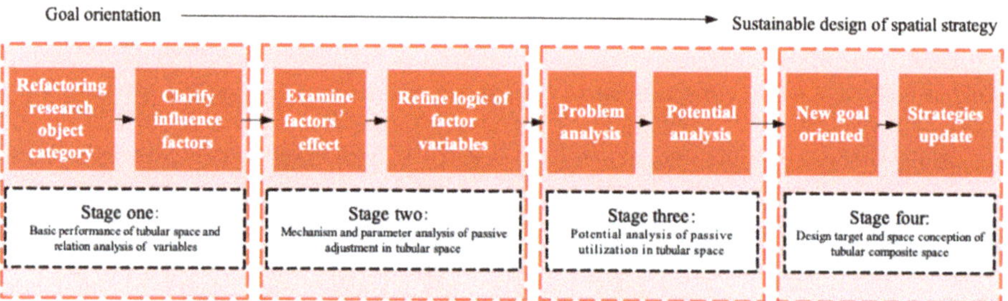

Figure 2. Research methodology route.

2.1. Stage One: Factor Analysis of the Effect of Passive Design on Tubular Space

Passive design belongs to the category of sustainable development in architecture, and is an essential part of addressing three factors: the environment, society, and the economy [35]. Tubular space is a typical passive spatial design strategy that coordinates contradictions among these three factors and architecture, leading building construction in a more positive direction. Architecture is a carrier of the climate and its human occupants; building space and outdoor climate conditions can be seen as the reasons for indoor physical environments and occupant satisfaction [36]. This research is based on an AHP (Analytic Hierarchy Process) methodology which decomposes complex issues into several group factors, and compares those factors to one another to determine their relative importance [37,38]. It adopted the method of factor quantification analysis to classify climate conditions, building spaces, physical environmental comfort, and occupancy satisfaction. This logical framework was established through measurements and simulations; the influences therein were determined by a correlation analysis, based on the acquired data. Therefore, the passive design of tubular space was divided into four factor groups: spatial parameters, climate parameters, the degree of physical environmental comfort, and occupancy satisfaction. (as shown in Figure 3)

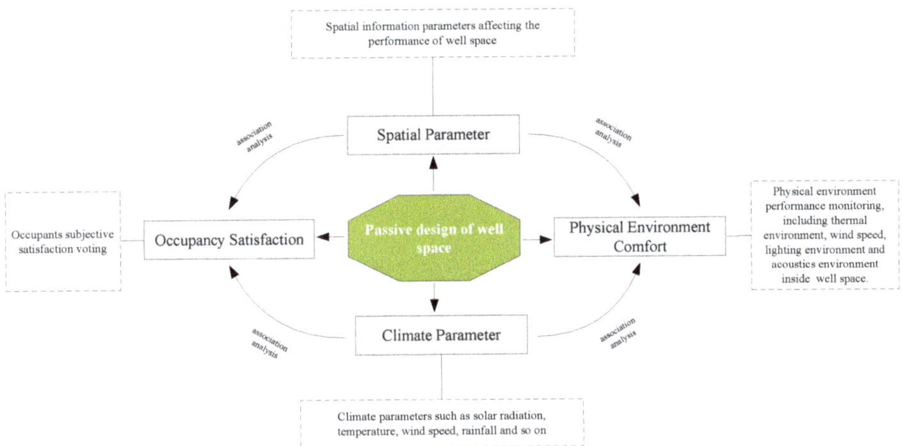

Figure 3. Analysis of the factors influential on tubular space in subway station building complexes.

According to authors pervious research in passive space design [36,39], the building space can be recognized as four categories including "shape", "mass", "quantity", and "connection". This research also adopts this research framework that the factor groups were further quantified and subdivided. The spatial parameters were again divided into four sub-factors: geometric dimensions, interface properties, and internal and external related categories (as shown in Table 1). Researchers have made useful contributions to indoor environment comfort research [40]. For example, Professor Fanger highlighted six elements that influence the comfort level of an indoor thermal environment (the mean radiation temperature, air temperature, relative humidity, air speed, clothing insulation, and metabolic rate) to form the Predicted Mean Vote (PMV) model [41]. LEED [42] in USGBC and China ESGB [43] lists lighting environment evaluation factors (including natural light, artificial light, view, and control) and offers clear standard indicators. As the tubular space can have a significant wind effect that affects the indoor air flow, while most tubular space exists underground, and the outdoor noise environment has little influence on the interior, therefore, outdoor climate parameters can be divided into four additional sub-factors: thermal environment, lighting environment, air quality, and wind speed. Corresponding to the outdoor climatic conditions, the indoor physical environment also includes four sub-factors: thermal environment, lighting environment, air quality, and air velocity. There are many factors involved in occupancy satisfaction. The method of semantic differential (SD) analysis [44] was used to divide the influencing factors into eight aspects: thermal conditions, humidity, light, air quality, air velocity, ease of use, cleanliness and maintenance, and overall environmental satisfaction (as shown in Table 1).

Table 1. Factor analysis of the passive adjustment effect of tubular space.

Factor Group	Factors	Parameter Acquisition Method	Parameter Unit
Spatial parameter	geometric dimensions (L:W:H)	distance measurement	m
	interface property (U-value)	material thermal performance calculation	$W/(m^2 \cdot K)$
	internal related categories (visitor flow rate)	statistics	N/h
	external related categories (outdoor, platform, commercial, none)	judgment	N/a
Climate parameter	thermal environment (temperature, relative humidity)	measurement	°C, %
	lighting (illuminance)		Lux
	air quality (PM2.5, PM10, CO_2)		$\mu g/m^3$, ppm
	wind speed		m/s
Degree of comfort with the physical environment	thermal environment (temperature, relative humidity)		°C, %
	lighting (illuminance)		Lux
	air quality (PM2.5, PM10, HCHO, CO_2)		$\mu g/m^3$, ppm
	air velocity, wind temperature		m/s, °C
Occupancy satisfaction	thermal comfort	occupant survey	Vote score [−3~3]
	humidity		
	air quality		
	lighting		
	ventilation		
	ease of use		
	cleanliness and maintenance		
	overall environmental quality satisfaction		

2.2. Stage Two: Field Survey

Corresponding with the public factors affecting the passive function of tubular space, the actual field investigation involved spatial drawings, monitoring the indoor and outdoor physical environments, and determining occupants' subjective levels of satisfaction. The relationship between buildings and people, especially in terms of the healthiness of the indoor environment, has a significant influence on human survival and sustainable development. Since the normal operating phase tends to be from 6:00 a.m. to 10:00 p.m., the opening time of building complexes (offices or businesses) is usually included within that time frame. Therefore, the survey had a clear research plan regarding a day cycle time, from 6:00 a.m. to 10:00 p.m., including two rush hours where there was peak human flow. The physical environment test called for the selection of a typical space, such as tube entrance, middle tubular space, connection point between a subway and complex building, station hall, or subway platform where the long-term physical environment could be monitored. The data were collected every five minutes. The physical quantities included nine parameters: temperature, humidity, illuminance, CO_2 concentration, PM2.5, PM10, HCHO, air velocity, and wind temperature. The test contents including long-term consecutive days of outdoor temperatures which measurement interval was 5 min, temperature measurements for each (selected) test point which measurement from 6:00 a.m. to 10:00 p.m. for each typical day; measurement interval was 5 min.

In addition, since piston wind can affect the interior tubular space and connection points of urban complexes during subway operation, and the aerodynamic forces of piston wind may be usable as a source of renewable energy [45], the fieldwork test also included instantaneous wind speed changes at the subway platform level. The observation frequency was 3 s, with a train cycle of arrival, stay, and departure (as shown in Table 2).

Table 2. Building the physical environment fieldwork test framework.

Measurement Items		Parameter Type	Test Content	Properties of the instruments
Thermal environment	outdoor temperature test	temperature	°C	Portable infrared temperature meter, Biaozhi GM700, Range: −50~700 °C, Resolution: 0.1 °C
	indoor temperature test for each (selected) test point			
	indoor humidity test for each (selected) test point	humidity	%	
Lighting	outdoor luminance test	luminance	lux/daylight factor %	Portable luminance meter, Reggiani DT-1301, Range: 0~50,000 Lux, Resolution: 1 Lux
	indoor luminance test for each (selected) test point			
IAQ	outdoor CO_3 concentration test	CO_2 concentration	ppm	Portable and self-record CO_2 meter, TJHY-EZY-1, Range: 0~5000 ppm, Resolution: 1 ppm
	indoor CO_2 concentration test for each (selected) test point			
	outdoor PM2.5/10 concentration test	PM2.5/10 concentration	$\mu g/m^3$	Portable air quality meter, tempopt LKC-1000S+, Range: 0~999 mg/m^3, Resolution: 0.01 mg/m^3
	indoor PM2.5/10 concentration test for each (selected) test point			
	indoor HCHO concentration test for each (selected) test point	HCHO	g/cm^3	
Ventilation	indoor air velocity test for each (selected) test point	air velocity	m/s	Self-record instrument for wind velocity and wind temperature, TJHY-FB-1, Range: 0~10 M/S, Resolution: 0.01 M/S
	indoor air temperature of each (selected) test point	air temperature	°C	Self-record instrument for environment, TJHY-HCZY-1, Range: 0~5000 ppm, Resolution: 1 ppm

The object of this investigation was the tubular spaces in a subway station building complexes, so the function and shape of the spaces were simpler and more explicit than other common building spaces. The design of the subjective questionnaire focused on the degree of occupancy comfort and the space's influence on human health during short stays. In general, the questionnaire included three categories: satisfaction with the physical environment; space satisfaction votes, such as ease of use, cleanliness, and maintenance; and overall satisfaction with the environment's quality.

In order to improve the efficiency of the investigation, spatial drawings, subjective satisfaction research, and studies of comfort related to the objective physical environment were all conducted. The characteristics of the subway space in utilization mode were special in that mostly the area was designed for a rapidly-passing crowd who would be present only for a short stay. The connection spaces often fluctuated in terms of the physical environment, and users' moods and physical health conditions likely varied considerably. Therefore, the method of subjective investigation also needed to be more diverse than in other similar types of research. The questionnaire was collected mainly by three means: website-based and on-site surveys, and on-site interviews. The purpose of the website-based questionnaire was to avoid misunderstandings related to temporally-subjective factors, and dispel individual elements through the long-term accumulation of memory. For the satisfaction and self-reported productivity questions, the survey used a 7-point semantic differential scale with endpoints of "very dissatisfied" and "very satisfied." For the purposes of comparison, the scale was assumed to be roughly linear, with ordinal values for each of the points that ranged from −3 (very dissatisfied) to +3 (very satisfied) and 0 as the neutral midpoint [46]. The on-site interviews were with specifically-selected occupants who stayed in the space for a long period of time, such as retail vendors, subway station operators, commercial building security, cleaning and maintenance staff, etc. The research methods were not rigidly obeyed for prescribed problems and formats, and were altered via conversations to help the researchers understand the respondent's age, cultural and economic background, space satisfaction, environmental problems, etc (as shown in Table 3).

Table 3. Occupancy satisfaction voting framework.

Test Items		Parameter Type	Test Content	Test Content
Physical environment satisfaction	thermal comfort	vote	web-based survey/ fieldwork-based survey/ human perception test	7-point scale $[-3,-2,-1,0,1,2,3]$ very dissatisfied to very satisfied
	humidity			
	air quality			
	lighting			
	ventilation			
Space satisfaction	ease of use	vote	web-based survey/ fieldwork-based survey/ human perception test	7-point scale $[-3,-2,-1,0,1,2,3]$ SD of feelings about space's atmosphere
	cleanliness and maintenance			
Overall environmental quality satisfaction		vote	web-based survey/ fieldwork-based survey/ human perception test	7-point scale $[-3,-2,-1,0,1,2,3]$ very dissatisfied to very satisfied

2.3. Stage Three: Problem and Analysis of the Spatial Potential

Stage two set up a framework for a comprehensive system that focused on three factors: architecture, humans, and the environment. Based on the conclusions made during that stage, the passive strategy factors affecting the research object were classified and recombined to analyze the functional characteristics of different positions of space that can be found throughout a subway station building complexes. According to the research by Margarita N. Assimakopoulos regarding the thermal environment in Greek subways [47], Teresa Moreno and colleagues' work on airborne particulate matter in the Barcelona subways, and John Burnett [48] and associates' investigation and analysis of the lighting environment in the Hongkong metro in China [49], obvious problems such as high humidity, low thermal comfort, and poor air quality and lighting environments in subway halls and platform spaces all emerged as worthy of further research. Many scholars also argued for energy-saving strategies in subway systems by means of passive ventilation designs for complete ventilation systems, for instance by developing ventilation systems in subway stations that could control indoor air pollutants [50,51]. The core of the third stage of this study includes two aspects. First, through an investigation and analysis of the status quo, the existing environmental problems were extracted and a design strategy put forward to resolve certain issues. Secondly, through statistical data, the researchers discovered the potential capacity of passive space, and identified design opportunities that could improve comfort, health, and energy efficiency.

2.4. Stage Four: Set up New Target Orientation and Space Update

The fourth step of this research put forward the spatial design goal of tubular space in subway station building complexes, from the perspective of sustainable development. Certain space assumptions and a particular design procedure for the compound tubular space were promoted to provide guidance for the optimized design.

Through data and problem analyses and potential excavation, this research searched typical models of complex tubular space complex systems that would be applicable to subway station building complexes. The key point was to determine an applicable and feasible space utilization model that could provide a design basis. The researchers put special emphasis on viable applications for potential natural resources in subway spaces, such as tunnel and piston wind, pull shafts, lighting, and landscape tubes, that could be further integrated into the design. Typical composite tubular spaces can be in the form of a tube tunnel (a solar chimney composite space system), combined wind tunnel (a displacement ventilation complex space system), combined active and passive ground source heat pump (wind tunnel composite tubular space system), hot air shaft ventilation, or lighting composite space system. Figure 3 offers an overall view of the study.

3. Results and Discussion

3.1. Building Space Information Factors

Based on the above-mentioned factors, a multi-criteria evaluation method for tubular space was proposed. The survey selected five typical subway station building complexes in Beijing, including Xizhimen (W1), Haidianhuangzhuang (W2), Guomao (W3), Dawanglu (W4), and Wangfujing Stations (W5) (as shown in Figure 4). The selected five stations involved six main subway lines: #1, #2, #4, #10, #13, and #14. All of the surveyed stations were urban subway transit hubs. Xizhimen Station (W1) is where the #2, #4, and #13 subway lines converge; it is also home to three high-rise commercial office buildings and the Beijing subway station. Haidianhuangzhuang (W2), Guomao (W3), and Dawanglu Stations (W4) all are points of convergence for more than two subway lines, and offer connections with urban complexes. Wangfujing Station (W5) is located in the middle of Beijing, and is connected to the largest integrated commercial building in Asia. As such, it features a substantial people flow rate; the location is also of geographical importance (as shown in Table 4).

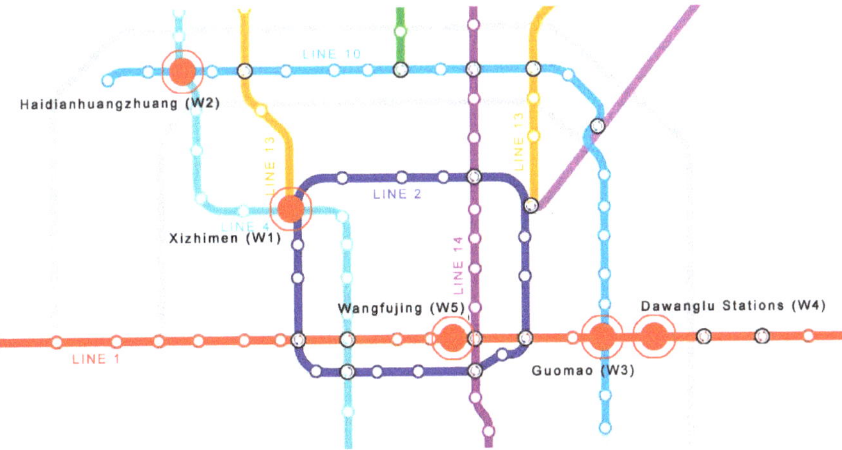

Figure 4. Five typical subway station building complexes in Beijing.

The test period was selected to be from 28th June 2017 to 20th July 2017, the highest-temperature time period in Beijing. It is a typical and continuous testing period of 3 weeks. The data excluded unstable factors which may conduct to instantaneous mutations data such as weather mutations, active equipment interference, people behavioral interference, misuse of testing instruments, etc, and used a mean value within the 3 weeks. The purpose of this research was to investigate the performance of the tubular space in the physical environment under the most unfavorable conditions in the summer climate.

Table 4. Survey object information.

No.	Station Name	City Complex	Building Function	Building Area (m^2)	Subway Line	Number of Test Points	Type of Test Point
W1	Xizhimen	Cade mall	Commercial, office	89,000	#2, #4, #13	3	middle tubular space, tube entrance, platform layer
W2	Haidianhuangzhuang	Gate City mall	Commercial	47,000	#4, #10	3	middle tubular space, tube entrance
W3	Guomao	Yintai mall	Commercial, office	350,000	#1, #10	4	middle tubular space, tube entrance, platform layer, station hall
W4	Dawanglu	China Trade Center mall	Commercial, office	710,000	#1, #14	4	middle tubular space, tube entrance, platform layer, station hall
W5	Wangfujing	Oriental Plaza mall	Commercial	120,000	#1	4	middle tubular space, tube entrance, platform layer, station hall

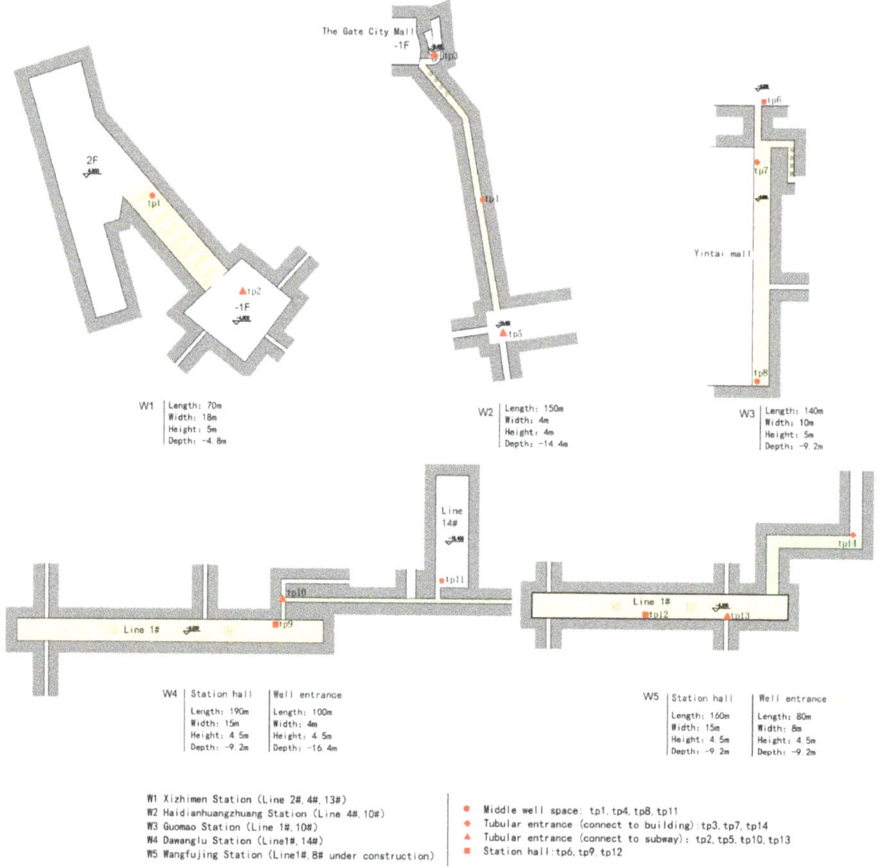

Figure 5. Test station space plan and test points location.

The first step in the fieldwork survey was to acquire the building information. This was performed in order to obtain the data supporting the tubular spaces. After the fieldwork test, the study first drew out five site plans and geometric scale parameters for the tubular space (as shown in Figure 5). The researchers then selected three or four test points for each site station; all tests contained 18 measuring points. Each test point had a certain representativeness. The W1 site contained three test points; tp1 was located in the middle of an above-ground glass corridor between the station and the complex, and was a middle tubular space.Tp4 and tp11 were located between the sites and underground associated tubes, which were all classified as middle tubular space. Tp8 was a link between a subway station and commercial building, which was also middle tubular space. Tp3, tp7, and tp14 belonged to the first kind of tube entrance space, and connected the commercial complex building at one end of the tube shaft. Tp2, tp5, tp10, and tp13 belonged to the second kind of tube entrance space, and connected to the station hall. Tp6, tp9, and tp12 were the test points at the subway station hall (as shown in Figure 5). Finally, tp15, tp16, tp17, and tp18 were the subway platform test points for W1, W3, W4, and W5, respectively.

Because the locations of the tubular spaces in each subway station building complexes were different, the environmental problems varied dramatically. Therefore, this research compared the physical environment parameters of the 18 measuring points, according to five types: middle tubular space (four test points), tube entrance (connection to building) (three test points), tube entrance (connection to subway) (four test points), station hall (four test points), and platform layer (four test points).

3.2. Field Survey Results and Analysis

3.2.1. Physical test results and analysis

Professor Fanger highlighted six elements that influence the comfort level of an indoor thermal environment (mean radiation temperature, air temperature, relative humidity, air speed, clothing insulation, and metabolic rate); all six are necessary to form the Predicted Mean Vote (PMV) model [41]. Tubular space in subway station building complexes is mostly underground, so the influence of radiation temperature can be ignored and people's metabolic rates can be set to the same level of 1.5 met [41]. During the test period, the hottest period in summer in Beijing was selected, so clothing level was chosen as 0.35 clo [41]. Data from this survey were collected according to eight physical parameters: environment temperature, humidity, illuminance, air velocity, PM2.5, PM10, and the HCHO and CO_2 concentrations at each point. The average result values are shown in Tables 5–9. According to the current national standards and norms, the typical range of thermal comfort is defined as between 16~28 °C [43], humidity comfort is between 30~60% [42], illumination should be no higher than 150 lux [52], and indoor air velocity in winter should be lower than 0.15 m/s and 0.25 m/s in summer. The concentrations of PM2.5 and PM10 should be lower than 75 μg/m^3 and 150 μg/m^3, respectively, according to the 24-hour average concentration limits of the two grades listed in the national standard [53]. The concentration of HCHO and CO_2 should be lower than 0.08 mg/m^3 [54] and 1000 ppm [55], respectively.

(1) Comfort analysis

Figures 6–9 are box diagrams of the physical test results for thermal conditions, lighting, IAQ, and ventilation in all five types of tubular space. The red zones show locations where the occupant comfort values were beyond the related comfort standard.

As regards the thermal environment, tp1 was a solar corridor space with a higher temperature than the other three points, which were at the boundary of the comfort zone (the point at which the human body would no longer enjoy thermal comfort). The tube lengths of tp4 and tp8 were 150 m and 140 m, respectively, and the humidity levels of the two test points significantly exceeded the standard. The highest reached 84.7%. The temperature in the space was lower than the human comfort level standard would deem acceptable, and the excessive humidity could easily cause mildew and affect users' health. As regards the lighting environment, the illuminance levels of tp4 and tp8 in the middle tubular space were not sufficient; the values did not reach the national standard requirement and thus could be hiding potential dangers. Due to the large amount of natural light at tp1, the average illuminance could reach 990 lux without artificial lighting. A better lighting environment would also improve the quality of the indoor environment. Past research results have shown that occupants become uncomfortable and can even experience headaches and chest tightness, causing their work ability to decline, when the CO_2 concentration is over 1000 ppm. In the test of CO_2 concentration for the middle tube, the two longer tube sat tp4 and tp8 had lower air quality; the highest CO_2 concentration was at tp4, which reached 1931 ppm. This is over two times the standard. The maximum PM2.5 concentration at tp8 reached 121 g/m^3; the outdoor concentration was 74.6 g/m^3. Thus, the disadvantages were greatly exacerbated. Although the air velocity values at tp4 and tp11 were slightly higher than the standard, moderate ventilation in a thermal environment can lower body surface temperature, taking away the sweat that collects on the human body's surface.

To sum up, the main problems as regards comfort in the middle tubular space were its high humidity, poor light environment, and low air quality.

Table 5. Average values for the physical test results in the middle tubular space.

Site No.	Test Point Number	Temperature (°C)	Humidity (%)	Illuminance (Lux)	Air Velocity (m/s)	PM2.5	PM10	HCHO	CO$_2$
W1	Tp1	32.1	47.7	990	0.196	47.9	67.2	0.022	681
W2	Tp4	28.1	70.2	72.0	0.265	45.4	64.3	0.030	1163
W3	Tp8	25.5	76.8	19.0	0.0	66.1	92.0	0.057	1000
W4	Tp11	28.3	57.4	153.6	0.307	52.0	73.1	0.044	892.1

Note: The data in the grey background indicate that it is beyond comfortable zone.

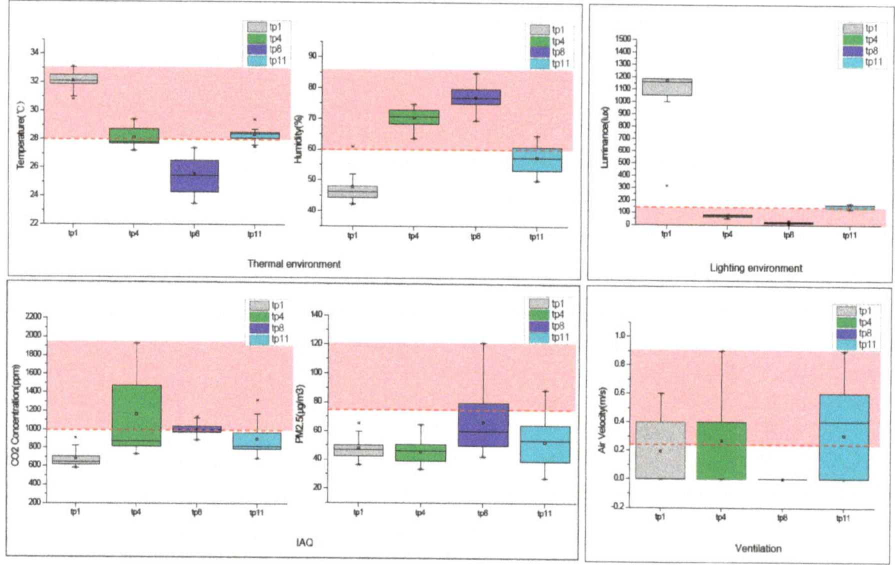

Figure 6. diagram of the physical test results from the middle tubular space.

Figure 7 shows the performance of the physical environment of the tube entrance space approach to one end of a shopping mall. All of the environment's temperatures and most of the humidity index values for the three test points exceeded the comfort zone. The maximum value of tp3 was 32.3 °C, while the mean outdoor temperature was 29 °C. The average illuminance levels of the three test points in the tube entrance (connection to building) did not reach the national standard; tp3 and tp7 were especially low. Both of the test points located at the indoor and outdoor junctions had light levels that would require human eyes a significant amount of time to adjust to the darkness. This can cause sensations of insecurity and discomfort when entering from a strong outdoor light environment. The concentration of CO_2 in these spaces was also high, with maximum values for tp3 and tp7 reaching 1761 ppm and 1690 ppm, respectively.

To sum up, because most of the tube entrance space was connected to a shopping mall and was close to the outdoors, the thermal environment and humidity levels were low, light problems were obvious, and concentrations of CO_2 were high.

Table 6. Average values for the physical test results from the tube entrance (connection to a building).

Site No.	Test Point Number	Temperature (°C)	Humidity (%)	Illuminance (Lux)	Air Velocity (m/s)	PM2.5	PM10	HCHO	CO_2
W2	Tp3	30.6	62.1	7.5	0.117	54.8	76.3	0.055	981.9
W3	Tp7	30.4	65.2	24.4	0.144	73.5	102.8	0.126	1308.2
W5	Tp14	28.9	52.9	147.9	0.604	32.2	45.0	0.047	928.9

Note: The data in the grey background indicate that it is beyond comfortable zone.

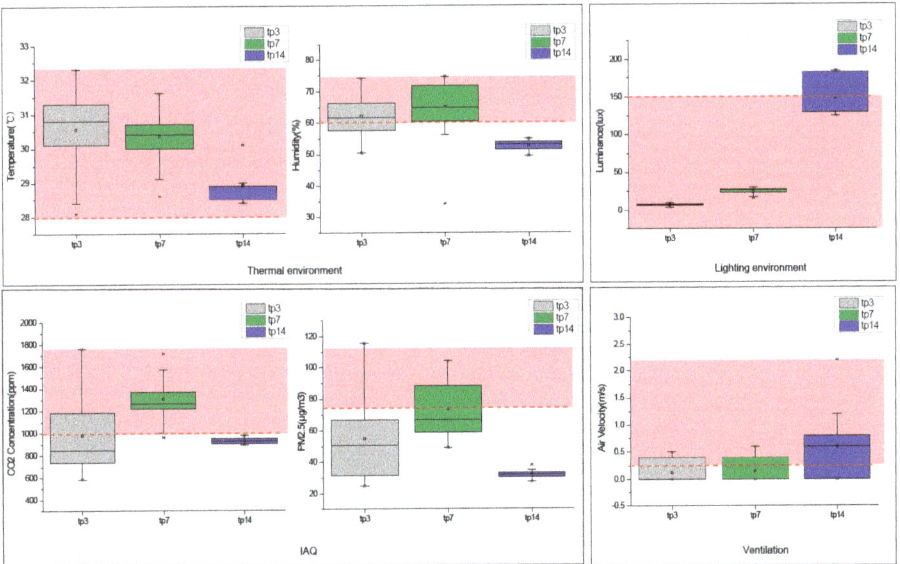

Figure 7. Box diagram of the physical test results from the tube entrance (connection to building).

According to Figure 8, the thermal and humidity environments at the tube entrance space near the subway were obvious. The temperatures at the four test points all exceeded the comfort level. The average temperature at tp5, with an elevation depth of 14.4 m, was 2.5 °C lower than that of tp10 and tp13, which were at an elevation depth of 9.2 m. The higher wind speed relieved the high humidity and CO_2 concentrations at the tube entrance space (connection to subway). Generally, the average humidity was lower than 70% and both the average and maximum values of CO_2 were greatly reduced. However, due to the influence of the subway piston wind, the wind speed presented a sinusoidal fluctuation and the direction of the wind speed changed periodically. Thus, the wind environment became the most unfavorable factor with regards to comfort.

To sum up, ventilation was one of the most detrimental comfort factors at the tube entrance space (connection to subway). It was affected by the piston wind so that cold air (from the air conditioning) was sometimes sucked out of the subway platform layer and hot air was pumped out of the tubular space. People who stayed at that location for long periods of time were frequently affected by two kinds of wind that had large temperature and direction differences. They expressed great discomfort and likely were experiencing threats to their health.

Table 7. Average values for the physical test results from the tube entrance (connection to subway).

Site No.	Test Point Number	Temperature (°C)	Humidity (%)	Illuminance (Lux)	Air Velocity (m/s)	PM2.5	PM10	HCHO	CO_2
W1	Tp2	30.8	50.0	154	0.365	53.5	75.9	0.019	688.4
W2	Tp5	28.6	66.0	32.5	0.859	52.7	73.6	0.019	840.5
W4	Tp10	31.3	68.3	107.3	1.73	107.6	151.1	0.021	920.8
W5	Tp13	31.1	70.0	91.2	0.683	149.2	201.7	0.026	802.4

Note: The data in the grey background indicate that it is beyond comfortable zone.

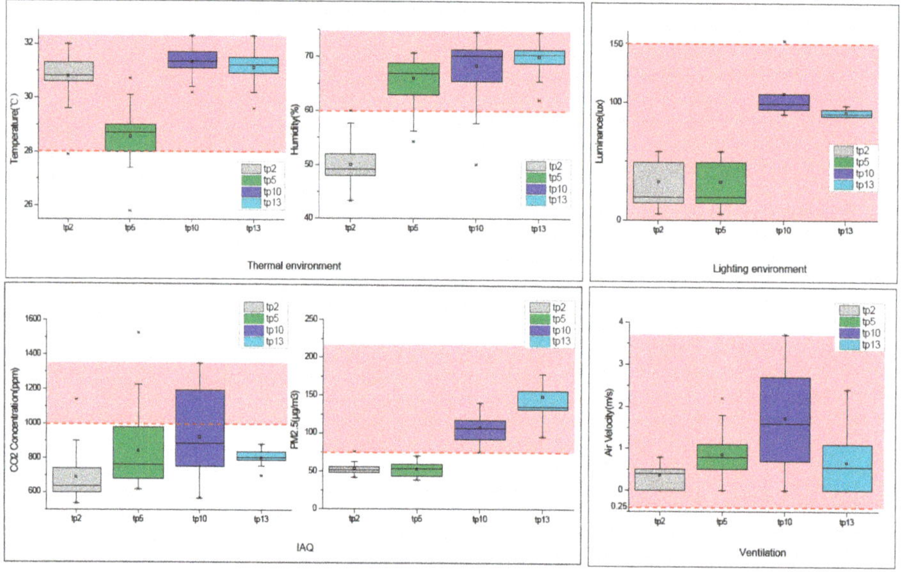

Figure 8. Box diagram of the physical test results at the tube entrance (connection to subway).

The physical environment test at the station hall focused on Beijing subway line #1, and included three test points. As can be seen from Figure 9, the environment temperature in the station hall was significantly higher than the comfort zone and 5 to 6 °C higher than the outdoor temperature at that time. The humidity in the station hall was also relatively high, but the light environment basically met the national standard. Due to a large number of people flowing through the middle of the station hall and the relatively longer tube lengths (130 m, 190 m, and 160 m), the CO_2 concentrations at the three points were correspondingly higher. The peak concentration of CO_2 at the tp6 measuring point reached 1833 ppm. Tp12 was located at a point with a smaller people flow rate, so its CO_2 concentration stayed within the standard range. However, due to the impact of the subway piston wind, there was a high level of PM pollution in the subway tunnel that was brought into the entrance hall, with a peak concentration of 160 g/m^3. Compared with the outdoor concentration 74.6 g/m^3, this was two times the outdoor concentration at that time. Thus, it can be seen that a large number of people were gathered at the station and hall levels, resulting in a higher concentration of CO_2. The station hall layer was directly connected to the train platform and had a higher PM concentration due to the influence of the piston wind from the subway.

In summary, influenced by the space size and flow rate, the physical environment of the hall at all stations was the worst, which is reflected in the high temperature and humidity, and poor air quality.

Table 8. Average values of the physical test results from the station hall.

Site No.	Test Point No.	Temperature (°C)	Humidity (%)	Illuminance (Lux)	Air Velocity (m/s)	PM 2.5	PM 10	HCHO	CO$_2$
W3	Tp6	33.5	62.2	144.1	0.724	68.6	96.0	0.038	1374.4
W4	Tp9	32.6	61.9	152.3	0.446	100.9	139.2	0.023	1153.6
W5	Tp12	31.2	71.4	92.0	0.172	119.6	167.2	0.021	922.5

Note: The data in the grey background indicate that it is beyond comfortable zone.

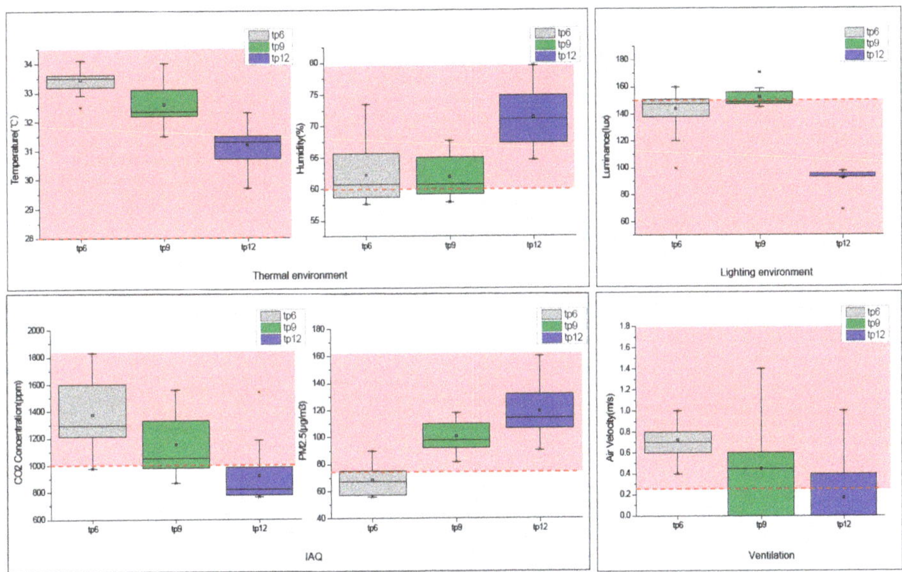

Figure 9. Box diagram of the physical test results from the station hall.

The physical environment of the platform layer is shown in Figure 10. Since the wind speed at the platform layer was significantly affected by the movement of the subway vehicles, it will be discussed in detail in the next chapter. The lighting environment met the national lighting standards. There were two key problems with comfort: the thermal environment and air quality. The temperature was generally too high; the highest value was from W4, which reached 34.5 °C. The values were 3.5 °C higher than the outdoor temperature at that time. For air quality, the most significant problem was the PM concentration. The PM2.5 data for almost all of the test sites were higher than the human body's comfort range; the PM10 concentration was too high at the W4 site as tube, reaching a maximum of 225.7 g/m^3, while the outdoor concentration was 83.2 g/m^3.

To sum up, the temperature and PM values were the key problems with comfort at the platform layer.

Table 9. Average values for the physical test results from the platform layer.

Site No.	Test Point No.	Temperature (°C)	Humidity (%)	Illuminance (Lux)	Air Velocity (m/s)	PM2.5	PM 10	HCHO	CO$_2$
W1	tp15	30.2	48	260.3	Instantaneous wind speed (as shown in Figure 12)	85.3	121.7	0.017	974.5
W3	tp16	31.6	64.3	263		98.5	138.3	0.019	867
W4	tp17	33.9	58.8	259.5		137.6	192.5	0.02	1005
W5	tp18	33.7	69.6	251.2		72.7	102.3	0.019	1283

Note: The data in the grey background indicate that it is beyond comfortable zone.

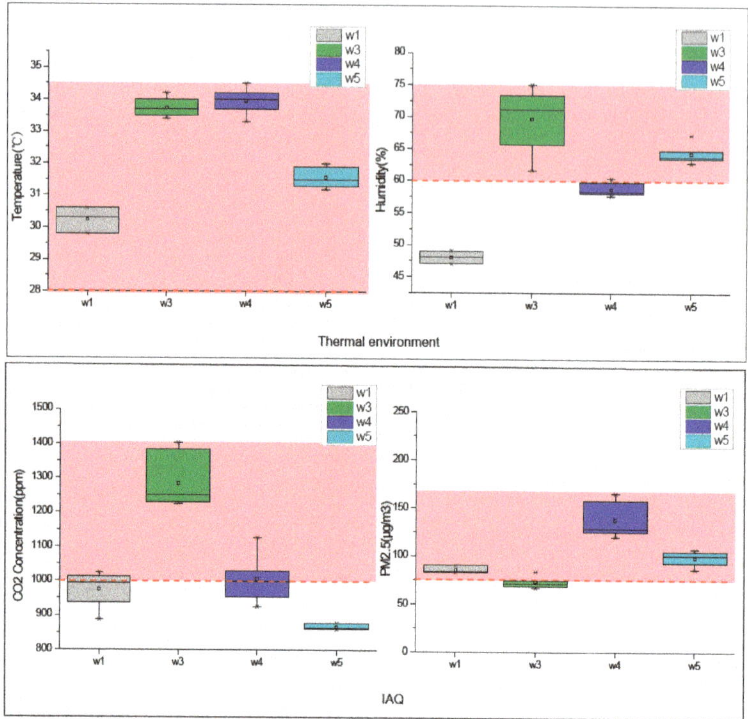

Figure 10. Box diagram of the physical test results from the platform layer.

(2) Changes in physical quantities with time and human flow

a: Temperature

The green curve in Figure 11 shows the fluctuation in outdoor temperature. Except for the aboveground space of tp1, the temperature at other points was not directly affected by solar radiation; therefore, the temperature curve hardly varied over time. The overall temperature environment at the middle of the tubular space was the best, followed by the tube entrance space near the shopping center. The third best was the tube entrance space near the subway measurement point. The thermal environment of the station hall was the worst, maintaining the highest temperature nearly the entire day. Changes in the flow rate had little influence on the thermal environment of the tubular space; the relative position was the decisive factor for the temperature there.

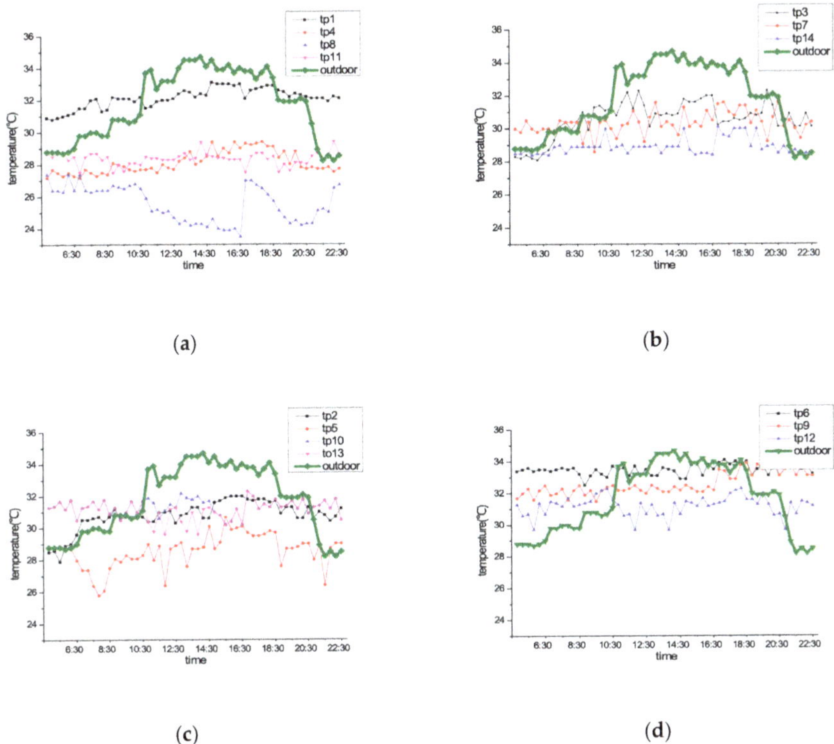

(a)

(b)

(c)

(d)

Figure 11. Comparison curves of hourly temperature data for each point during the monitoring day. (**a**) Middle tubular space; (**b**) Tube entrance (connection to building); (**c**) Tube entrance (connection to subway); (**d**) Station hall.

b: CO_2 concentration

In the diurnal variation curve for CO_2 concentration, there was almost no consistent fluctuation pattern for the same type of tubular space. From the overall curvilinear relation in Figure 12, it can be seen that there were two main trends. The first was the stable value for the whole day; the range of change was not large, as illustrated by test points tp8, tp13, and tp14. The second was the dramatic changes during two time periods, 8:00~9:00 a.m. and 4:30~7:30 p.m., which indicated that the concentration of CO_2 was significantly affected by indoor people flow, and the elevated concentration occurred during morning and evening peak hours. In addition, tp3, tp4, tp5, tp6, tp7, tp9, tp10, and tp12 exposed a lack of air adjustment capacity when a large number of people gathered together.

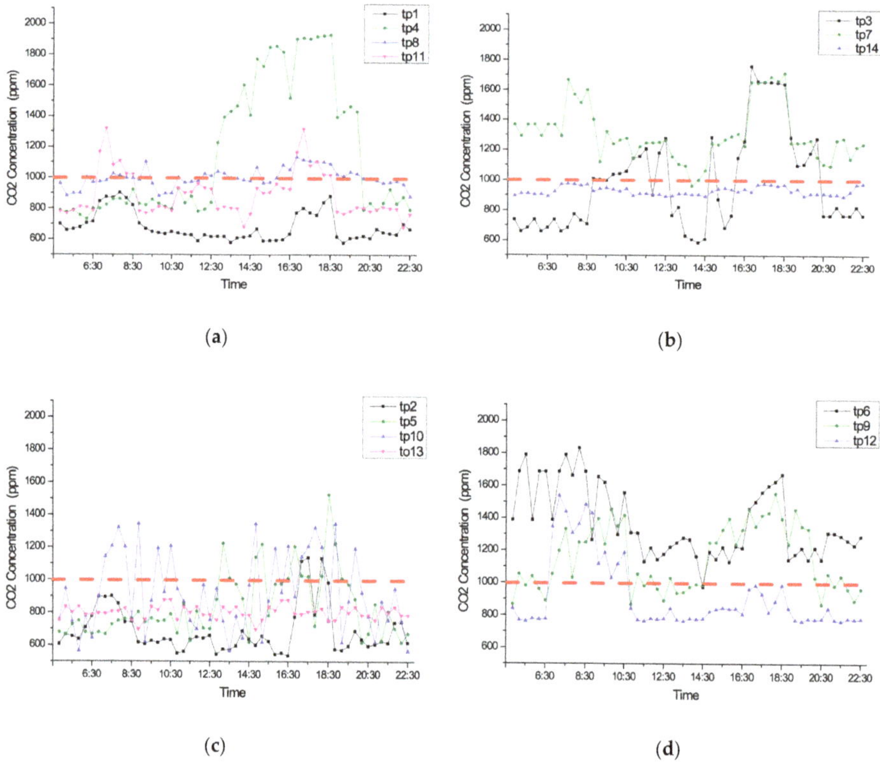

Figure 12. Comparison curves of hourly CO_2 concentration data for each point during the monitoring day. (**a**) Middle tubular space; (**b**) Tube entrance (connection to building); (**c**) Tube entrance (connection to subway); (**d**) Station hall.

c: PM2.5/PM10

As shown in Figure 13, the PM2.5 and PM10 data for each test point varied with time. This study took the middle of the tubular space to be the object of analysis. There were change rules over time for four test points: tp1, tp4, tp8, and tp11. The survey found that three points—tp1, tp4, and tp8—had a sudden increase in PM concentration at 11:00 a.m. and 6:00 p.m., almost two times that of other times. Tp11 was located in the tubular space near the #14 subway line, and was affected by the piston wind from the subway platform. The variations in PM concentration throughout the day were large, and the regularity presented was closely related to the subway's operation time.

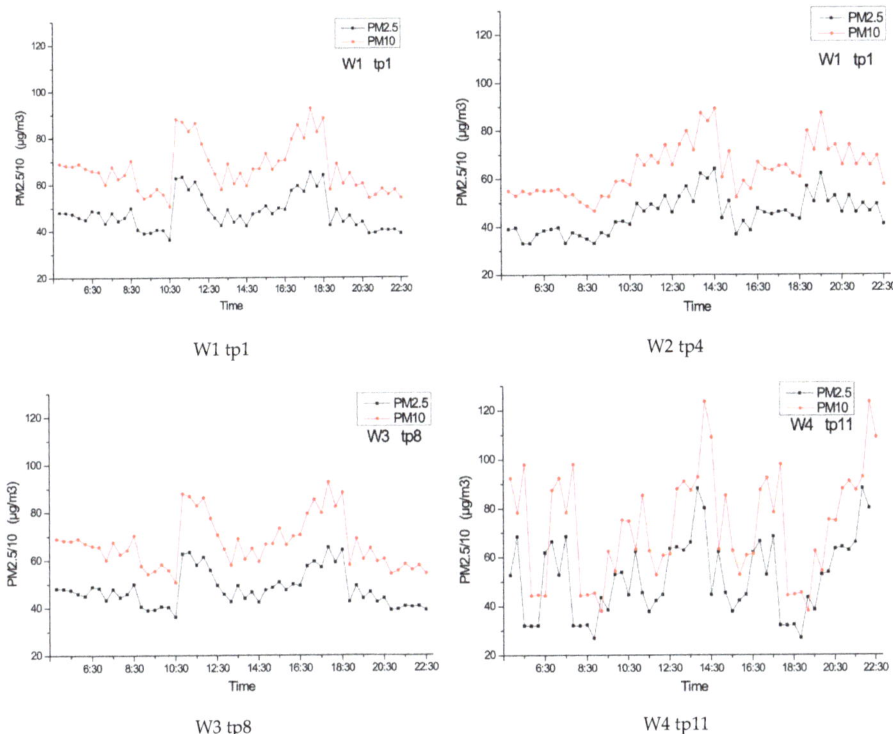

Figure 13. Comparison curves of hourly particulate matter 2.5 (PM2.5)/PM10 concentration data for each point during the monitoring day.

d: Wind speed

The test point height was 1.5 m (tp1, as shown in Figure 14a). One operating cycle for a train (arrival, stay, and departure) was around three minutes. Due to the piston effect caused by the train's operation, the tunnel wind reached remarkable levels. The wind speed when the train was arriving lasted for 15 s, with an average wind speed of 1.2 m/s. The stage during which the train stayed at the station lasted for 40~50 s (at the test point it was 48 s), with an average wind speed of 0.84 m/s. The train's departure stage lasted longer, about 120 s, with a maximum instantaneous wind speed of up to 3.6 m/s and an average wind speed of 1.79 m/s. During each three-minute cycle, the maximum wind speed was 3.6 m/s during the departure stage, while the minimum wind speed was 0 m/s during the period when the train stayed at the station (as shown Figure 14b). The wind direction was opposite during the arrival and departure stages. These two wind directions were offset when the train stayed at the station, presenting a brief state of calm. According to the coupling experiment in the Beijing subway conducted by Mingliang Ren and others, the maximum wind speed near the train could reach 7.5 m/s at the test point height of 2 m [56] (tp2, as shown in Figure 14a).

3.2.2. Occupancy satisfaction survey results and analysis

The subjective occupancy survey was comprised of two parts: on-site questionnaires and on-site interviews. The subjective questionnaire adopted a 7-point scale, where −3 corresponded with "Very Dissatisfied", 3 referred to the respondent being "Very Satisfied", and 0 was neutral (as shown in Table 10). The subjective questionnaire was issued at each test point (tp1 to tp14), 40 copies each, for a total of 560 copies. There was a total of 551 valid questionnaires completed, for a recovery rate of 98%. The test period was selected to be the same as the physical environment test, the highest temperature period for the Beijing area, from 28 June 2017, to 20 July 2017. The purpose of this research was to investigate the performance of the tubular space in terms of occupancy satisfaction under the most unfavorable conditions in the summer climate. In addition, the survey also included on-site interviews. Researchers talked with subway staff, retail traders, security guards, cleaners, and passers-by for a substantial period of time.

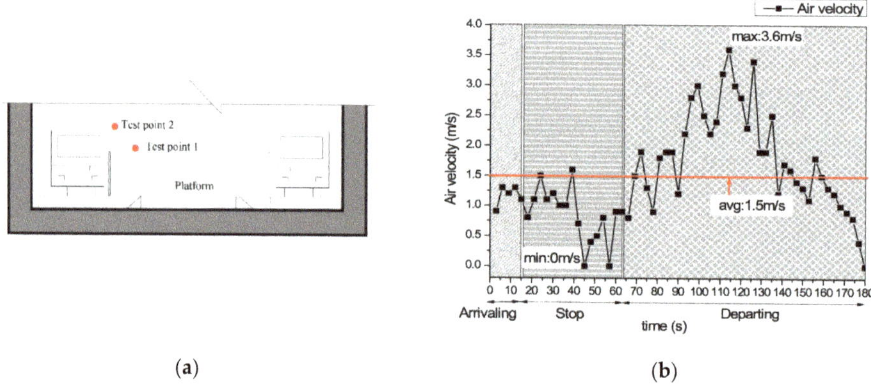

(a)   (b)

Figure 14. Wind speed curve for one operating cycle of a train. (**a**) Test point location; (**b**) Curve line.

Table 10. Questionnaire for the occupancy satisfaction survey.

Occupancy Satisfaction	Very Dissatisfied		Neutral		Very Satisfied		
Thermal comfort							
Humidity							
Air quality							
Lighting							
Ventilation	−3 ☐	−2 ☐	−1 ☐	0 ☐	1 ☐	2 ☐	3 ☐
Ease of use							
Cleanliness and maintenance							
Overall environmental quality satisfaction							

Table 11 shows the average data collected from the questionnaire. Corresponding with the analysis of the physical environment test results, this research also classified the 14 test points according to their location. There were a total of four types: middle tubular space, tube entrance (connection to building), tube entrance (connection to subway), and station hall.

Based on a histogram analysis of Figure 15, the researchers found that occupants had higher levels of satisfaction in the middle tubular space, and most of the collected data were positive. The problems with higher temperatures and poor lighting were clear in the physical environment test, but the subjective feelings of the users were not obvious. However, in terms of humidity, both the subjective questionnaire and the interviews reflected the occupants' discomfort.

The satisfaction results at the tube entrance (connection to building) fell into two categories. The first type was in relation to where the tube entrance connected the building to the outdoor space (tp3 and tp7). All indexes of satisfaction were low, especially in terms of thermal comfort, humidity, air quality, and convenience. The other type related to where the tube entrance connected to the underground building (tp14), which demonstrated positive advantages for all indicators.

Table 11. Results data from occupancy satisfaction survey.

Occupancy Satisfaction Vote	Tp1	Tp2	Tp3	Tp4	Tp5	Tp6	Tp7	Tp8	Tp9	Tp10	Tp11	Tp12	Tp13	Tp14
Thermal Comfort	−0.23	1.1	−0.9	1.3	−1.1	−1.8	−2.3	2.1	−2.1	−1.2	0.9	−2.2	−2.5	1.9
Humidity	−0.25	0.2	−1.5	−1.6	−2.5	−2.2	−2.2	−0.6	−2.4	−1.5	0.8	−2.3	−2.1	1.5
Air Quality	−0.15	−0.3	−1.2	0.2	−1.9	−2.6	−1.9	1.1	−1.5	−1.2	1.1	−2.3	−0.5	1.3
Lighting	2	−0.9	0.5	0.3	−0.8	0.2	−0.6	1.2	0.8	0.8	1.2	−0.5	0.2	1.5
Wind Comfort	−0.9	1.1	−0.6	0.5	−2.6	−0.5	−0.5	1.4	0.5	−1.6	1.2	−1.1	−2.4	−0.1
Ease of Use	−0.15	0.6	−2.1	1.1	0.2	0.9	1.1	0.9	0.6	−0.2	1.1	0.3	0.2	0.2
Cleaning and Maintenance	0.6	0.5	1.6	0.5	1.8	1.2	−1.6	1.3	0.7	0.2	1.3	0.2	0.3	0.3
Overall Environmental Quality	0.5	0.6	1.1	0.5	−0.9	0.4	−1.2	1.4	0.5	0.3	1.3	−1.2	−0.9	0.8
Overall	0.18	0.36	−0.39	0.35	−0.98	−0.55	−1.15	1.10	−0.36	−0.55	1.11	−1.14	−0.96	0.93

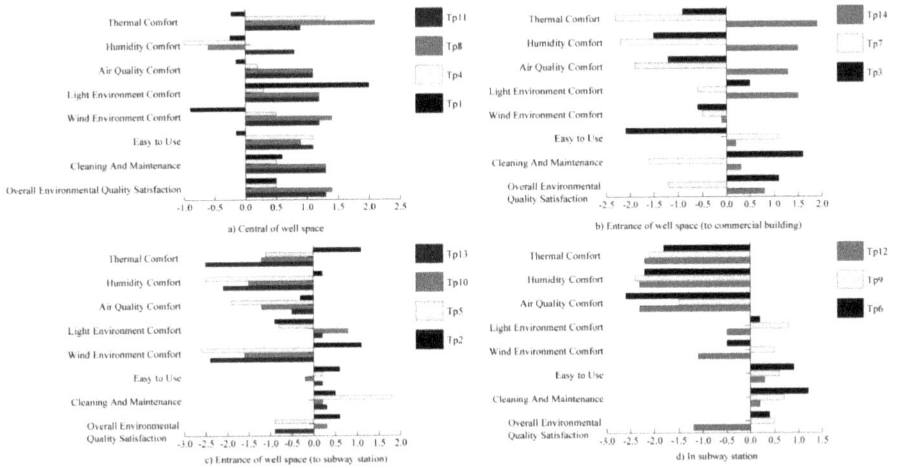

Figure 15. Results analysis of the occupancy satisfaction survey.

The most complaints from occupants were from the area near the subway. Both the space entrance near the subway and the subway platform level received lower occupancy satisfaction scores. The average thermal comfort of the tube entrance space near the subway was −1.85, the humidity score was −1.48, and air velocity comfort score was −1.38. Compared with the physical environment test results, the thermal environment of the tube entrance space near the subway had the greatest influence on the physiology and psychology of the human body. It was also the weakest point of the inner tubular space. In the physical environment test, the performance of the station hall was the worst, which is reflected in its high temperature, humid environment, and poor air quality. The averages of the three categories from the subjective survey data—thermal environment, humidity, and air quality—were as follows. Tp6 was −2.2, Tp9 was −2, and tp12 was −2.3, which indicates extreme occupant dissatisfaction.

All of the questionnaire items can be sorted as follows: tp11 (middle tubular space) > tp8 (middle tubular space) > tp14 (tube entrance type1) > tp2 (tube entrance type2) > tp4 (middle tubular space) > tp1 (middle tubular space) > tp9 (station hall) > tp3 (tube entrance type1) > tp6 (station hall) = tp10 (tube entrance type2) > tp13 (tube entrance type2) > tp5 (tube entrance type2) > tp12 (station hall) > tp7 (tube entrance type 1).

3.2.3. Satisfaction–Comfort Matrix Results and Analysis

Figure 16 shows the results of the Satisfaction–Comfort Matrix in the overall 14 target test points. The horizontal axis of the matrix corresponds to the level of each comfort (thermal, lighting, ventilation, and air quality) with the physical environment inside of the tubular space. The data was calculated into a comfort percentage obtained from the test results, and divided into six classes from 0 to 100%. The vertical axis of the matrix corresponds to the subjective analysis of occupancy space satisfaction, following with the −3~3 score scale outlined in the above research method.

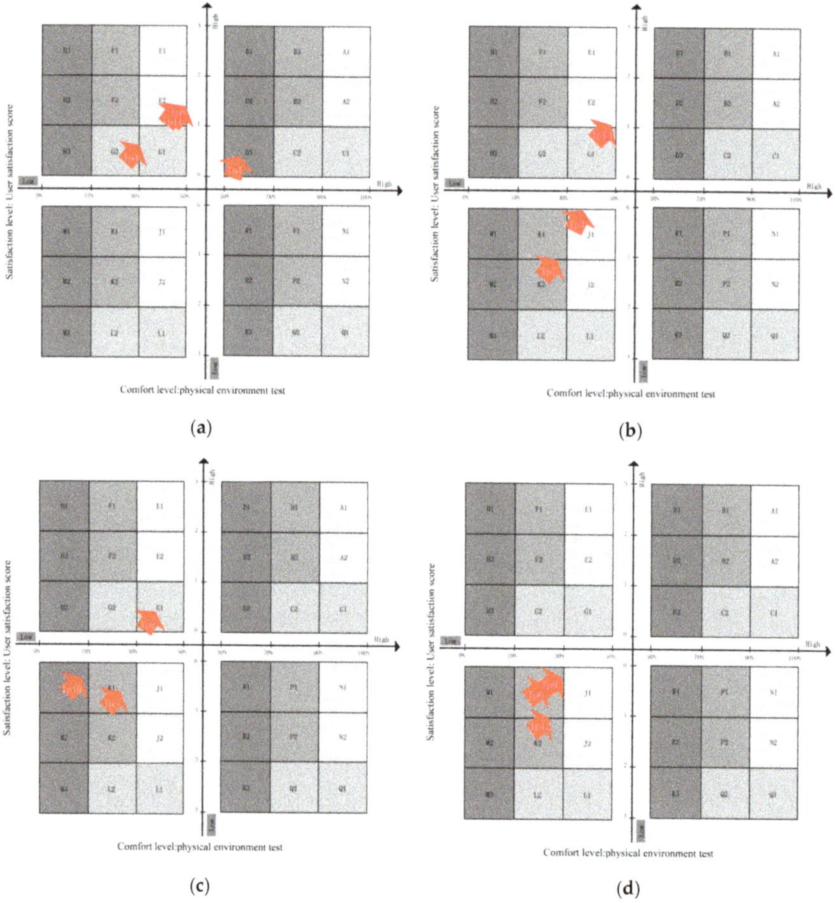

Figure 16. Test Point Satisfaction-Comfort Matrix results. (**a**) Middle tubular space; (**b**) Tube entrance (connection to building); (**c**) Tube entrance (connection to subway); (**d**) In subway station.

The Satisfaction–Comfort Matrix is divided into four quadrants. The results located in the first quadrant indicate that the animate space has a positive effect in terms of both satisfaction and comfort. The results located in the second quadrant mean it has a positive effect in terms of satisfaction but a negative effect in terms of comfort. The results located in the third quadrant imply neither satisfaction nor comfort. The results located in the fourth quadrant mean it has a negative effect in terms of satisfaction but a possible positive effect in terms of comfort [39]. Each quadrant's evaluation is divided into four grades (as shown in Figure 16).

3.3. Problem Analysis and Potential

According to the investigation of the objective physical environment and subjective comfort feelings, it can be concluded that the potential for passive utilization of tubular space includes two aspects: Improvement in comfort and utilization of climate.

3.3.1. Comfort Improvement from the Perspective of Passive Space Design

Tube entrance space plays a role in cohesion and intermediary conversion. However, it was revealed that in the Beijing subway station building complex in the underground tube entrance convergence space, the instantaneous wind speed was too high in winter and transition seasons; this seriously affected user comfort and the long-term health of security personnel. Combining a design for the complex building's tube entrance transitional buffer space with setting a reasonable wind barrier along the narrow tube entrance area, as tube as identifying transfer space in the underground connected tubular space, would help to avoid the local wind outlet issue (as shown in Figure 17).

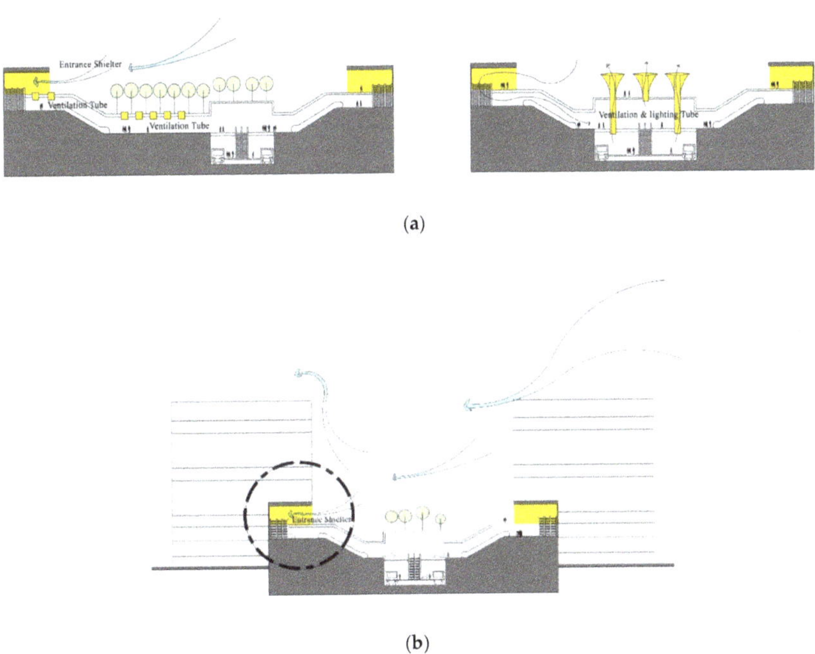

(a)

(b)

Figure 17. Passive strategy for relieving excessive wind velocity at the tube entrance. (a) Setting conversion space in the underground interconnected tubular space; (b) Transition buffer space at the tube entrance that combines with the complex building design.

The tubular space in the subway station area is always long and narrow. People in these locations experience high traffic flow, and the space lacks connection with the outdoor natural environment; this often results in poor air quality. Because it is a pass-through space, it is regularly neglected. Considering the complex building shape, setting aside areas such as air shafts, light tubes, atriums, and sunken courtyards would be an effective way of improving the air quality and indoor environment through space-based strategies.

3.3.2. Climate Utilization from the Perspective of Passive Space Design

However, tubular space as a means of connecting the external and internal environments has the advantage of transmitting natural resources. Effective natural resources can be transferred to the ground and underground spaces in subway transportation areas to solve comfort-based shortcomings in tubular spaces. These spaces can improve the quality of the area and reduce possible energy consumption during operation. In subway station building complexes, using aerodynamic, piston, and mixed ventilations, and exporting hot air to the outside through ventilation shafts or entrances can reduce energy consumption from air conditioning equipment, as tube as the area's levels of heat and humidity. Research has explored new modes for composite space systems, such as tubes with wind tunnels and solar chimneys [57], air supply tunnels that displace wind for better ventilation, active and passive combinations of wind tunnels and ground source heat pumps, and lighting from hot-press ventilation channels. These are all typical spatial patterns that can improve the utilization efficiency of passive tubular space. In addition, the shape of the tubular space has the advantage of introducing natural light. Natural light can improve the comfort level of the light environment and reduce lighting energy consumption. According to the actual test data, in a summer outdoor temperature of 28 °C, the air temperature in a space covered by four meters of soil can be maintained at 10 °C. When the outdoor temperature drops to −5 °C, the air temperature in that same underground area will remain stable at 10 °C [58]. In subway station building complexes, using geothermal energy with tubular spaces will not only improve the thermal environment quality and passive space design, but also transfer comfortable temperatures to other functional areas surrounding the tubular space, improving comfort and reducing the cost of operation.

3.4. Design Target and Space Conception of Tubular Composite Spaces from the Perspective of Sustainable Development

3.4.1. Passive Design for High Performance and Low Energy Consumption-Oriented Tubular Spaces in Subway Station Building Complexes

It was determined that there is often low levels of comfort and user satisfaction problems with tubular space. To design a new integrated subway station building complex both at home and abroad, one must continuously attempt to optimize the indoor environment and further coordinate the design strategy with the setting. Thus, this research studied tubular spaces in subway station building complex integration designs, forming a passive design method for tubular spaces that would offer high levels of performance and low amounts of energy consumption. This high level of performance is expressed in greater comfort and improved human health related to the indoor thermal environment, light and air quality, and user experience. Low energy consumption comes in the form of an energy-saving role in the building, such as by providing passive cooling, a fresh air supply, natural light, and wind energy utilization via the functional and shape advantages of tubular space (as shown in Figure 18).

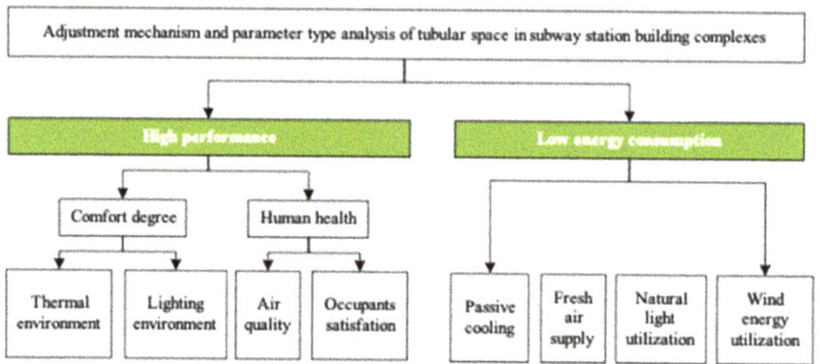

Figure 18. Type analysis of the passive parameters of high performance and low energy consumption-oriented tubular space in subway station building complexes.

3.4.2. Three Typical Design Concepts for Composite Tubular Space

An integrated design concept was presented in the study; it included two key aspects: the integration of an independent original design division that considered the organization of a single building, the underground station, and station space in the subway station building complex design process; and the integration of the space system with the goals of high performance and low energy consumption, to optimize the comprehensive performance of the composite space. Based on these issues, three design concepts for composite spaces are described below. The verification of their performance will be analyzed in detail in future research.

(a) Tubular complex space system for a wind tunnel and solar chimney (building atrium) for air filtration

The subway station building complex is a relatively complicated building space. Because of its large area, functional space coordination, and significant number of users, atriums are usually employed to organize the functional relationship. Vertically high atrium spaces can serve as a draft for solar chimneys. Combining them with a horizontal tunnel shaft design will help improve the quality of the indoor space's thermal environment and reduce the building's amount of energy consumption. In terms of the thermal environment, tunnel air temperature tends to be low, and combining the low temperature of tunnel air with air conditioning will reduce energy consumption in summer. Combining wind tunnels with the construction of solar chimneys and using the air dynamics principle of cold air sinking and hot air rising will benefit passive air circulation in buildings, exhausting high-temperature air and sucking low-temperature air from the tunnels. In addition, moderate thermal pressure ventilation in buildings is beneficial to the circulation of air and can reduce surface body temperature in summer, thus improving thermal comfort. However, due to the wind tunnels the air quality will remain low. Therefore, the composite tubular space system should be combined with an air filtration system to improve the air quality inside the tube and air circulation throughout the subway station building complex (as shown in Figure 19).

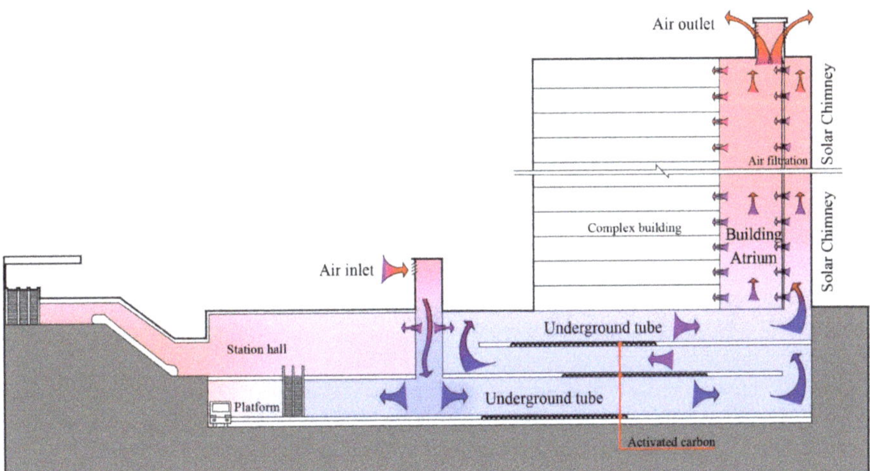

Figure 19. Sketch map of a complex tubular space system for wind tunnel and solar chimney air filtration.

(b) Complex tube path space system for lighting using hot air ventilation

The types of tubular space in a building can be divided into vertical, horizontal, and mixed. The advantage of a vertical tubular space is that it penetrates space in a vertical direction, which is beneficial for energy flowing from the top to the bottom, or vice versa. Natural lighting is a scarce resource in underground space, and thus underground areas consume much more energy for lighting than do general ground-level buildings. In most subway station building complex buildings, atrium spaces or pass heights are the most common organizational structures. The integrated design idea connects building complexes and subway station halls to provide natural lighting in tubular space, forming light tubes with the atrium space. In addition, vertical tubes in atriums can offer useful hot air ventilation. The low air temperature of tunnel wind can result in a substantial temperature difference from the higher air temperatures at the tops of atriums, which can lead low-temperature air into the building. This composite space system can also integrate the composite tubular space system of a wind tunnel with a solar chimney (atrium) for air filtration, and integrate natural light with the air filtration system, ventilation, and other functions resulting in a composite function space. This yields further improvement in the passive space adjustment effect (as shown in Figure 20).

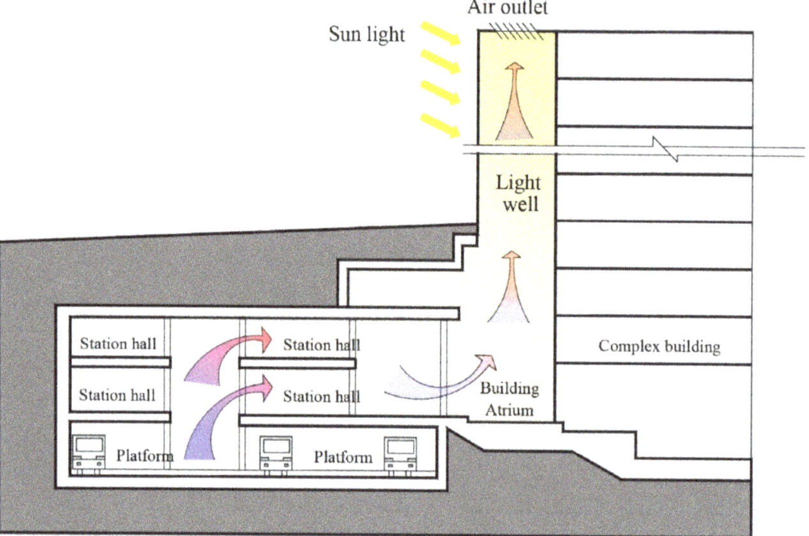

Figure 20. Sketch map of lighting through hot pressing the ventilation in a composite space system.

(c) Piston wind power generation for an auxiliary lighting system

According to past research, if a large number of fans are installed in subway tunnels, the blades produce air resistance, resulting in resistance to the train operation. This can lead to more energy consumption. However, if a few small micro-sized wind power generators are laid out in the subway station, the machines will generate renewable energy for the platform, station hall layer, and artificial lighting for the complex connection space. This is a useful way of using wind energy in subway tunnels. According to the survey distributed for this research, the average wind speed at a height of 1.5 m above the platform is 1.5 m/s, and the average wind speed near the upper end of the metro vehicle can reach 7.5 m/s. Previous research has shown that large, lightweight twist blades are the best choice for subway tunnels. In micro-sized wind turbines, the starting wind speed for vertical axis force generator equipment is 1 m/s, and the rated wind speed is 11 m/s. The safe velocity is 45~60 m/s, and the rated power is 200 W, 300 W, or 400 W. If the subway's daily schedule runs from 6:00 a.m. to 10:00 p.m., the system will be in operation for 16 hours a day. If the 400 W fan is selected, two units can be installed at the corner of each platform floor (for a total of eight units). This would result in 51.2 KW·h of electricity sent per day for artificial lighting demands. According to the general industrial and commercial electricity fee collection standards in Beijing, the annual payback period for each of these typhoon machines would be two to three years, with an average value of 1 Rmb/degree (as shown in Table 12).

Table 12. Economic benefit calculation for piston wind power generation for an auxiliary lighting system.

Rated Power	Impeller Diameter	Lump Sum Investment	Additional Investment	Service Life	Annual Maintenance Cost	Daily Power Output	Annual Power Output	Payback Period of Investment
200	0.47	2000	1200	10	100	3.2	1100	2.99
300	0.66	2100	1500	10	150	4.8	1750	2.24
400	0.66	2400	2000	10	300	6.4	2300	2.16

Note		
	1.	A one-time investment would be the total cost of the system's installation.
	2.	Additional investments would be used to replace two battery charges for 10 years of life.

4. Conclusions

This study focused on the various tubular space forms in subway station building complexes; the goal was to identify those that would have a moderating effect on passive potential space. The result of this investigation, research, and analysis was an improvement of the indoor environment in terms of comfort and energy consumption. This work followed four key pathways.

(1) It attempted to establish a scientific and logical method for verifying the value of tubular space by establishing causal relationships among the parameterized building space information factors, occupancy satisfaction elements, physical environment comfort aspects, and climate conditions. This research adopted an analytical hierarchical process methodology for classifying each quantized building information factor, and then compared this information to the actual fieldwork physical environment test and occupant satisfaction vote data, in order to discern the key advantages and weaknesses in tubular space design in subway station building complexes.

(2) Based on the actual field investigation, a database of physical environment performance data and users' subjective satisfaction information was established. The results showed that 59% and 57.4% time in in middle tubular space was out of thermal and humidity comfort zone, and the uncomfortable temperature and humidity at the tube entrance measuring point reached 100%. Air quality was the worst in middle tube space, 65% of the time is unhealthy which exposing significant environmental problems. By analyzing correlations therein, the comfort and health problems found in different locations that were related to tubular spaces, as tube as the potential for passive utilization, were able to be discussed.

(3) According to the investigation of the objective physical environment and subjective comfort feelings, it can be concluded that the potential for passive utilization of tubular space includes two aspects: improvement in comfort level itself and utilization of climate between natural or artificial. Based on the measured data, an integrated design method for tube path spaces exhibiting high levels of performance and low amounts of energy consumption in subway station building complexes was put forward.

(4) The research described three typical composite tubular space designs, including wind tunnels/solar chimneys (atriums), which are composite tubular spaces with air filtration systems; composite space systems that provide lighting by hot-pressing the ventilation; and piston wind power generation for use in auxiliary lighting systems. These provide methods and ideas for future research and design.

Author Contributions: Conceptualization, J.L.; methodology, J.L.; software, S.L.; validation, Q.W.; formal analysis, J.L.; investigation, J.L., S.T., Y.J.; data curation, J.L.; writing—original draft preparation, J.L.; writing—review and editing, S.L.; visualization, Q.W.; project administration, J.L.; funding acquisition, J.L.

Funding: This work was supported by the Fundamental Funds for Beijing Natural Science Foundation of China (Grant No. 8182043), and the National Natural Science Foundation of China (Grant No. 51708019).

Acknowledgments: Bejing Jiaotong University; Beijing MTR Construction Administration Corporation.

Conflicts of Interest: The authors declare no conflict of interest.

References

1. Wey, W.; Zhang, H.; Chang, Y. Alternative transit-oriented development evaluation in sustainable built environment planning. *Habitat Int.* **2016**, *55*, 109–123. [CrossRef]
2. Yang, J.; Chen, J.; Le, X.; Zhang, Q. Density-oriented versus development-oriented transit investment: Decoding metro station location selection in Shenzhen. *Transp. Policy* **2016**, *51*, 93–102. [CrossRef]
3. Papa, E.; Bertolini, L. Accessibility and Transit-Oriented Development in European metropolitan areas. *J. Transp. Geogr.* **2015**, *47*, 70–83. [CrossRef]
4. Peng, Y.; Li, Z.; Choi, K. Transit-oriented development in an urban rail transportation corridor. *Transp. Res. Part B Methodol.* **2017**, *103*, 269–290. [CrossRef]
5. Yang, Y.; Zhang, P.; Ni, S. Assessment of the Impacts of Urban Rail Transit on Metropolitan Regions Using System Dynamics Model. *Transp. Res. Procedia* **2014**, *4*, 521–534. [CrossRef]
6. Liu, J. *Rail Transit Complex Design Based on TOD*; Beijing Jiaotong University: Beijing, China, 2013.
7. Hu, Y.; Zhao, C. Brief Discussion on Architectural Design of Green Subway Station. *Build. Sci.* **2014**, *6*, 132–138.
8. Song, Y.; Wang, J.; Zhu, N. Pondering over the Passive Design Strategy for Native Green Buildings of China. *Archit. J.* **2013**, *7*, 94–99.
9. Wang, Z.; Chen, F.; Shi, Z. Prediction on Medium and long term energy consumption of urban rail transit network in Beijing. *China Railw. Sci.* **2013**, *34*, 133–136.
10. Chen, J.; Gao, G.; Wang, X.; Wang, X. Calculation method of whole life-cycle energy consumption for urban rail transit. *J. Traffic Transp. Eng.* **2014**, *14*, 89–97.
11. Gao, G.; Guan, W.; Li, J.; Dong, H.; Zou, X.; Chen, W. Experimental investigation of an active–passive integration energy absorber for railway vehicles. *Thin-Walled Struct.* **2017**, *117*, 89–97. [CrossRef]
12. Marzouk, M.; Abdelaty, A. Monitoring thermal comfort in subways using building information modeling. *Energy Build.* **2014**, *84*, 252–257. [CrossRef]
13. Li, X.; Wang, Z.; Ma, C.; Liu, L.; Liu, X. Energy Consumption Test and Analysis of Large Public Buildings Based on Gray Box Model. *Procedia Eng.* **2016**, 146–150. [CrossRef]
14. He, J.; Yan, Z.F.; Liu, H.J. On the Current Energy Consumption and Countermeasures of Large Public Buildings in China. *Adv. Mater. Res.* **2012**, 840–842. [CrossRef]
15. Lam, J.C.; Yang, L.; Liu, J. Development of passive design zones in China using bioclimatic approach. *Energy Convers Manag.* **2006**, *47*, 746–762. [CrossRef]
16. Badescu, V.; Laaser, N.; Crutescu, R.; Crutescu, M.; Dobrovicescu, A.; Tsatsaronis, G. Modeling, validation and time-dependent simulation of the first large passive building in Romania. *Renew. Energy* **2011**, *36*, 142–157. [CrossRef]
17. Sadineni, S.B.; Madala, S.; Boehm, R.F. Passive building energy savings: a review of building envelope components. *Renew. Sustain Energy Rev.* **2011**, *15*, 3617–3631. [CrossRef]
18. Feist, W.; Schnieders, J.; Dorer, V.; Haas, A. Re-inventing air heating: convenient and comfortable within the frame of the Passive House concept. *Energy Build.* **2005**, *37*, 1186–1203. [CrossRef]
19. Zhang, H.; Li, J.; Dong, L.; Chen, H. Integration of sustainability in Net-zero House: Experiences in Solar Decathlon China. *Energy Procedia* **2014**, *57*, 1931–1940. [CrossRef]
20. Yiing, C.F.; Yaacob, N.M.; Hussein, H. Achieving Sustainable Development: Accessibility of Green Buildings in Malaysia. *Procedia Soc. Behav. Sci.* **2013**, *101*, 120–129. [CrossRef]
21. Russell-Smith, S.V.; Lepech, M.D.; Fruchter, R.; Littman, A. Impact of progressive sustainable target value assessment on building design decisions. *Build. Environ.* **2015**, *85*, 52–60. [CrossRef]
22. Olgyay, V. *Design with Climate*, New ed.; John Wiley & Sons Inc.: Hoboken, NJ, USA, 1992.
23. Li, J. *Passive Adjustment Performance of Intermediary Space in Buildings*; Tsinghua University: Beijing, China, 2016.
24. Jiang, Y. *The Study on Ventilation Design of Deep Plan Buildings with Vertical Sapce*; Shenyang Architecture University: Shenyang, China, 2011; Volume 12, p. 15.
25. Brown, G.Z.; Dekey, M. *Sun, Wind & Light: Architecture Design Strategies*; John Wiley & Sons Inc.: Hoboken, NJ, USA, 2006.
26. Pieter, D.W.; Marinus, V.D.V. Providing computational support for the selection of energy saving building components. *Energy Build.* **2004**, *36*, 749–758.

27. Kenji, M.; Keishi, D.; Li, M. Tokyo Tokyu Toyoko line Shibuya Station. *Archit. J.* **2009**, *4*, 40–45.

28. Katavoutas, G.; Assimakopoulos, M.N.; Asimakopoulos, D.N. On the determination of the thermal comfort conditions of a metropolitan city underground railway. *Sci. Total Environ.* **2016**, *566*, 877–887. [CrossRef] [PubMed]

29. Li, Y.; Geng, S.B.; Zhang, X.S.; Zhang, H. Study of thermal comfort in underground construction based on field measurements and questionnaires in China. *Build. Environ.* **2017**, *116*, 45–54. [CrossRef]

30. Kim, M.J.; Richard, D.B.; Kim, J.T.; Yoo, C.K. Indoor air quality control for improving passenger health in subway platforms using an outdoor air quality dependent ventilation system. *Build. Environ.* **2015**, *92*, 407–417. [CrossRef]

31. *Chip Sullivan.Garden and Climate*; China Building Industry Press: Beijing, China, 2005.

32. Wu, C.; Wang, Z.; Lu, Z. Strategy of urban development from the view of Shibuya Hikarie. *Archit. Tech.* **2015**, *11*, 40–47.

33. Zhang, F.; Zhou, X. From the utilization of underound space to the conformity of underground space and the city: a case study of the two renovations of Les Hasses. *Mod. Urban Res.* **2014**, *29*, 29–38.

34. Li, J. Multi-criteria Approach to Impact Evaluation of Passive Adjustment Performance of Intermediary Space in Buildings. *Archit. J.* **2016**, *2*, 50–55.

35. Song, Y.; Li, J.; Wang, J.; Hao, S.; Zhu, N.; Lin, Z. Multi-criteria approach to passive space design in buildings: Impact of courtyard spaces on public buildings in cold climates. *Build. Environ.* **2015**, *89*, 295–307. [CrossRef]

36. Li, B. *The Research on Climatic-Active Design Strategy of Building Skin in Hot-Summer and Cold-Winter Zone*; Tsinghua University: Beijing, China, 2004; pp. 127–281.

37. Qiu, J. *Evaluation Science: Theory Method Practice*; Science Press: Beijing, China, 2010. (In Chinese)

38. Haas, R.; Meixner, O. An Illustrated Guide to the Analytic Hierarchy Process. Available online: https://mi.boku.ac.at/ahp/ahptutorial.pdf (accessed on 22 February 2019).

39. Li, J.; Song, Y.; lv, S.; Wang, Q. Impact evaluation of indoor environmental performance of animate space in buildings. *Build. Environ.* **2015**, *94*, 353–370. [CrossRef]

40. McMulla, R. *Environmental Science in Building*; Macmillan: Basingstoke, UK, 2007.

41. Fanger, P.O. Fundamentals of thermal comfort. *Adv. Sol. Energy Technol.* **1988**, *4*, 3056–3061.

42. Leadership in Energy and Environmental Design (LEED). The U.S. Green Building Council (USGBC): Washington, DC, USA, 2001. Available online: http://www.usgbc.org/ (accessed on 22 February 2019).

43. MOHURD. *Assessment Standard for Green Building of China*; GB/T50378-2014; China Building Industry Press: Beijing, China, 2014.

44. Kang, J.; Zhang, M. Semantic differential analysis of the soundscape in urban open public spaces. *Build. Environ.* **2010**, *45*, 150–157. [CrossRef]

45. Sun, Z.; Wei, D.; Xie, J. Research on subway tunnel wind power system. *Renew. Energy Resour.* **2016**, *9*, 1333–1341.

46. Hummelgaard, J.; Juhl, P.; Sabjornsson, K.O.; Clausen, G.; Toftum, J.; Langkilde, G. Indoor air quality and occupant satisfaction in five mechanically and four naturally ventilated open-plan office buildings. *Build. Environ.* **2007**, *42*, 4051–4058. [CrossRef]

47. Assimakopoulos, M.N.; Katavoutas, G. Thermal comfort conditions at the platforms of the Athens Metro. *Procedia Eng.* **2017**, *180*, 925–931. [CrossRef]

48. Moreno, T.; Reche, C.; Minguillón, M.C.; Capdevila, M.; de Miguel, E.; Querol, X. The effect of ventilation protocols on airborne particulate matter in subway systems. *Sci. Total Environ.* **2017**, *584*, 1317–1323. [CrossRef] [PubMed]

49. Burnett, J.; Pang, A.Y. Design and performance of pedestrian subway lighting systems. *Tunn. Undergr. Space Technol.* **2004**, *19*, 619–628. [CrossRef]

50. Han, L.; Feng, L.; Yuan, Y. Numerical Analysis of Cooling Effect of Tunnel Ventilation System in Subway Station. *Refrig. Air Cond.* **2016**, *2*, 1–4.

51. Yan, L. *The Effect of Piston Wind on Subway Environment and Energy Saving Character*; Beijing University of Technology: Beijing, China, 2015.

52. *Standard for Urban Rail Transit Lighting*; GB/T16275-2008; Ministry of Housing and Urban-Rural Development of the People's Republic of China: Beijing, China, 2008.

53. *Indoor Air Quality Standard GB/T18883-2002*; Ministry of Housing and Urban-Rural Development of the People's Republic of China: Beijing, China, 2002.

54. *Environemt Air Quality Standard GB3095-2012*; Ministry of Housing and Urban-Rural Development of the People's Republic of China: Beijing, China, 2012.
55. ASHRAE (American Society of Heating, Refrigerating, and Air Conditioning Engineers). *ASHRAE Standard 62.1—2013.Ventilation for Acceptable Indoor Air Quality*; ASHRAE: Atlanta, GA, USA, 2013.
56. Ren, M.; Guo, C.; Guo, Q.; Yang, Y.X.; Kang, G.Q.; Luo, H.L. Numerical Analysis and Effectively Using of Piston-effect in Subway. *J. Shanhaijiaotong Univ.* **2008**, *8*, 1376–1380, 1391.
57. Hu, Y. An Analysis of the Transitional Space of Traffic Hubs Case Studies of Traffic Hubs in Japan. *Archit. J.* **2014**, *6*, 109–113.
58. Wang, S.; Zhu, X.; Jiang, Y. Measurement and analysis of wall heat flow in thermal environment of Beijing Subway. *J. Undergr. Work. Tunn.* **1997**, *3*, 32–37.

Article

Application of Airborne Microorganism Indexes in Offices, Gyms, and Libraries

Pietro Grisoli [1,*], **Marco Albertoni** [1] and **Marinella Rodolfi** [2]

1   Department of Drug Sciences, Laboratory of Microbiology, University of Pavia, 27100 Pavia, Italy; marco.albertoni@unipv.it
2   Department of Earth and Environmental Sciences, Mycology Section, University of Pavia, 27100 Pavia, Italy; marinella.rodolfi@unipv.it
*   Correspondence: pietro.grisoli@unipv.it; Tel.: +39-0382-987397

Received: 19 December 2018; Accepted: 9 March 2019; Published: 15 March 2019

Abstract: The determination of microbiological air quality in sporting and working environments requires the quantification of airborne microbial contamination. The number and types of microorganisms, detected in a specific site, offer a useful index for air quality valuation. An assessment of contamination levels was carried out using three evaluation indices for microbiological pollution: the global index of microbiological contamination per cubic meter (GIMC/m^3), the index of mesophilic bacterial contamination (IMC), and the amplification index (AI). These indices have the advantage of considering several concomitant factors in the formation of a microbial aerosol. They may also detect the malfunction of an air treatment system due to the increase of microbes in aeraulic ducts, or inside a building compared to the outdoor environment. In addition, they highlight the low efficiency of a ventilation system due to the excessive number of people inside a building or to insufficient air renewal. This study quantified the levels of microorganisms present in the air in different places such as offices, gyms, and libraries. The air contamination was always higher in gyms that in the other places. All examined environments are in Northern Italy.

Keywords: airborne microorganisms; bacteria; fungi; gyms; indoor air quality; libraries; offices

---

1. Introduction

The evaluation of microbiological air contamination is an important aspect of applied microbiology and industrial hygiene [1–3]. In Italy, several legislative decrees, in particular the Legislative Decree 81/2008, impose that employers are responsible for the assessment of risks of biological origin arising from the activities in the workplace; moreover, they must adopt suitable measures so that the air in the workplace be healthy [4]. Therefore, the realization of measuring methods of airborne contamination is a matter of prime importance in the most complex procedure of risk evaluation that must involve, necessarily, different scientific skills [5–8]. Air contamination, both in confined environments and outdoors, can be caused not only by chemical agents but also by biological sources such as pollen, dust, mites, insects, pet allergens, bacteria, fungi, and viruses [9]. The monitoring of bioaerosols represents an important tool both for assessing the risk in the working environment [10] and for the evaluation of the environmental impact of certain activities that take place outdoors, such as biopurification or waste disposal [11]. Several investigations have been directed to study the disease called Sick Building Syndrome (SBS), which is characterized by varied and non-specific symptoms observed in workers employed in confined environments [12–14]. The World Health Organization has defined SBS as "an increase in frequency in the occupants of non-industrial buildings, of not specific acute symptoms (irritation of eyes, nose, throat, headache, fatigue, nausea) that improve when building is left" [15]. The American Conference of Governmental Industrial Hygienists (ACGIH) does not suggest threshold limit values (TLV) for the environmental concentration of biological agents, because the existing information does not allow to establish a scientifically acceptable dose–response relationship [16].

However, the assessment of culturable microbial agents is currently the easiest and most practical measure to determine changes in the air quality of confined workplaces. The purpose of these controls may be to set guide values, technologically attainable by adopting containment measures of contamination established on the basis of repeated samplings in specific locations. With this aim, this study examined structures destined for different types of work and recreational activity such as offices, gyms, and libraries, measuring air-diffused contamination to identify and apply indices useful for the microbiological classification of the air quality [17].

The indoor air quality was evaluated by applying the following indices: the global index of microbiological contamination per cubic meter (GIMC/m^3), the index of mesophilic bacterial contamination (IMC), and the amplification index (AI). The GIMC considers different microbial types and emphasizes their ability to grow in a wide temperature range. GIMC may be a simple method to evaluate a potential biological risk in indoor and outdoor environments and to monitor sources of microbiological contamination. The IMC reveals the presence of obligated mesophilic bacteria, organisms of probable human origin in the indoor air. This index represents a practical instrument to underline bacterial growth caused by hypoventilation and overcrowding. The AI is important as it reveals microbial pollution in indoor ventilation systems. These indices showed their applicability in other types of environments [18,19].

2. Materials and Methods

2.1. Sampling Method

Environmental monitoring was performed in Northern Italy in 10 diverse offices, gyms, and libraries situated in buildings equipped with a ventilation system able to work in the following modes: heating, air conditioning, and simple ventilation. The offices are placed in bank edifices, the gyms in fitness centers, while the libraries belong to university buildings. The samplings were realized every month for one year, inside the various environments, during normal people activity, and outside the building. Quantitative data were determined in triplicate by means of an orthogonal impact Microflow Air Sampler (AQUARIA, Lacchiarella, Italy), kept 1.5 m above ground level, in the center of the room for offices, in the middle of the weight-training room for gyms, and in the reading rooms for libraries. Air samplers worked at a fixed speed of 1.5 Ls^{-1}, collecting a volume of 200 L. For the different types of environments, the number of persons (n. p.) and the air speed (m/s) were recorded, so that in three sampling periods, the following average values were obtained: offices (n. p. 4.7; 0.03 m/s), gyms (n. p. 21.3; 0.060 m/s), and libraries (n. p. 16.2; 0.05 m/s).

2.2. Microorganisms Assessed

Bacteria were collected using Tryptone Soya Agar (TSA, Oxoid, Basingstoke, UK), and the cultures were incubated at 37 °C for 48 h for mesophilic bacteria and at 20 °C for 6 days for psychrophilic bacteria. Fungi were collected on Sabouraud Dextrose Agar (SAB Oxoid, Basingstoke, UK), and the cultures were incubated at 20 °C for 6 days. The outdoor air quality used as control was analyzed following the same criteria. All total microbial counts are indicated as the number of colony-forming units per cubic meter of air (CFU/m^3), calculated as an average of three determinations from three samples collected serially. The assessment of microbial contamination was effected by using bacterial and fungal counts, and the following indexes were calculated: GIMC per cubic meter (GIMC/m^3), which is the sum of the values of the total microbial counts determined for mesophilic bacteria, psychrophilic bacteria, and fungi in all sampled areas; IMC, derived from the ratio between the values of CFU per cubic meter measured for mesophilic and psychrophilic bacteria at the same sampling point; AI, resulting from the ratio between the GIMC/m^3 values measured inside the building and those measured outdoor.

2.3. Data Analysis

The air microbial contamination values were expressed as colony-forming units (CFUs), and the limit of quantification was 1 CFU/m^3. The number of CFUs for contact plate after appropriate incubation was corrected using the positive-hole correction table provided by the supplier. Statistical analysis of the data was performed by comparing the results obtained in the different environments by means of analysis of variance one-way (post hoc test) values transformed into logarithms (natural log). The analyses were conducted using Prism 3.0. The significance level was $p < 0.05$.

3. Results

Table 1 shows the results of microbiological contamination of 10 different offices located in buildings with centralized systems for ventilation. The mean values were higher for mesophilic bacteria during the air conditioning phase compared to heating ($p = 0.034$) and during simple ventilation compared to heating ($p = 0.023$). The contamination of psychrophilic bacteria was greater during simple ventilation compared to air conditioning ($p = 0.016$) and to heating ($p = 0.032$). The mean values of CFU/m^3 for the fungi were not significantly different during the different periods of air treatment. Although the average contamination was low, there were high maximum values of CFU/m^3 for mesophilic bacteria in the periods of air conditioning and simple ventilation and for psychrophilic bacteria during simple ventilation.

Table 1. Total microbial concentrations measured in the offices during heating, air conditioning, and simple ventilation.

Functioning Modes	N	Mesophilic Bacteria CFU/m^3		Psychrophilic Bacteria CFU/m^3		Fungal Count CFU/m^3	
		M ± SD	Min–max	M ± SD	Min–max	M ± SD	Min–max
Heating	10	191.20 ± 144.53	8–450	176.70 ± 117.21	7–366	136.00 ± 32.80	22–654
Simple ventilation	10	667.60 ± 1602.60	6–5200	761.90 ± 1519.7	10–5000	131.50 ± 72.30	5–190
Air Conditioning	10	1089.70 ± 2392.20	19–7800	495.60 ± 307.20	76–860	119.60 ± 48.84	28–199

The number of samples (N), mean (M), standard deviation (SD), and range of values are indicated for each sampling period. CFU: colony-forming units.

The calculation of $GIMC/m^3$ confirmed the presence of higher values of contamination during conditioning ($p = 0.01$) and simple ventilation ($p = 0.01$) compared to heating. The average value of IMC during heating amounted to 7.30 and was higher than the values of the other modes of operation, corresponding to 0.80 during the simple ventilation ($p = 0.04$) and 1.60 during air conditioning (Table 2).

Table 2. Global index of microbial contamination per cubic meter ($GIMC/m^3$) and index of mesophilic bacterial contamination (IMC) measured in offices during heating, air conditionining and simple ventilation.

Functioning Modes	N	GIMC/m³		IMC	
		M ± SD	Min–max	M ± SD	Min–max
Heating	10	503.90 ± 282.22	124–1111	7.30 ± 20.03	0.20–64.30
Simple ventilation	10	1561.00 ± 3154.80	21–10,450	0.80 ± 0.40	0.10–1.30
Air Conditioning	10	1704.90 ± 2537.57	229–8720	1.60 ± 2.67	0.02–9.10

The number of samples (N), mean (M), standard deviation (SD), the range of values are indicated for each sampling period.

The data regarding the microbiological contamination found in 10 different gyms with centralized ventilation systems are summarized in Table 3. The results show highest mean values for mycetic contamination during the air conditioning phase compared to simple ventilation ($p = 0.04$) and heating ($p = 0.01$). As regards bacteria, the average contamination values for psychrophilic bacteria were superior to those for mesophilic bacteria in the three air treatment modalities and were significantly different during conditioning ($p = 0.02$).

Table 3. Total microbial concentrations measured in gyms during heating, air conditioning, and simple ventilation.

Functioning Modes	N	Mesophilic Bacteria CFU/m^3		Psychrophilic Bacteria CFU/m^3		Fungal Count CFU/m^3	
		M ± SD	Min–max	M ± SD.	Min–max	M ± SD	Min–max
Heating	10	1096.60 ± 924.40	140–2850	1446.40 ± 1356.82	180–4650	187.50 ± 167.02	20–568
Simple ventilation	10	393.00 ± 257.00	120–980	819.00 ± 432.40	330–1800	1208.90 ± 1459.60	90–4848
Air Conditioning	10	666.50 ± 477.76	200–1800	1509.00 ± 1354.02	440–4400	2430.90 ± 3678.31	8–10,848

The number of samples (N), mean (M), standard deviation (SD), and range of values are indicated for each sampling period.

There was a notable increase in fungal count when the central heating was switched off. In fact, the highest maximum values of CFU/m^3 were found for fungi during simple ventilation and conditioning, in contrast to what observed for bacteria ($p = 0.016$).

Table 4 shows the results of microbial contamination based on the GIMC/m^3 and on the IMC measured in the gyms. Because of the elevated fungal count observed in this period, the mean of GIMC/m^3 was higher during air conditioning ($p = 0.019$), with a maximum value of 15,248. The mean of the IMC values was always <1, which implies that the counts of the psychrophilic bacteria were almost always higher than the counts of the mesophilic bacteria; however, no significant differences in the index were observed when central heating was on or off.

Table 4. GIMC/m^3 and IMC measured in gyms during heating, air conditioning, and simple ventilation.

Functioning Modes	N	GIMC/m^3		IMC	
		M ± SD	Min–max	M ± SD	Min–max
Heating	10	2703.50 ± 2144.84	1160–7090	0.90 ± 0.74	0.12–2.40
Simple ventilation	10	2420.90 ± 1645.70	910–6548	0.50 ± 0.30	0.17–1.20
Air Conditioning	10	4606.40 ± 4428.70	1480–15,248	0.60 ± 0.39	0.20–1.40

The number of samples (N), mean (M), standard deviation (SD), and range of values are indicated for each sampling period.

Similar sampling and monitoring methods were adopted for the evaluation of the air quality of libraries. The average levels of contamination detected during air sampling showed small oscillations with very restricted values of CFU/m^3 both for bacterial loads (mesophilic and psychrophilic) and for fungi. The results of airborne bacteria and fungi recovered from the libraries showed that the CFU/m^3 values were higher during heating than during simple ventilation or air conditioning. However, no significant differences in the concentrations of all microbial counts were evidenced in the three functioning modes of the ventilation system (Table 5).

Table 5. Total microbial concentrations in libraries during heating, air conditioning, and simple ventilation.

Functioning Modes	N	Mesophilic Bacteria CFU/m^3		Psychrophilic Bacteria CFU/m^3		Fungal Count CFU/m^3	
		M ± SD	Min–max	M ± SD	Min–max	M ± SD	Min–max
Heating	10	171.20 ± 72.43	112–320	301.10 ± 237.26	125–725	123.90 ± 54.52	62–220
Simple ventilation	10	65.10 ± 29.40	38–110	152.40 ± 90.70	81–380	66.00 ± 31.70	32–110
Air Conditioning	10	71.80 ± 30.50	30–120	200.60 ± 127.65	97–400	76.60 ± 36.55	35–120

The number of samples (N), mean (M), standard deviation (SD), and range of values are indicated for each sampling period.

In addition, the transformation of total microbial count values into the indices evidenced a general reduction of the GIMC/m^3 values and of the IMC values in air conditioning compared to heating, but the statistical analysis showed no significant differences. The IMC index assumed values lower than 1 and showed that air contamination was mainly due to environmental bacteria. All values were limited to a small range in every period of the monitoring campaign (Table 6).

Table 6. GIMC/m^3 and IMC measured in libraries during heating, air conditioning, and simple ventilation.

Functioning Modes	N	GIMC/m^3		IMC	
		M ± SD	Min–max	M ± SD	Min–max
Heating	10	596.20 ± 335.08	300–1250	0.70 ± 0.22	0.40–0.90
Simple ventilation	10	283.50 ± 131.70	153–578	0.50 ± 0.20	0.25–0.70
Air Conditioning	10	349.00 ± 176.43	182–652.50	0.40 ± 0.18	0.20–0.60

The number of samples (N), mean (M), standard deviation (SD), and range of values are indicated for each sampling period.

In indoor environments, microbiological air contamination can also be described by the Amplification Index of microbial contamination (AI). To calculate this index, it was necessary to use the GIMC values measured outside the buildings in the different sampling periods (Table 7). These values were significantly different in the three sampling points; in particular, the widest variations were detectable outside gyms in the air conditioning period ($p < 0.0001$).

Table 7. GIMC/m^3 measured outside gyms, libraries, and offices during heating, air conditioning, and simple ventilation periods.

Sampling Points	N	Heating GIMC/m^3		Simple Ventilation GIMC/m^3		Air Conditioning GIMC/m^3	
		M ± SD	Min–max	M ± SD	Min–max	M ± SD	Min–max
Gyms- outdoor	10	714.40 ± 294.00	328–1268	703.10 ± 226.40	484–1236	3586.10 ± 2326.42	1325–9240
Libraries outdoor	10	410.65 ± 187.81	266–880	596.55 ± 167.19	359–892	508.40 ± 141,01	364–830
Offices-outdoor	10	421.70 ± 179.62	248–890	619.90 ± 169.30	392–933	703.80 ± 226.20	485–1068

The number of samples (N), mean (M), standard deviation (SD), and range of values are indicated for each sampling period.

AI is determined by calculating the ratio between the GIMC/m^3 values measured inside a building and those measured outside. This index is greater than 1 when the microbial contamination inside a building is higher than the outdoor contamination. In most cases, the values of microbial contamination of indoor air are higher than the outdoor values. In the heating phase, AI showed a worsening of air contamination especially in gyms compared the other environments considered (3.7). On the contrary, during the conditioning period, the highest values were recorded in offices (3.0) (Figure 1).

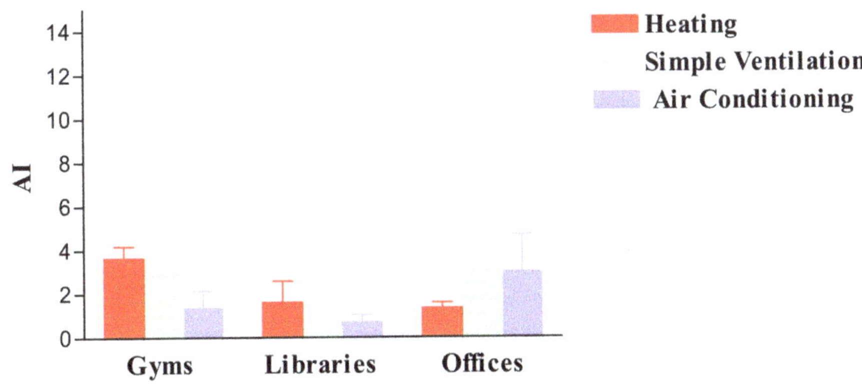

Figure 1. Amplification Index of microbial contamination (AI) measured in the three types of environments during heating, air conditioning, and simple ventilation.

4. Discussion

It can be reasonably assumed that the microflora existing in recreational and working indoor environments has a concentration lower than or equal to the external one detected in the same location and in the same climatic conditions [20,21]. In buildings equipped with centralized ventilation systems there is a reduction of microbial contamination compared to the outside when air filtration and treatment systems are maintained in appropriate conditions [22]. The air inside buildings may be contaminated by the growth of microorganisms on floors, walls, aeraulic plants or because of inadequate air changes [23]. The ventilation systems can have very different structural characteristics, ranging from the simple aspiration of exhaust air to integrated heating, cooling, and humidification. It is, therefore, clear that the air introduced from these systems can be considered as a matrix whose microbiological quality can be evaluated and classified, as for any other product that may have effects on human health. In addition, the activities carried out in a confined environment may be responsible for the spread of microorganisms.

Dissenting opinions have been expressed on the possibility to monitor microbiological environmental contamination in workplaces. Information on the dose–response relationship related to the exposure to microorganisms is still not available today. The ACGIH does not indicate TLV for biological agents. Furthermore, different analytical methods do not allow to recover and identify all the microorganisms present in the air. In fact, the percentage of viable, culturable bacteria recovered with normal sampling systems oscillates in a range of values that varies between 0.1% and 10% of the total bacteria present in the air [24–26]. The presence of microorganisms should also be sought in working environments because of their toxigenic potential and the possibility of spreading cell fragments and volatile organic compounds into the environment. The lack of viable, culturable cells does not necessarily indicate a healthy environment. Some studies, carried out in diversified working environments, report environmental contamination values referable to the number of viable cells belonging to a single class of microorganisms [27–29]. For example, a microbiological classification of the air quality in non-industrial environments and homes considers the contamination from bacteria that develop at 20–25 °C. In this classification, for non-industrial environments, a level <100 CFU/m^3 corresponds to the category of low contamination, while a value >2000 CFU/m^3 corresponds to the category of very high contamination [30]. It is clear that such assessment is incapable not only to describe all the factors that determine the accumulation and spread of microorganisms in the environment but also to identify possible risks for workers.

In this research, GIMC was used in order to include in the same data several categories of microorganisms. This index considers microorganisms proliferating at different temperatures, such as mesophilic and psychrophilic bacteria, and fungi capable of adapting to different types of environments. These characteristics make this quantitative measure of microbiological contamination significant and allow to evaluate the salubrity of a work environment; in fact, it considers microorganisms that can develop in a wide range of temperatures, including ambient temperatures, typical of saprophytic life, and 37 °C, which is the temperature of development of pathogens. This index is particularly useful because it is able to highlight even anomalous situations in indoor environmental microbiological contamination: in fact, in the offices we analyzed, the mesophilic bacteria during the conditioning phase were higher than the psychrophilic ones. This value represents an exception because it is usually reasonable to assume that psychrophilic bacteria, which might grow well in air conditioning systems, are more numerous than mesophilic bacteria during the cooling season, as shown in the results for gyms and libraries buildings and in previous researches [22]. In particular, GIMC attributes importance to bacteria that can proliferate in a wide range of temperatures. It is true that the two incubation temperatures (20 °C and 37 °C) do not differentiate the two categories of bacteria completely. Nevertheless, it is useful to determine the two total counts during environmental monitoring because they have different significances and allow a more complete evaluation of airborne bacterial contamination. In fact, the purpose of the index is to provide a measure of biological risk and to aggregate several "environmental indicators". IMC is mainly an index of anthropic contamination; it highlights the share of mesophilic bacteria in the microbial population examined. This index derives from the ratio between the value of CFU/m^3 at 37 °C and that at 20 °C. This value, determined outdoor, is always very close to 1, whereas in indoor environments, it may be higher, also depending on the number of people present. In fact, mesophilic bacteria derive from the normal bacterial flora of humans and can therefore constitute the predominant population in confined environments, as already verified in other works [9]. Finally, the amplification index is fundamental to detect the accumulation and proliferation of microorganisms in ventilation systems or in buildings. AI describes global indoor aerial modification. Generally, there are no relations between indoor and outdoor fungal and bacterial counts. However, it is important to judge Indoor Air Quality (IAQ) not only by the measurement of single parameters, but also using the global value of microbial contamination. High AI values may only be indicative of IAQ deterioration when caused by high total fungal and bacterial counts [10,22]. Moreover, AI is essential to determine the environmental impact of outdoor work activities with potential spread of pathogenic and non-pathogenic microorganisms; it must be calculated by referring to the contamination detected in a control point that must provide the background value [18]. AI and the other two indices could provide criteria to evaluate indoor environmental quality more uniformly [31]. In fact, these indices have different advantages: they consider several factors which contribute to the development of different types of microorganisms (determination of GIMC), they may reveal an inadequate functioning of the air conditioning system due to an excessive number of people in a building or insufficient ventilation (IMC determination), and they may suggest a malfunction of the air system due to the increase of microbial contamination in the ventilation ducts or in the building compared to the outside environment (AI measures). The utilization of these indices can enable the determination of threshold values as a function of the structures (buildings) analyzed. In previous research in other environments, such as university classrooms, a reference value of $GIMC/m^3 = 1000$ was connected to a correct maintenance of the aeraulic system [19]. In this work, the measured GIMC values were higher than 1000, with the exception of libraries that presented very low contamination. In particular, both offices and gyms had the highest values of pollution during the conditioning phase, corresponding to $GMIC/m^3$ of 1704.90 and 4606.40, respectively. GIMC exceeding 1000 does not necessarily indicate a health risk; however, it is appropriate to analyze the contamination levels through the calculation of IMC and AI. During air conditioning in the offices, IMC and AI showed that mesophilic bacteria underwent a real amplification due to few air changes or the accumulation of microorganisms in the air system. In fact, as reported by other authors, variations and fluctuations in indoor humidity and temperature have significant effects on microbial diffusion and growth [32]. For gyms, on the contrary, the GIMC values were higher in summer because the proportion of mesophilic bacteria decreased passing from the heating phase to the conditioning phase, but the portion of psychrophilic bacteria and mycetes increased consistently, as evidenced by the IMC.

It is important to note that the rise of the microbial load inside the gyms coincided with an increase of the microorganisms outdoors, as shown by the calculation of AI. Therefore, the presence of a greater microbial contamination during summer could depend on the incorrect functioning or poor controls of the ventilation systems, which resulted in the accumulation of microorganisms from the outside [33].

In conclusion, our research results show that the microbial contamination changes depending on the environment analyzed and highlight the easy applicability of the proposed indices. GIMC, IMC, and AI complete the information provided by the single classes of airborne microorganisms and, together, they allow the analysis of indoor air quality. On the basis of sufficient data, these indices classify environments in function of their microbiological contamination and identify guide values to be adopted for routine monitoring and for the implementation of containment and remediation measures.

Author Contributions: Conceptualization, P.G.; methodology, P.G.; formal analysis, P.G., M.R.; investigation, P.G., M.A., M.R.; funding acquisition, P.G., M.A.; writing—original draft preparation, P.G. and M.R.

Funding: This work was financially supported by a research fund of the Laboratory of Microbiology, Department of Drug Sciences, University of Pavia.

Conflicts of Interest: The authors declare no conflict of interest.

References

1. Anderson, A.M.; Weiss, N.; Raine, F.; Salkinoja-Salonen, M.S. Dust-borne bacteria in animal sheds, schools and children's day care centers. *J. Appl. Microbiol.* **1999**, *86*, 622–634. [CrossRef]
2. Bernasconi, C.; Rodolfi, M.; Picco, A.; Grisoli, P.; Dacarro, C.; Rembges, D. Pyrogenic activity of air to characterize bioaerosol exposure in public buildings: A pilot study. *Lett. Appl. Microbiol.* **2010**, *50*, 571–577. [CrossRef] [PubMed]
3. Maroni, M.; Seifert, B.; Lindvall, T. Indoor Air Quality: A comprensive reference book. In *Air Quality Monographs*; Elsevier: Amsterdam, The Netherlands, 1995; Volume 3, ISBN 9780444816429.
4. Testo unico in materia di tutela della salute e della sicurezza nei luoghi di lavoro. D.Lgs n. 81/2008. In *Gazzetta Ufficiale della Repubblica Italiana*; n. 101 (30 Aprile 2008); Ministero della Giustizia, Ufficio Pubblicazione Leggi e Decreti: Rome, Italy.
5. Brief, R.S.; Bernath, T. Indoor pollution: guidelines for prevention and control of microbiological respiratory hazards associated with air conditioning and ventilation systems. *Appl. Ind. Hyg.* **1988**, *3*, 5–10. [CrossRef]
6. Jensen, P.A.; Schafer, M.P. Sampling and characterization of bioaerosols. In *NIOSH Manual of Analytical Methods Niosh/DPSE*; NIOSH Publications: Washington, DC, USA, 1998; Volume 2, pp. 82–112.
7. Švajlenka, J.; Kozlovská, M.; Pošiváková, T. Assessment and biomonitoring indoor environment of buildings. *Int. J. Environ. Health Res.* **2017**, *27*, 427–439. [CrossRef] [PubMed]
8. Burge, P.S. Sick building syndrome. *Occup. Environ. Med.* **2004**, *61*, 185–190. [CrossRef] [PubMed]
9. Cabo Verde, S.; Almeida, S.M.; Matos, J.; Guerreiro, D.; Meneses, M.; Faria, T.; Botelho, D.; Santos, M.; Viegas, C. Microbiological assessment of indoor air quality at different hospital sites. *Res. Microb.* **2015**, *166*, 557–563. [CrossRef] [PubMed]
10. Dacarro, C.; Grisoli, P.; Del Frate, G.; Villani, S.; Grignani, E.; Cottica, D. Micro-organisms and dust exposure in an Italian grain mill. *J. Appl. Microbiol.* **2005**, *98*, 163–171. [CrossRef] [PubMed]
11. Prazmo, Z.; Krysinska-Traczyk, E.; Skorska, C.; Sitkowska, J.; Cholewa, G.; Dutkiewicz, J. Exposure to bioaerosols in a municipal sewage treatment plant. *Ann. Agric. Environ. Med.* **2003**, *10*, 241–248. [PubMed]
12. Mendell, M.J.; Lei-Gomez, Q.; Mirer, A.G.; Seppänen, O.; Brunner, G. Risk factors in heating, ventilating, and air-conditioning systems for occupant symptoms in US office buildings. *Ind. Air* **2008**, *18*, 301–316. [CrossRef] [PubMed]
13. Qian, J.; Hospodsky, D.; Yamamoto, N.; Nazaroff, W.W.; Peccia, J. Size-resolved emission rates of airborne bacteria and fungi in an occupied classroom. *Ind Air.* **2012**, *22*, 339–351. [CrossRef] [PubMed]
14. Wolff, C.H. Innate immunity and the pathogenicity of inhaled microbial particles. *Int. J. Biol. Sci.* **2011**, *7*, 261–268. [CrossRef] [PubMed]
15. World Health Organization. *Indoor Air Pollutants: Exposure and Health Effects*; Report on a WHO Meeting, EURO Reports and Studies no. 78; WHO Regional Office for Europe: Copenhagen, Denmark, 1983.
16. ACGIH. On-site investigation, pp. 1–8; Fungi, pp. 1–10; Bacteria, pp. 1–7. In *Guidelines for the Assessment of Bioaerosols in the Indoor Environment*; Committee on Bioaerosols, Ed.; American Conference of Governmental Industrial Hygienists: Cincinnati, OH, USA, 1989.
17. Traistaru, E.; Moldovan, R.C.; Menelaou, A.; Kakourou, P.; Georgescu, C. A comparative study on the quality of air in offices and homes. *J. Environ. Sci. Health A* **2013**, *48*, 1806–1814. [CrossRef] [PubMed]
18. Grisoli, P.; Rodolfi, M.; Villani, S.; Grignani, E.; Cottica, D.; Berri, A.; Picco, A.; Dacarro, C. Assessment of airborne microorganism contamination in an industrial area characterized by an open composting facility and a wastewater treatment plant. *Environ. Res.* **2009**, *109*, 135–142. [CrossRef] [PubMed]
19. Grisoli, P.; Rodolfi, M.; Chiara, T.; Zonta, L.A.; Dacarro, C. Evaluation of microbiological air quality and of microclimate in university classrooms. *Environ. Monit. Assess.* **2012**, *184*, 4171–4180.
20. Heildelberg, J.F.; Shahamat, M.; Levin, M.; Rahman, I.; Stelma, G.; Grim, C.; Colwell, R.R. Effect of aerosolization on culturability and viability of Gram-Negative bacteria. *Appl. Environ. Microbiol.* **1997**, *63*, 3585–3588.
21. Macher, J.M.; Chatigny, M.A.; Burge, H.A. Sampling airborne microorganisms and aeroallergens. In *Air Sampling Instruments for Evaluation of Atmospheric Contaminants*, 8th ed.; Cohen, B.S., Hering, S.V., Eds.; American Conference of Governmental Industrial Hygienists Inc.: Cincinnati, OH, USA, 1995; pp. 589–617.
22. Dacarro, C.; Grignani, E.; Lodola, L.; Grisoli, P.; Cottica, D. Proposta di indici microbiologici per la valutazione della qualità dell'aria degli edifici. *G. Ital. Med. Lav. Ergon.* **2000**, *XXII*, 229–235.

23. Gołofit-Szymczak, M.; Górny, R.L. Microbiological Air Quality in Office Buildings Equipped with Different Ventilation Systems. *Ind. Air* **2018**, *28*, 792–805. [CrossRef] [PubMed]

24. Jouzaitis, A.; Willeke, K.; Grinshpun, S.A.; Donnelly, J. Impaction onto a Glass Slide or Agar versus Impingement into a Liquid for the Collection and Recovery of Airborne Microorganisms. *Appl. Environ. Microbiol.* **1994**, *60*, 861–870.

25. Stewart, S.L.; Grinshpun, S.A.; Willeke, K.; Terzieva, S.; Ulevicius, V.; Donnelly, J. Effect of Impact Stress on Microbial Recovery on an Agar Surface. *Appl. Environ. Microb.* **1995**, *61*, 1232–1239.

26. Walter, M.V.; Marthi, B.; Fieland, V.P.; Ganio, L.M. Effect of Aerosolization on Subsequent Bacterial Survival. *Appl. Environ. Microb.* **1990**, *56*, 3468–3472.

27. Flannigan, B. Indoor microbiological pollutants—Sources, species, characterization and evaluation. In *Chemical, Microbiological, Health and Comfort Aspects of Indoor Air Quality—State of the Art in SBS*; Knöppel, H., Wolkoff, P., Eds.; EDSC, EEC, EAEC: The Netherlands, 1992; pp. 73–98.

28. Miller, J.D. Fungi as contaminants in indoor air. *Atmos. Environ.* **1992**, *6A*, 2163–2172. [CrossRef]

29. Rao, C.Y.; Burge, H.A. Review of quantitative standards and guidelines for fungi in indoor air. *J. Air Waste Manag. Assoc.* **1996**, *46*, 899–908. [CrossRef] [PubMed]

30. Commission of European Communities. *Indoor Air Quality & Its Impact on Man*; Report n.12; Commission of European Communities: Luxembourg, 1993.

31. Sofuoglu, S.C.; Moschandreas, D.J. The link between symptoms of office building occupants and in-office air pollution: The Indoor Air Pollution Index. *Ind. Air* **2003**, *13*, 332–343. [CrossRef]

32. Flannigan, B.; Miller, J.D. Microbial growth in indoor environments. In *Microorganisms in Home and Indoor Work Environments: Diversity, Health Impacts, Investigation and Control*; Flannigan, B., Samson, R.A., Miller, J.D., Eds.; Taylor & Francis: Abingdon, UK, 2001; pp. 35–67.

33. Dacarro, C.; Picco, A.M.; Grisoli, P.; Rodolfi, M. Determination of aerial microbiological contamination in scholastic sports environments. *J. Appl. Microbiol.* **2003**, *95*, 904–912. [CrossRef] [PubMed]

Article

How Working Tasks Influence Biocontamination in an Animal Facility

Anna M. Marcelloni [1], Alessandra Chiominto [1], Simona Di Renzi [1], Paola Melis [1], Annarita Wirz [2], Maria C. Riviello [2,3], Stefania Massari [4], Renata Sisto [1], Maria C. D'Ovidio [1] and Emilia Paba [1,*]

[1] Department of Occupational and Environmental Medicine, Epidemiology and Hygiene, Italian Workers' Compensation Authority (INAIL), Via Fontana Candida 1, Monte Porzio Catone, 00078 Rome, Italy; a.marcelloni@inail.it (A.M.M.); a.chiominto@inail.it (A.C.); s.direnzi@inail.it (S.D.R.); p.melis@inail.it (P.M.); r.sisto@inail.it (R.S.); m.dovidio@inail.it (M.C.D.O.)

[2] Santa Lucia Foundation IRCCS, Via Ardeatina 306, 00142 Rome, Italy; a.wirz@hsantalucia.it

[3] Institute of Cell Biology and Neurobiology, National Research Council (CNR), Via E. Ramarini 32, Monterotondo, 00015 Rome, Italy; cristina.riviello@cnr.it

[4] Department of Occupational and Environmental Medicine, Epidemiology and Hygiene, Italian Workers' Compensation Authority (INAIL), Via Stefano Gradi 55, 00143 Rome, Italy; s.massari@inail.it

* Correspondence: e.paba@inail.it; Tel.: +39-06-9418-1432

Received: 30 April 2019; Accepted: 27 May 2019; Published: 29 May 2019

Abstract: The exposure to biocontaminants in animal facilities represents a risk for developing infectious, allergic and toxic diseases. The aim of this study was to determine what factors could be associated with a high level of exposure to biological agents through the measure and characterization of airborne fungi, bacteria, endotoxin, $(1,3)$-β-D-glucan and animal allergens. Airborne microorganisms were collected with an air sampler and identified by microscopic and biochemical methods. Endotoxin, $(1,3)$-β-D-glucan, Mus m 1, Rat n 1, Can f 1, Fel d 1, Equ c 4 allergens were detected on inhalable dust samples by Kinetic LAL, Glucatell, and ELISA assays, respectively. Our data evidenced that changing cages is a determinant factor in increasing the concentration of the airborne biocontaminants; the preparation of bedding and distribution of feed, performed in the storage area, is another critical working task in terms of exposure to endotoxins (210.7 EU/m^3) and $(1,3)$-β-D-glucans (4.3 ng/m^3). The highest concentration of Mus m 1 allergen (61.5 ng/m^3) was observed in the dirty washing area. The detection of expositive peaks at risk of sensitization (>2 µg/g) by Fel d 1 in animal rooms shows passive transport by operators themselves, highlighting their role as vehicle between occupational and living environments.

Keywords: allergens; endotoxin; biological agents; laboratory animal allergy; environmental monitoring; occupational exposure

---

1. Introduction

The exposure to biocontaminants is well documented and studied in several occupational settings, and few data are available in animal facilities, although more attention has been addressed to animal allergens. These biological agents are aeroallergens, mainly lipocalins, derived by different biological fluids and/or tissue (saliva, serum, urine, dander, hair, fur) that remain in suspension for different times in relation to meteorological conditions and factors influencing their dispersion [1,2]. In recent years, the topic regarding animal allergen exposure has been treated both in pets [3–5] and experimental animals [6–9], and attention has been also addressed to co-exposure of allergens and endotoxins in these workplaces, although in a more restricted way [10–15]. Exposure to biocontaminants in animal facilities in USA and in the United Kingdom has received great attention [16–19] as demonstrated by the NIOSH publication in 1998 [20], where the so-called LAA (Laboratory Animal Allergy) was considered an occupational risk, and by Gordon in 2001 [21]. Various papers have been published in scientific literature both as a review [22–24] and experimental studies [25–27] while in Italy the topic of LAA is not studied carefully yet.

Health effects reported on laboratory animal workers (LAWs) are mainly represented by asthma, rhinitis, conjunctivitis, dermatitis and anaphylaxis [8,19,28]. Discordant percentages of LAA have been reported in the LAWs and these differences could be attributed to different evaluation methods (questionnaires, analytical test, etc.). In this context a relevant aspect is represented by environmental exposure to animal allergens, and although a lot of papers report the monitoring of these biocontaminants, there is some difficulty in comparing results because of the differences in sampling and analytical methods [29–32]. In addition to allergens, there are several other biological agents in animal facilities with possible health consequences such as airborne bacteria, fungi and microbial indicators (endotoxin and (1,3)-β-D-glucan) which may originate from animals (fur, epidermal material), their waste (feces and urine) and materials used in their maintenance (food, bedding) [33]. Airborne endotoxin, an outer membrane component of Gram-negative bacteria, has been identified as the major risk factor for nasal, chest and skin symptoms also contributing to allergy diseases [11,34]. Some authors report that fecal bacteria in soiled bedding may produce increased concentrations of endotoxin compared to other settings and this may be an occupational health concern. The authors concluded that the percentage of dust and endotoxin in different types of rodent bedding could be an important factor affecting the occupational exposure of personnel working with laboratory animals causing airway inflammation, hypersensitivity pneumonitis, organic dust toxic syndrome (ODTS), chronic obstructive pulmonary disease (COPD) and asthma-like syndrome, as well as having harmful effects on the animals themselves [33].

Pathogenic moulds, yeast and their products may also exist in the air of animal units, causing further risk for personnel working there. In particular, (1,3)-β-D-glucans are non-allergenic and hardly soluble in water glucose polymers, which consist of a part of the cellular wall of most fungi, but also many plants and some bacteria. Due to their presence in both viable and dead cells, they may be considered a good indicator of exposure to fungi [35]. Although little information is available on their health effects in animal facilities, however, their association with dry cough, cough associated with phlegm, hoarseness, and atopy, has been reported in different studies [36–38]. Very few and not recent investigations dealing with airborne biological agents in laboratory animal settings, other than allergens, can be found [12–15,39] to date. These issues have received little attention in our country. The objective of this study was to determine factors (working tasks, cage changing frequency and animal strains) were associated with the greatest level of exposure to biological agents of the personnel working with rodents through the measure and characterization of airborne viable fungi, bacteria, endotoxin, (1,3)-β-D-glucan and animal allergens.

2. Materials and Methods

2.1. Animal Facility

The study was conducted in a Biomedical Research Institute where animals are maintained in agreement with Italian Legislative decree 26/2014 [40]. The facility includes 14 conventional animal rooms (3 housing rats and 11 housing mice) where animals are housed in stainless steel wire cages (on average 2–3 animals/cage) with perforated floor and trays underneath; filters on the top prevent the distribution of airborne particles. Each cage is stacked in a rack that can hold up to 30 cages for mice and 24 for rats. The facility houses about 7000 rodents.

The racks are washed every 4 months. The cages and their equipment (grids containing bottles of water and feed, the bottles and soiled bedding) are changed once or twice per week depending on experimental protocols. When cages are changed, a clean rack is brought into the room, animals and accessories are moved to new cages and dirty cages are moved to the cage-washing area where they are emptied and washed (dirty washing area). Subsequently, the cages are allowed to dry and restacked for future use (clean washing area). There is also a storage area where preparation of the bedding and distribution of feed are carried out; these activities are performed only in the afternoon. The staff is divided into laboratory technicians and researchers; laboratory technicians perform shifts of 8 h and 30 min, and are engaged in changing cages, bedding and feeding animals, cleaning rooms, washing cages; researchers are engaged in several experimental procedures such as collecting blood and urine specimens, surgery and sacrifice. Personal protective equipment is required for all personnel working in the facility and includes surgical cap, mask, gloves, shoe cover, and disposable coat.

Animal rooms and the storage area have a ventilation system separate from the rest of the facility. The outside air is collected and treated by an Air Handling Unit (AHU) located on the coverage plan. Another AHU treats the air entering the two washing areas. In each work environment the air enters through vents located on the ceiling (two in each room), while the exhaust air is removed through grilles located on the walls near the floor (four in each rooms) or on the ceiling in the dirty and clean washing areas. In animal rooms 15–20 air changes per hour are guaranteed; the values of temperature and relative humidity are maintained in a controlled range (T: 20 °C ± 1 °C; RH: 45–55%).

2.2. Sampling and Analysis of Airborne Bacteria and Fungi

Triplicate air samples were collected using a portable microbiological sampler (SAS Super ISO, PBI International, Milan, Italy) at a flow rate of 100 L min^{-1}, from 10 workplaces: 7 animal rooms randomly selected (3 housing rats and 4 housing mice) (see Table 1), 2 washing areas (dirty and clean), and the storage area. Two offices were investigated: the technician's and manager's offices. All measurements were taken in the middle. In order to evaluate the influence of changing cages on environmental contamination, in animal rooms air samples were taken over three consecutive days: the day of changing cages, the one before, and the one after. Regarding the storage area, air sampling was carried out in the morning (without activity) and in the afternoon.

Table 1. Animals housed in each investigated room during the study and frequency of changing cages.

Rooms	Strains	Total Number	Male	Female	Pups	Changing Cages
A	Rats	149	63	86	0	Bi-weekly
B	Mice	494	328	166	0	Weekly
C	Mice	682	311	267	104	Weekly
D	Mice	613	225	283	105	Bi-weekly
E	Rats	155	85	60	10	Weekly
F	Mice	723	316	380	27	Bi-weekly
G	Rats	377	37	85	255	Bi-weekly

Air volumes of 300 L were sampled. Total cultivable and Gram-negative bacteria were impacted on Tryptone Soy Agar (TSA) and Mac Conkey 3 (McC$_3$) plates, respectively, fungi on Malt Extract Agar (MEA) and Dichloran Glycerol Agar (DG18) (all from Oxoid S.p.A, Milan, Italy). TSA and McC$_3$ plates were subsequently incubated for 1–2 days at 37 °C (mesophilic bacteria), DG18 and MEA at 25 °C for 7–10 days.

The bacterial isolates were identified with microscopic and biochemical methods by API 20E and NE tests (bioMerieux, Marcy l'Etoile, France) and Microstation ID instrument (Biolog Inc., Hayward, CA, USA). Identification of moulds and yeasts were accomplished via macroscopic and microscopic examination, referring to the manual Medically Important Fungi: A guide to Identification [41], while biochemical identifications were performed using the Microstation ID instrument. Data are expressed as CFU/m^3. The limit of detection (LOD) was 2 CFU/m^3. The uncertainty assessment results are presented in Table 2.

2.3. Inhalable Dust Sampling and Analysis of Endotoxin, (1,3)-β-D-glucan and Allergens

Stationary inhalable dust samples were collected using airChek2000 pumps (SKC Inc., Eighty Four, PA, USA), at a sampling flow rate of 2 L·min^{-1}, equipped with stainless IOM sampler and glass (GF, pore size 1.6 μm) and polycarbonate (PC, pore size 0.8 μm) for respectively endotoxin and (1,3)-β-D-glucan analysis.

Closed-face cassette with support pad and mixed cellulose ester (MCE, diameter 33 mm, 0.8 μm porosity), at a flow rate of 2.5 L·min^{-1}, were used for allergen sampling. Samplers were placed in the middle of the room at 1.5 m above the floor to simulate human breathing zone. A field blank was deployed on each sampling. The sampling strategy was the same as environmental monitoring (three consecutive days in animal rooms, in the morning and in the afternoon inside the storage area). The sampling time was 3 h.

GF extracts (5 mL of 0,05% Tween 20 in Pyrogen Free Water) were centrifuged at 1000× g rpm for 20 min. and analysed in duplicate with Kinetic LAL method (QCL-LAL assay Lonza Walkersville, MD, USA). Concentrations are reported in EU/m^3. The LOD was 0.005 EU/mL.

For (1,3)-β-D-glucan, PC extracts (5 mL of 0.85% NaCl and 0.05% tween 80) were vigorously shaken for 30′ at room temperature (250 rpm) and analyzed with the (1,3)-β-D-glucan specific Kinetic Chromogenic LAL assay (Glucatell, Associates of Cape Cod), as previously described [42,43]. (1,3)-β-D-glucan concentrations are presented in ng/m^3. The LOD in suspension was 2.53 pg/mL.

MCE filters were extracted by submerging in 1 mL phosphate saline buffer (PBS plus 0.01% Tween 20) in a 15 mL centrifuge tube, vortexed at room temperature and shaken for 2 h. The eluate was collected by compressing the filter in a plastic syringe, 1% of human serum albumin was added for protein stabilization and stored at −20 °C until assay. ELISA kits for Mus m 1, Rat n 1, Can f 1, Fel d 1, Equ c 4 were performed following the protocols of analysis (Indoor Biotechnologies Ltd., Manchester, UK.) with read of the plates at 405 nm optical density. Results are expressed as ng/m^3. The uncertainty assessment results are reported in Table 2.

(*) Type A standard uncertainty with a confidence level 1-α > 0.95 using the Student's T to find the appropriate coverage factor based on degrees of freedom. (**) Measures are detected in the morning (without activity) and in the afternoon (during activity).

Table 2. Uncertainty measurement (*) for sampled biological agents by working areas.

Workplace	Mesophilic Bacteria (UFC/m³)	Fungi (UFC/m³)	Endotoxin (EU/m³)	(1,3)-β-D-glucan (ng/m³)	Rat n 1 (ng/m³)	Mus m 1 (ng/m³)
Animal rooms	±52.26	±0.58	±1.35	±0.59	±0.16	±4.56
Washing area clean	±21.18	±16.91	±9.25			
Washing area dirty	±434.69	±66.18	±6.54			
Storage area (**)	±44.77	±94.89	±163.54	±4.57	±0.00	±8.32

(*) Type A standard uncertainty with a confidence level 1-α > 0.95 using the Student's T to find the appropriate coverage factor based on degrees of freedom. (**) Measures are detected in the morning (without activity) and in the afternoon (during activity).

2.4. Statistical Analysis

One way ANOVA and Students't-test were considered to evaluate significant differences in environmental biocontaminant concentrations in various work tasks and areas. Univariate and multivariate, both fixed and mixed effect linear regression models were used, in which the concentration of the different biological agents acts as outcome variable.

Data were analyzed using the statistical software R (R Foundation for Statistical Computing, Vienna, Austria—ISBN 3-900051-07-0, URL http://www.R-project.org).

3. Results

Table 1 shows strains and number of animals (divided into males, females and pups) housed in investigated rooms (A–G) and the frequency of changing cages. Concentrations (mean value and standard deviation) of mesophilic bacteria, fungi, endotoxin, (1,3)-β-D-glucan, Rat n 1 and Mus m 1 detected in each working areas are reported in Table 3. Regarding animal rooms, data are also presented by sampling room and the number of rodents is indicated.

3.1. Airborne Bacteria and Fungi

A total of 306 air samples were analysed. Mean mesophilic bacteria levels ranged from 33.8 to 480 CFU/m^3 with a peak value of 885 CFU/m^3 found in an animal room during the changing cages. Gram-negative concentrations were below the LOD (2 CFU/m^3) except for animal rooms, during the changing cages, and for the storage area where a mean value of 3 CFU/m^3 was recorded. Among a total of 18 genera and species, the most frequently isolated were *Staphylococcus warneri* (23.9%), *Staphylococcus xylosus* (21.4%) *Micrococcus* spp. (14.5%), *Acinetobacter schindleri* (10.4%).

Colonies of *Escherichia coli*, *Enterobacter cloacae* and *Enterococcus faecalis* (risk group 2, according to Annex XLVI of the Italian Legislative Decrees [44,45], were isolated in four animal rooms during the changing cages.

Mean concentrations of moulds and yeasts were low (flora was dominated by moulds such as *Alternaria*, *Ulocladium*, *Cladosporium*, *Penicillium*, *Eurotium*, *Aspergillus*, *Scopulariopsis*, *Phoma*, with occasional isolates of yeasts. No pathogen was identified.

3.2. Endotoxin, (1,3)-β-D-glucan and Allergens

The mean concentrations of endotoxin, (1,3)-β-D-glucan and allergens (See Table 3) were obtained on 89 samples taken in different occupational environments. Statistically significant differences were found between offices and animal rooms, these last having a higher concentration of endotoxin ($p = 0.00616$) and between the animal rooms and the storage area in the concentration of (1,3)-β-D-glucan ($p = 0.038$).

The evaluation of allergens was performed on 181 air samples for ELISA tests. Mus m 1 allergens were detected in all working areas except for offices, with the highest value (61.5 ng/m^3) in the dirty washing area while Rat n 1 allergens were absent except for in animal rooms. Can f 1 and Equ c 4 allergens were absent while Fel d 1 allergen was detected in rat rooms with the highest levels during the changing cages activity (range 0.4–3 ng/m^3).

Table 3. Concentrations (mean value and standard deviation) of mesophilic bacteria, fungi, endotoxin, (1,3)-β-D-glucan, Rat n 1 and Mus m 1 by working areas (number of animals is indicated for each room).

Workplace	Mesophilic Bacteria (UFC/m^3)	Fungi (UFC/m^3)	Endotoxin (EU/m^3)	(1,3)-β-D-glucan (ng/m^3)	Rat n 1 (ng/m^3)	Mus m 1 (ng/m^3)
Animal rooms	130.7 ± 207.5	1.6 ± 3.04	5.9 ± 3.6	1.5 ± 1.3	0.1 ± 0.4	4.3 ± 11.0
Room A (149)	63.1 ± 92.7	2.1 ± 4.4	7.8 ± 0.7	1.8 ± 1.2	0.0 ± 0.0	0.0 ± 0.1
Room B (494)	122.9 ± 110.2	2.0 ± 1.7	7.2 ± 4.2	3.6 ± 2.3	0.0 ± 0.0	6.1 ± 9.9
Room C (682)	112.4 ± 115.7	1.1 ± 2.4	3.6 ± 3.5	0.9 ± 0.6	0.1 ± 0.3	2.7 ± 4.2
Room D (613)	64.8 ± 83.5	1.3 ± 2.7	0.1 ± 0.0	1.9 ± 0.8	0.5 ± 0.9	0.3 ± 0.2
Room E (155)	317.0 ± 427.0	3.3 ± 5.3	7.3 ± 3.2	0.1 ± 0.0	0.3 ± 0.5	0.1 ± 0.2
Room F (723)	66.7 ± 75.8	0.4 ± 1.1	11.1 ± 4.8	1.3 ± 0.8	0.0 ± 0.0	20.6 ± 24.1
Room G (377)	220.5 ± 232.5	1.7 ± 1.7	5.2 ± 2.6	0.9 ± 0.3	0.0 ± 0.0	0.1 ± 0.2
Washing area clean	78.3 ± 2.4	19.3 ± 13.6	7.4 ± 1.8	1.3 ± 0.7		0.4 ± 0.0
Washing area dirty	211.1 ± 175.0	107.8 ± 63.1	16.7 ± 4.7	1.6 ± 0.9		61.5 ± 0.0
Storage area *	266.7 ±36.1	112.5 ± 76.4	124.2 ± 102.8	3.9 ± 1.7		1.5 ± 0.8
Offices	308.0 ± 91.2	60.0 ± 25.9	2.4 ± 0.8	1.4 ± 1.1		

* Mean concentrations measured in the storage area refer to samplings performed the morning and in the afternoon.

3.3. Effect of the Working Activities

One of our main questions was whether or not specific working tasks may cause an increased risk of biological exposure for personnel working in animal facilities. The hypothesis that the airborne concentration of biocontaminants could increase during the changing cages was tested. To point out the increase in level of (1,3)-β-D-glucan the time was considered a two-level factor: sampling during the changing cages was compared to the sampling before and after. The β-glucans concentration "during" increases on average by a factor of 1.3 with respect to the concentration "before" and "after" and the difference is statistically significant (p = 0.04). For endotoxins, a mixed effect linear regression model was used in which animal room was introduced as a random effect variable. The concentrations increased on average by a factor of 1.8 with respect to the levels before or after (p = 0.04). The highest mean concentrations of endotoxin (210.7 EU/m^3) and (1,3)-β-D-glucan (4.3 ng/m^3) were measured in the afternoon during the preparation of bedding and distribution of feed (storage area) (See Table 4).

The allergenic concentration was treated as a continuous variable. As in the case of endotoxin, to increase the statistics, the time was considered a two-level factor and a mixed effect linear regression model was used with the animal room treated as a random effect variable. The concentration level of the mouse allergenic contaminant increases on average by a factor of 9.3 with respect to the levels before or after the changing cages (p = 0.037).

Table 4. Concentrations (mean value and standard deviation) of inhalable endotoxin, (1,3)-β-D-glucan, Rat n 1 and Mus m 1 by working activities.

Workplace	Endotoxin (EU/m^3)	(1,3)-β-D-glucan (ng/m^3)	Rat n 1 (ng/m^3)	Mus m 1 (ng/m^3)
Animal rooms				
Before	5.2 ± 3.7	0.8 ± 0.5		0.1 ± 0.1
During	7.1 ± 3.8	2.4 ± 1.9	0.4 ± 0.6	10.5 ± 17.4
After	5.4 ± 3.5	1.4 ± 0.6		2.2 ± 5.3
Storage area				
Morning	37.8 ± 37.3	3.5 ± 2.4		0.9 ± 0.0
Afternoon	210.7 ± 20.0	4.3 ± 1.6		2.2 ± 0.0

3.4. Effect of Animal Strains and Changing Cages Frequency

In order to determine the effect of the animal strains on environmental biocontamination, the concentration measured in animal rooms was normalized according to the number of animals. This normalized quantity was related to the strain treated as a two level factor. In the case of the mesophilic bacteria and endotoxin, the average level was lowered in the case of mice with respect to the case of rats by a factor of −0.5070 (p = 0.020) and by a factor of −0.032 (p = 1.5·10-8) respectively. The result was not statistically significant for Gram-negative bacteria, fungi and (1,3)-β-D-glucan.

To test if animal species have an effect on the allergenic pollutants concentration, a Pearson's Chi-squared with Yates' continuity correction test was applied. The test was statistically significant in the case of the Mus m 1 and Fel d 1 (p = 0.004 and p = 0.045 respectively). The positive result of the Mus m 1 is statistically associated with the mouse species whilst the Fel d 1 is statistically associated with the rat.

In regards to the cage changing frequency, the concentrations normalized to the number of animals have been compared with once per week and twice per week cage changing frequency, treated as a two-level factor. A statistically significant difference has been found only in the case of mesophilic bacteria concentration. The twice per week frequency has on average a bacteria concentration reduced by a factor of 0.13 with respect to the level of rooms with a once per week frequency (p = 0.019).

4. Discussion

To the best of our knowledge, this is the first Italian study which assesses the exposure to biological agents in a conventional animal facility, taking into account the biocontamination of different occupational areas in conjunction with the effect of specific working tasks, cage changing frequency and animal strains on it.

The most frequent genera identified in this study are common skin inhabitants and are thus easily shed into the environment on desquamated epithelial skin cells. Some bacterial species have been reported to cause a variety of diseases, ranging from pneumonia (*Acinetobacter*) to intestinal infections (*E. coli*) (https://www.cdc.gov). Other microorganisms, such as *Enterobacter* and *Pseudomonas*, exhibit a high degree of biological activity, are producers of endotoxin, and often show strong allergenic properties that increase the risk of disease in workers [46,47]. However, colonies of bacterial species belonging to risk group 2, according to the Italian Legislative Decrees 81/2008 and 106/2009 [44,45], have been isolated in some animal rooms during the changing cages. With regard to endotoxin, no regulatory value is currently available, but various recommendations are available. Among them, Dutch Expert Committee on Occupational Standards recommends an occupational exposure limit of 90 EU/m^3 [48]. Endotoxin concentrations measured in this study were below this standard in all settings except for the storage area during activities where a mean value of 210.6 EU/m^3 was observed. Our data confirm what was reported by Kaliste et al., 2002 [33], that bedding material could be a significant source of accumulation and release of bacterial endotoxins. This evidence was demonstrated also by Whiteside et al., 2010 [49] who measured a range of endotoxins between 3121 and 5401 EU/g in hardwood rodent bedding.

Endotoxin concentrations measured in this study are higher than those reported by Hwang et al., 2016 [50], who found a mean value of 0.14 EU/m^3 (range = 0.03–0.60 EU/m^3) in an animal laboratory housing mice, where the main tasks were feeding/weighing with air exchange rates comparable to ours, and then those reported by Pacheco et al., 2006 [13] who measured a mean endotoxin concentration of 315 pg/m^3 (about 3.15 EU/m^3) with a peak value of 678 pg/m^3 (corresponding about 6.78 EU/m^3) during the changing cages in a mouse facility.

Our data are also slightly higher than those reported by Ooms et al., 2008 [39] (mean value = 4.7 EU/m^3; range = 3.4–6.3 EU/m^3) in rabbit rooms; this difference may be attributed to higher mechanical air-change rates (30.3 air changes/h) in the animal housing space. Finally, results of the present study are quite lower than those shown by Lieutier-Colas et al., 2001 [15], who found the highest concentration (15.4 ng/m^3 = 154.7 EU/m^3) associated with cage cleaning and feeding in animal rooms housing rats.

The detection of endotoxin in the technician's office (mean value = 2.3 EU/m^3) suggests that these components may be carried on clothes and shoes or on airborne particles like allergens.

The highest levels of (1,3)-β-D-glucan were found in the storage area, indicating that activities performed in this environment were also associated with an increased exposure to fungi, some of which belong to strongly allergenic and mycotoxin producers genera (*Alternaria* spp., *Penicillium* spp.).

Regarding animal allergens, no occupational environmental limits are available, partly due to the lack of standardized measurement methods and partly because of complex exposure-response relationships and the influence of genetic susceptibility; it is important to keep the allergen exposure as low as possible.

We found the highest concentration for Mus m 1 in the dirty washing area (mean value = 61.5 ng/m^3) and its concentrations increased significantly during the changing cages on average by a factor of 9.3 with respect to the levels before or after ($p = 0.037$). Straumfors et al., [16] in a recent paper report that cage emptying and cage washing in the cage washroom represented the highest exposure with a median value of 3.0 ng/m^3 for Mus m 1. The lower concentration found by these authors could be explained by the use of individually ventilated cages in respect to cage-rack systems and by the different sampling methodologies (stationary versus personal monitoring).

Our data regarding the highest values of Mus m 1 in respect to Rat n 1 in animal facilities are in agreement with the ones reported in literature (range: 17–564 ng/m^3 for Mus m 1; 0.43–27.36 ng/m^3 for Rat n 1) [1,15,51].

For cat allergens (Fel d 1), limit values between 2 and 8 μg/g Fel d 1 for sensitization and above 8 μg/g for the development of acute asthma attacks, respectively, have been proposed [52]. In this work we measured concentrations at risk of sensitization (>2 μg/g) in two rat rooms.

The detection of cat allergens (Fel d 1) in animal rooms is unexpected, along with the presence of rat and mouse allergens (Figure 1), showing a cross-contamination likely due to the passive transport by operators themselves; this highlights their role as vehicle between occupational areas and living environments [53–56].

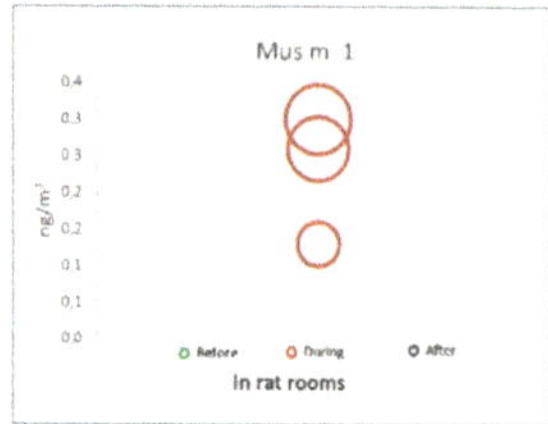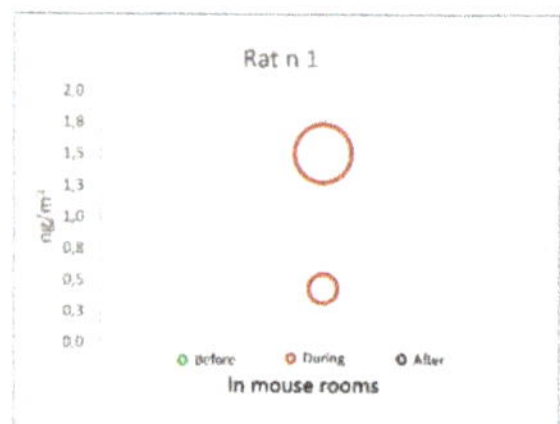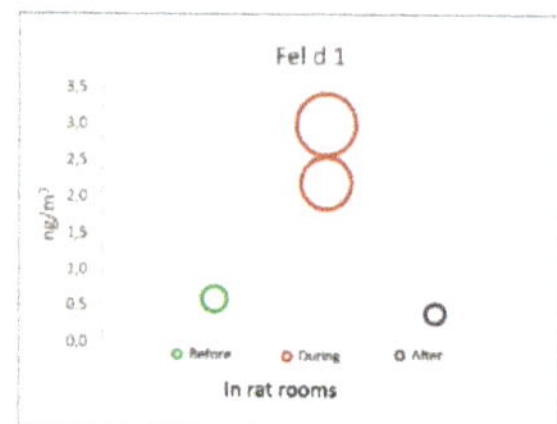

Figure 1. Aeroallergen levels over three consecutive days in mouse and rat rooms: before, during and after changing cages (Each circle represents the observed value and dimension/diameter indicates the concentration).

5. Conclusions

The environmental monitoring of airborne biocontaminants has been confirmed as a valid tool to identify the working tasks more critical in terms of exposure to biological agents in animal facilities.

Statistical analysis proves that the changing of cages is a determinant factor in increasing the concentration of all the airborne biocontaminants and in releasing biological agents that could risk the health of workers. It can also be pointed out that the concentrations of the biocontaminants increase significantly during this task with respect to the other days, but return to the background levels indicating the effectiveness of the ventilation system. Our data also identify the preparation of bedding and distribution of feed performed in the storage area as critical working tasks in terms of exposure to endotoxins, fungi and (1,3)-β-D-glucans.

The frequency of the changing of cages does not seem to affect the microbiological contamination; a statistical difference has been found only in the case of mesophilic bacteria levels. In regard to the animal strains, rats seem to influence only mesophilic bacteria and endotoxin levels.

Our data can be particularly useful for increasing knowledge about the biological risk in this occupational sector and to support all actors of prevention, employers, health and safety representatives and workers in all phases of risk management.

Simple preventive measures must be adopted to control the biological risk in this occupational setting.

Author Contributions: Environmental monitoring and laboratory procedures, A.M.M. and A.C.; Laboratory procedures, S.D.R.; Data set built, P.M.; Logistics support A.W. and M.C.R.; Statistical analysis, writing-review and editing, S.M. and R.S.; Conceived and designed the research topic, writing original draft, revising, editing and supervision, M.C.D. and E.P.

Funding: This research received no external funding.

Acknowledgments: The authors are grateful to the workers who collaborated in our experimental phases of the study.

Conflicts of Interest: The authors declare no conflict of interest.

References

1. Zahradnik, E.; Raulf, M. Respiratory allergens from furred mammals: Environmental and occupational exposure. *Vet. Sci.* **2017**, *4*, 38. [CrossRef] [PubMed]

2. Zahradnik, E.; Raulf, M. Animal allergens and their presence in the environment. *Front. Immunol.* **2014**, *3*, 76. [CrossRef] [PubMed]

3. Liccardi, G.; Calzetta, L.; Baldi, G.; Berra, A.; Billeri, L.; Caminati, M.; Capano, P.; Carpentieri, E.; Ciccarelli, A.; Crivellaro, M.A.; et al. Italian Allergic Respiratory Diseases Task ForceAllergic sensitization to common pets (cats/dogs) according to different possible modalities of exposure: An Italian Multicenter Study. *Clin. Mol. Allergy* **2018**, *16*, 3. [CrossRef]

4. Liccardi, G.; Salzillo, A.; Piccolo, A.; Calzetta, L.; Rogliani, P. Dysfunction of small airways and prevalence, airway responsiveness and inflammation in asthma: Much more than small particle size of pet animal allergens. *Ups. J. Med. Sci.* **2016**, *121*, 196–197. [CrossRef] [PubMed]

5. Patelis, A.; Dosanjh, A.; Gunnbjörnsdottir, M.; Borres, M.P.; Högman, M.; Alving, K.; Janson, C.; Malinovschi, A. New data analysis in a population study raises the hypothesis that particle size contributes to the pro-asthmatic potential of small pet animal allergens. *Ups. J. Med. Sci.* **2016**, *121*, 25–32. [CrossRef]

6. Stave, G.M. Occupational animal allergy. *Curr. Allergy Asthma Rep.* **2018**, *18*, 11. [CrossRef] [PubMed]

7. Mason, H.J.; Willerton, L. Airborne exposure to laboratory animal allergens. *AIMS Allergy Immunol.* **2017**, *1*, 78–88. [CrossRef]

8. Simoneti, C.S.; Ferraz, E.; Menezes, M.B.; Bagatin, E.; Arruda, L.K.; Vianna, E.O. Allergic sensitization to laboratory animals is more associated with asthma, rhinitis, and skin symptoms than sensitization to common allergens. *Clin. Exp. Allergy* **2017**, *47*, 1436–1444. [CrossRef]

9. Bush, R.K.; Stave, G.M. Laboratory animal allergy: An update. *ILAR J.* **2003**, *44*, 28–51. [CrossRef]

10. Lai, P.S.; Allen, J.G.; Hutchinson, D.S.; Ajami, N.J.; Petrosino, J.F.; Winters, T.; Hug, C.; Wartenberg, G.R.; Vallarino, J.; Christiani, D.C. Impact of environmental microbiota on human microbiota of workers in academic mouse research facilities: An observational study. *PLoS ONE* **2017**, *12*, e0180969. [CrossRef] [PubMed]

11. Oppliger, A.; Barresi, F.; Maggi, M.; Schmid-Grendelmeier, P.; Huaux, F.; Hotz, P.; Dressel, H. Association of endotoxin and allergens with respiratory and skin symptoms: A descriptive study in laboratory animal workers. *Ann. Work. Expo. Health* **2017**, *61*, 822–835. [CrossRef] [PubMed]

12. Samadi, S.; Heederik, D.J.; Krop, E.J.; Jamshidifard, A.R.; Willemse, T.; Wouters, I.M. Allergen and endotoxin exposure in a companion animal hospital. *Occup. Environ. Med.* **2010**, *67*, 486–492. [CrossRef] [PubMed]

13. Pacheco, K.A.; McCammon, C.; Thorne, P.S.; O'Neill, M.E.; Liu, A.H.; Martyny, J.W.; Vandyke, M.; Newman, L.S.; Rose, C.S. Characterization of endotoxin and mouse allergen exposure in mouse facilities and research laboratories. *Am. Occup. Hyg.* **2006**, *50*, 563–572.

14. Platts-Mills, J.; Custis, N.; Kenney, A.; Tsay, A.; Chapman, M.; Feldman, S.; Platts-Mills, T. The effects of cage design on airborne allergens and endotoxin in animal rooms: High-volume measurements with an ion-charging device. *Contemp. Top. Lab. Anim. Sci.* **2005**, *44*, 12–16. [PubMed]

15. Lieutier-Colas, F.; Meyer, P.; Larsson, P.; Malmberg, P.; Frossard, N.; Pauli, G.; de Blay, F. Difference in exposure to airborne major rat allergen (Rat n 1) and to endotoxin in rat quartes according to tasks. *Clin. Exp. Allergy* **2001**, *31*, 1449–1456. [CrossRef] [PubMed]

16. Straumfors, A.; Eduard, W.; Andresen, K.; Sjaastad, A.K. Predictors for increased and reduced rat and mouse allergen exposure in laboratory animal facilities. *Ann. Work Expo. Health* **2018**, *62*, 953–965. [CrossRef]

17. Feary, J.; Cullinan, P. Laboratory animal allergy: A new world. *Curr. Opin. Allergy Clin. Immunol.* **2016**, *16*, 107–112. [CrossRef]

18. Simoneti, C.S.; Freitas, A.S.; Barbosa, M.C.; Ferraz, E.; de Menezes, M.B.; Bagatin, E.; Arruda, L.K.; Vianna, E.O. Study of risk factors for atopic sensitization, asthma, and bronchial hyperresponsiveness in animal laboratory workers. *J. Occup. Health* **2016**, *58*, 7–15. [CrossRef]

19. Kampitak, T.; Betschel, S.D. Anaphylaxis in laboratory workers because of rodent handling: Two case reports. *J. Occup. Health* **2016**, *58*, 381–383. [CrossRef]

20. National Institute for Occupational Safety and Health (NIOSH). *Preventing Asthma in Animal Handlers*; NIOSH: Washington, DC, USA, 1998; pp. 97–116.

21. Gordon, S. Laboratory animal allergy: A British perspective on a global problem. *ILAR J.* **2001**, *42*, 37–46. [CrossRef]

22. Corradi, M.; Ferdenzi, E.; Mutti, A. The characteristics, treatment and prevention of laboratory animal allergy. *Lab. Anim.* **2012**, *42*, 26–33. [CrossRef] [PubMed]

23. D'Ovidio, M.C.; Martini, A.; Melis, P.; Signorini, S. Value of the microarray for the study of laboratory animal allergy (LAA). *G. Ital. Med. Lav. Ergon.* **2011**, *33*, 109–116. [PubMed]

24. Folletti, I.; Forcina, A.; Marabini, A.; Bussetti, A.; Siracusa, A. Have the prevalence and incidence of occupational asthma and rhinitis because of laboratory animals declined in the last 25 years? *Allergy* **2008**, *63*, 834–841. [CrossRef]

25. D'Ovidio, M.C.; Wirz, A.; Zennaro, D.; Masssari, S.; Melis, P.; Peri, V.M.; Rafaiani, C.; Riviello, M.C.; Mari, A. Biological occupational allergy: Protein microarray for the study of laboratory animal allergy (LAA). *AIMS Public Health* **2018**, *5*, 352–365. [CrossRef] [PubMed]

26. Larese Filon, F.; Drusian, A.; Mauro, M.; Negro, C. Laboratory animal allergy reduction from 2001 to 2016: An intervention study. *Respir. Med.* **2018**, *136*, 71–76. [CrossRef] [PubMed]

27. Tafuro, F.; Selis, L.; Goldoni, M.; Stendardo, M.; Mozzoni, P.; Ridolo, E.; Boschetto, P.; Corradi, M. Biomarkers of respiratory allergy in laboratory animal care workers: An observational study. *Int. Arch. Occup. Environ. Health* **2018**, *91*, 735–744. [CrossRef]

28. Bhabha, F.K.; Nixon, R. Occupational exposure to laboratory animals causing a severe exacerbation of atopic eczema. *Austr. J. Dermatol.* **2012**, *53*, 155–156. [CrossRef] [PubMed]

29. Jones, M. Laboratory animal allergy in the modern era. *Curr. Allergy Asthma Rep.* **2015**, *15*, 73. [CrossRef]

30. Muzembo, B.A.; Eitoku, M.; Inaoka, Y.; Oogiku, M.; Kawakubo, M.; Tai, R.; Takechi, M.; Hirabayashi, K.; Yoshida, N.; Ngatu, N.R.; et al. Prevalence of occupational allergy in medical researchers exposed to laboratory animals. *Ind. Health* **2014**, *52*, 256–261. [CrossRef]

31. Acton, D.; McCauley, L. Laboratory animal allergy: An occupational hazard. *AAOHN J.* **2007**, *55*, 241–244. [CrossRef]

32. Elliott, L.; Heederik, D.; Marshall, S.; Peden, D.; Loomis, D. Incidence of allergy and allergy symptoms among workers exposed to laboratory animals. *Occup. Environ. Med.* **2005**, *62*, 766–771. [CrossRef] [PubMed]

33. Kaliste, E.; Linnainmaa, M.; Meklin, T.; Nevalainen, A. Airborne contaminants in conventional laboratory rabbit rooms. *Lab. Anim.* **2002**, *36*, 43–50. [CrossRef]

34. Freitas, A.S.; Simoneti, C.S.; Ferraz, E.; Bagatin, E.; Brandão, I.T.; Silva, C.L.; Borges, M.C.; Vianna, EO. Exposure to high endotoxin concentration increases wheezing prevalence among laboratory animal workers: A cross-sectional study. *BMC Pulm. Med.* **2016**, *16*, 69. [CrossRef] [PubMed]

35. Douwes, J. (1–>3)-Beta-D-glucans and respiratory health: A review of the scientific evidence. *Indoor Air* **2005**, *15*, 160–169. [CrossRef] [PubMed]

36. Paba, E.; Chiominto, A.; Marcelloni, A.M.; Proietto, A.; Sisto, R. Exposure to airborne culturable microorganisms and endotoxin in two Italian poultry slaughterhouses. *J. Occup. Environ. Hyg.* **2014**, *11*, 469–478. [CrossRef]

37. Rylander, R.; Lin, R.H. (1–>3)-beta-D-glucan-relationship to indoor air-related symptoms, allergy and asthma. *Toxicology* **2000**, *152*, 47–52. [CrossRef]

38. Rylander, R.; Norhall, M.; Engdahl, U.; Tunsater, A.; Holt, P.G. Airways inflammation, atopy, and (1–>3)-beta-D-glucan exposure in two schools. *Am. J. Respir. Crit. Care Med.* **1998**, *158*, 1685–1687. [CrossRef]

39. Ooms, T.G.; Artwohl, J.E.; Conroy, L.M.; Schoonover, T.D.; Fortman, J.D. Concentration and emission of airborne contaminants in a laboratory animal facility housing rabbits. *J. Am. Assoc. Lab. Anim. Sci.* **2008**, *47*, 39–48. [PubMed]

40. Implementation of the Directive 2010/63/EU on the protection of animals used for scientific purposes, Italian Legislative Decree 4 march 2014, n. 26. *Off. J.* **2014**, *60*.

41. Larone, D.H. *Medically Important Fungi: A guide to Identification*, 3rd ed.; ASM: Washington, DC, USA, 1995.

42. Lee, T.; Grinshpun, S.A.; Kim, K.Y.; Iossifova, Y.; Adhikari, A.; Reponen, T. Relationship between indoor and outdoor airborne fungal spores, pollen, and (1→3)-β-D-glucan in homes without visible mold growth. *Aerobiologia* **2006**, *22*, 227–236. [CrossRef]

43. Iossifova, Y.Y.; Reponen, T.; Bernstein, D.I.; Levin, L.; Kalra, H.; Campo, P.; Villareal, M.; Lockey, J.; Hershey, G.K.; LeMasters, G. House dust (1-3)-beta-D-glucan and wheezing in infants. *Allergy* **2007**, *62*, 504–513. [CrossRef]

44. Implementation of the article 1 of the law 3 august 2007, no. 123 concerning the protection of health and safety in the workplaces, Italian Legislative Decree no. 81/2008. *Ordinary Suppl. No.108 Off. J.* **2008**, *101*.

45. Supplementary and corrective provisions of the legislative decree 9 April 2008, no. 81, concerning the protection of health and safety in the workplaces, Italian Legislative Decree no. 106/2009. *Ordinary Suppl. No. 142 Off. J.* **2009**, *180*.

46. Duchaine, C.; Grimard, Y.; Cormier, Y. Influence of building maintenance, environmental factors, and seasons on airborne contaminants of swine. *Am. Ind. Hyg. Assoc. J.* **2000**, *61*, 56–63. [CrossRef]

47. Mandryk, J.; Alwis, K.U.; Hocking, A.D. Effects of personal exposures on pulmonary function and work-related symptoms among sawmill workers. *Ann. Occup. Hyg.* **2000**, *44*, 281–289. [CrossRef]

48. Health Council of the Netherlands. *Endotoxins-Health-Based Recommended Occupational Exposure Limit*; Publication no. 2010/04OSH. 010; Health Council of the Netherlands: The Hague, The Netherlands, 2010.

49. Whiteside, T.E.; Thigpen, J.E.; Kissling, M.G.; Grant, M.G.; Forsythe, D.B. Endotoxin, coliform, and dust levels in various types of rodent bedding. *J. Am. Assoc. Lab. Anim. Sci.* **2010**, *49*, 184–189. [PubMed]

50. Hwang, S.H.; Park, D.J.; Park, W.M.; Park, D.U.; Ahn, J.K.; Yoon, C.S. Seasonal variation in airborne endotoxin levels in indoor environments with different micro-environmental factors in Seoul, South Korea. *Environ. Res.* **2016**, *145*, 101–108. [CrossRef]

51. Ohman, J.L.; Hagberg, K.; MacDonald, M.R.; Jones, R.R.; Paigen, B.J.; Kacergis, J.B. Distribution of airborne mouse allergen in a major mouse breeding facility. J. Allergy. *Clin. Immunol.* **1994**, *94*, 810–817.

52. Liccardi, G.; Triggiani, M.; Piccolo, A.; Salzillo, A.; Parente, R.; Manzi, F.; Vatrella, A. Sensitization to common and uncommon pets or other furry animals: Which may be common mechanisms? *Transl. Med. UniSa* **2016**, *14*, 9–14.

53. Liccardi, G.; Salzillo, A.; Piccolo, A.; D'Amato, M.; D'Amato, G. Can the levels of Can f 1 in indoor environments be evaluated without considering passive transport of allergen indoors? *J. Allergy Clin. Immunol.* **2013**, *131*, 1258–1259. [CrossRef]

54. Krop, E.J.; Doekes, G.; Stone, M.J.; Alberse, R.C.; van der Zee, J.S. Spreading of occupational allergens: Laboratory animal allergens on hair-covering caps and in mattress dust of laboratory animal workers. *Occup. Environ. Med.* **2007**, *64*, 267–272. [CrossRef] [PubMed]

55. Pereira, F.L.; Silva, D.A.; Sopelete, M.C.; Sung, S.S.; Taketomi, E.A. Mite and cat allergen exposure in Brazilian public transport vehicles. *Ann. Allergy Asthma Immunol.* **2004**, *93*, 179–184. [CrossRef]

56. Ferrari, M.; Perfetti, L.; Moscato, G. A case of indirect exposure to cat at school. *Monaldi Arch. Chest. Dis.* **2003**, *59*, 169–170. [PubMed]

MDPI

St. Alban-Anlage 66

4052 Basel

Switzerland

Tel. +41 61 683 77 34

Fax +41 61 302 89 18

www.mdpi.com

Applied Sciences Editorial Office

E-mail: applsci@mdpi.com

www.mdpi.com/journal/applsci

www.ingramcontent.com/pod-product-compliance
Lightning Source LLC
LaVergne TN
LVHW071521170726
843515LV00008B/2035